ERGEBNISSE DER MATHEMATIK UND IHRER GRENZGEBIETE

UNTER MITWIRKUNG DER SCHRIFTLEITUNG DES „ZENTRALBLATT FÜR MATHEMATIK“

HERAUSGEGEBEN VON

L. V. AHLFORS · R. BAER · R. COURANT · J. L. DOOB · S. EILENBERG
P. R. HALMOS · T. NAKAYAMA · H. RADEMACHER
F. K. SCHMIDT · B. SEGRE · E. SPERNER

NEUE FOLGE · HEFT 12

REIHE:

ALGEBRAISCHE GEOMETRIE

BESORGT
VON

B. SEGRE

SPRINGER-VERLAG
BERLIN · GÖTTINGEN · HEIDELBERG
1956

ALGEBRAIC VARIETIES

BY

M. BALDASSARRI

SPRINGER-VERLAG
BERLIN · GÖTTINGEN · HEIDELBERG
1956

ISBN 978-3-642-52763-0 ISBN 978-3-642-52761-6 (eBook)
DOI 10.1007/978-3-642-52761-6

BRÜHLSCHE UNIVERSITÄTSDRUCKEREI GIESSEN

Preface

Algebraic geometry has always been an eclectic science, with its roots in algebra, function-theory and topology. Apart from early researches, now about a century old, this beautiful branch of mathematics has for many years been investigated chiefly by the Italian school which, by its pioneer work, based on algebro-geometric methods, has succeeded in building up an imposing body of knowledge. Quite apart from its intrinsic interest, this possesses high heuristic value since it represents an essential step towards the modern achievements. A certain lack of rigour in the classical methods, especially with regard to the foundations, is largely justified by the creative impulse revealed in the first stages of our subject; the same phenomenon can be observed, to a greater or less extent, in the historical development of any other science, mathematical or non-mathematical. In any case, within the classical domain itself, the foundations were later explored and consolidated, principally by SEVERI, on lines which have frequently inspired further investigations in the abstract field.

About twenty-five years ago B. L. VAN DER WAERDEN and, later, O. ZARISKI and A. WEIL, together with their schools, established the methods of modern abstract algebraic geometry which, rejecting the classical restriction to the complex groundfield, gave up geometrical intuition and undertook arithmetisation under the growing influence of abstract algebra. In this way was perfected one of the most refined tools of present-day mathematics, which was then applied to various subjects in the classical domain, such as linear systems, geometric and arithmetic genera and, in an outstanding manner, to the theory of abstract Abelian varieties.

In the meantime the already familiar ties between transcendental theory and algebraic geometry, due chiefly to M. NOETHER, POINCARÉ and SEVERI, were strengthened by W. V. D. HODGE and K. KODAIRA through the new theory of differential forms; these researches, by extending classical properties of algebraic surfaces to higher varieties, and passing from these to KÄHLER's analytic varieties, enriched the field of algebraic geometry proper with fresh insight and viewpoints.

Finally the transcendental investigations, whose topological nature was thoroughly explored by H. CARTAN and his school, had the effect of introducing quite naturally into algebraic geometry such powerful methods as the theory of fibre spaces and stacks. From this vantage point it is possible to give a purely topological construction for the canonical systems of higher algebraic varieties over the complex field; such systems had previously been defined algebraically by M. EGER, B. SEGRE and J. A. TODD on the basis of SEVERI's work on rational equivalence. This is one of the most fascinating topics of the new algebraic geometry, in which algebra and topology, transcendental theory and differential geometry merge into a single stream.

The aim of the present work is to describe — perforce briefly, but as completely as space allows — modern algebraic geometry in its most important aspects. Generally speaking, the various methods are here surveyed from a strictly utilitarian point of view, and not for their own interest; the special algebraic, transcendental and topological techniques required for the purpose are taken for granted; while the short introductions to the different sections are intended to recall definitions and fix the terminology. For the more algebraic part of the work we have often referred to the exhaustive monograph of P. SAMUEL in this collection, and we regret that it has proved impossible to do the same for the topological part since F. HIRZEBRUCH's monograph, also in this collection, appeared after our work was completed.

It must be obvious that such a programme cannot be free from imperfections due to sheer lack of space, not to mention those for which the author is more directly to blame. Thus we have been obliged to restrict the exposition to fundamentals, to be very concise throughout, to sketch many proofs and to omit others; we can only hope that the work has not been condensed beyond the point of intelligibility. The relative amounts of space devoted to theories or proofs should not be scrutinised too closely; and the historical and bibliographical notes and lists are intended solely to help the reader and are not claimed to be complete.

In conclusion I take pleasure in thanking Prof. B. SEGRE for his valuable assistance during the compilation, and also Prof. L. ROTH, who has read the English text and proof sheets, and suggested corrections and improvements: our debt to both is great. To all friends and colleagues who have interested themselves in the work I express the hope that this book may prove useful in diffusing the knowledge of algebraic geometry beyond the confines of any personal fashion or school.

MARIO BALDASSARRI

Padova, May 1956

Table of Contents

ALGEBRAIC VARIETIES

I. A Survey of Foundations

This chapter is chiefly modelled on SAMUEL's account [c], where the reader may find almost all the proofs which are here omitted; however, the salient facts will be quoted explicitly. Our purpose is merely to agree on the language and to recall, as briefly as possible, some of the fundamentals: more elementary arguments, as for instance projections in affine or projective spaces, are not expounded; they may be found in SAMUEL [c] or in HODGE-PEDOE [a], the latter being restricted to the case of characteristic zero.

For the algebraic background VAN DER WAERDEN [a], SAMUEL [a, b] and NORTHCOTT [a] may be consulted. On the other hand, the reader who needs or wishes to know how classical algebraic geometry can supply a satisfactory basis, will find the best treatment in the critical works of SEVERI [20, 29, 31, 32] or in the treatise [c] of SEVERI.

The treatises [b] of VAN DER WAERDEN and [a] of HODGE-PEDOE are intermediate between classical and abstract geometry. The deepest synthesis of both classical and modern tendencies is to be found in the treatise [a] of A. WEIL which, in spite of some later simplifications, remains the highest stage achieved in our subject.

1. Algebraic Varieties

Let k be a subfield of an algebraically closed field k^* and $\mathbf{A}^n(k^*)$ (or $\mathbf{P}^n(k^*)$) the affine (or projective) n-dimensional space over k^*. A subset H of $\mathbf{A}^n(k^*)$ is called an algebraic set over k or a *k-set* if the points of H are the zeros of a set $\mathfrak{F}$ of polynomials in $k[\xi_1, \xi_2, \ldots, \xi_n] = k[\xi]$, where $\xi_1, \xi_2, \ldots, \xi_n$ are indeterminates: we shall then write $H = V_k(\mathfrak{F})$ {criticism in ZARISKI [b], p. 79}.

The set of all polynomials in $k[\xi]$ null over H is an ideal in $k[\xi]$ to be denoted by $\mathfrak{J}_k(H)$.

We recall that conversely an arbitrary ideal $\mathfrak{J}$ in $k[\xi]$ is not associated with some algebraic k-set H: a n. a. s. c. for this is that $\mathfrak{J}$ coincides with its own *radical*, i. e. with the set $\mathfrak{J}^*$ of elements $f \in k[\xi]$ such that if $f \in \mathfrak{J}^*$, then $f^m \in \mathfrak{J}$, where m is a positive integer. The necessity is obvious and the sufficiency is the famous HILBERT "*Nullstellensatz*", which, by a remark of RABINOWITSCH [1], is equivalent to the fact that if an ideal $\mathfrak{J}$ in $k[\xi]$ is free from zeros, then it is the unit ideal {see SAMUEL [c], p. 4}.

This theorem, in its homogeneous form, may be used as a foundation for the elementary theory of algebraic varieties: see ZARISKI [b], p. 81, HODGE-PEDOE [a_2], Ch. X. Other simple proofs of HILBERT's theorem have been given by BRAUER [1], GOLDMAN [1] and ZARISKI [13].

The k-sets of $\mathbf{A}^n(k^*)$ form a system closed both by intersection and union provided the latter be repeated finitely: a k-set is said irreducible or a *k-variety* V if it is not a proper union of two k-sets. A reducible k-set is always the union of k-varieties which are called the *components* and are uniquely determined by the set.

A k-set V is irreducible if and only if $\mathfrak{J}_k(V)$ is a prime ideal $\mathfrak{p}$. Then if $k[\xi]/\mathfrak{p} \cong k[v]$ where v is the reduced class mod $\mathfrak{p}$ of ξ, $k[v]$ is an integral ring: the *ring of the affine coordinates or of the polynomials on* V. The quotient field $k(v)$ of $k[v]$ is *the field* $R_k(V)$ *or also* $k(V)$ *of the rational functions on* V*:* it is a finite transcendental extension of $k[v]$ whose degree is called *the dimension of* $V(\dim V)$.

This strictly algebraic notion of dimension coincides with the topological definition, which has, however, a sense only when k is the complex field {see VAN DER WAERDEN [b], p. 123}. For other purely algebraic definitions of the dimension see: ABELLANAS [1] and ZARISKI [a], p. 1.

If k^* is of infinite degree of transcendence over k, the point $P=(x)$, where x is the transform of v in the k-isomorphism of $k[v]$ into k^*, is called *a generic point of* V *over* k. Such a point determines over k the ideal $\mathfrak{p}$: it is determined by V to within a k-isomorphism and satisfies the relation $\dim_k(x) = \dim V$.

Defining as *specialisation of* P *over* k each point P' such that every polynomial in $k[\xi]$ null at P is null also at P' (then $P' \in V$), we can define a generic point of a k-set H as a point P such that each point P' of H is a specialisation of P over k*:* a set H having such a generic point P *is a* k*-variety* and P to within a k-isomorphism coincides with a generic point as already defined.

If V and W are k-varieties with $W \subseteq V$, then $\dim W \leq \dim V$, these dimensions being equal if and only if $W = V$*:* in particular if $\dim W = \dim V - 1$, W is called a *k-hypersurface* or a *k-primal* of V. A k-primal in $\mathbf{A}^n(k^*)$, which will be called a *k-form,* can be represented by *a single* equation {see, for proofs, SAMUEL [c], p. 6}.

The previous definitions and results can be extended to the *projective* k-sets in $\mathbf{P}^n(k^*)$ by using homogeneous polynomials in $k[\xi_0, \xi_1, \ldots, \xi_n] = k[\xi]$, where $(\xi_0, \xi_1, \ldots, \xi_n)$ are indeterminates related to the homogeneous coordinates $(x_0, x_1, \ldots, x_n)$ of a point P in $\mathbf{P}^n(k^*)$. Denoting by $k(P)$ the subfield of k^* generated by adding to k the ratios of the coordinates (x) of P to one which is non-null, the system of homogeneous coordinates (x) in $\mathbf{P}^n(k^*)$ will be called *strictly homogeneous* if $k(x)$ is a simple transcendental extension of $k(P)$: such systems certainly exist if k^* is of infinite degree of transcendency over k.

The point P will be called *a projective zero* of a polynomial $F(\xi_0, \xi_1, \ldots, \xi_n)$ over k if $F(x) = 0$ for each system of homogeneous coordinates (x) of P, and for this it is sufficient that $F(x) = 0$ for a strictly homogeneous system (x). The question is thus reduced to the study of the zeros of the homogeneous polynomials.

The affine and projective k-sets may be related to each other by identifying $\mathbf{P}^n(k^*)$ with a star of rays in $\mathbf{A}^{n+1}(k^*)$ so that the projective k-sets in $\mathbf{P}^n(k^*)$ are biunivocally associated with the conical affine k-sets in that star: in particular the intersection and union properties are immediately transferred to projective k-sets, and hence also the notion of irreducibility and *projective k-variety*. Consequently we can define *the ring of the homogeneous coordinates* $k[v_0, v_1, \ldots, v_n] = k[v]$ for a projective k-variety V and *the field* $R_k(V)$ *of the rational functions on* V, the latter being restricted to homogeneous fractions. They are coincident with the same sets affinely defined for the cone $C(V)$ obtained by projecting V into $A^{n+1}(k^*)$.

Again taking $\dim V = [R_k(V):k]$, we find that $\dim C(V) = \dim V + 1$. The extension of the notion of *generic point with respect to k* and of *specialisation with respect to k* can be effected in an obvious manner: we point out that projective specialisations can always be finitely extended.

This result, stated in purely algebraic language, is as follows: "Let (x) and (y) be given sets of elements of a field $k^* \supset k$: then every specialisation (x') of (x) over k can be extended, in at least one way, to a specialisation (x', y') of (x, y) over k, if we admit specialisation to ∞ (generalised specialisations)", where ∞ must be considered as an element of k^*. This has been proved by VAN DER WAERDEN in [2] {see also [b], p. 164} by means of elimination theory and later, independently of this theory, by the American school {see A. WEIL [a], p. 30}. For comments see ZARISKI [b], p. 81 and VAN DER WAERDEN [12].

It is well known that a projective space $\mathbf{P}^n(k^*)$ becomes an affine space $\mathbf{A}^n(k^*)$ by simply removing from $\mathbf{P}^n(k^*)$ the hyperplane at infinity. Thus is defined an application f of $\mathbf{A}^n(k^*)$ in $\mathbf{P}^n(k^*)$: the smallest k-set of $\mathbf{P}^n(k^*)$, let it be $\overline{H}$, containing a k-set H of $\mathbf{A}^n(k^*)$, is called *the projective closure* of H. It is easily seen that $\overline{H}$ can be represented by the homogeneous system associated with any system of equations over k representing H. The application f preserves dimensions and irreducibility over k; the inverse f^{-1} is obviously defined for every k-set in $\mathbf{P}^n(k^*)$ not completely at infinity {see SAMUEL [c], p. 12}.

2. Absolute and Relative Varieties

The definitions given up to now are related to a fixed reference field k and so they apply only to sets in $\mathbf{A}^n(k^*)$ (or in $\mathbf{P}^n(k^*)$) which can be represented by a system of equations over k. To avoid this, without taking $k = k^*$, which would not allow either the definition of the generic point or the use of reference fields not algebraically closed, we can, following A. WEIL, take a fixed "*universal domain*": i. e. a field $\mathbf{K}$ of infinite degree of transcendence over each other field to be considered. It is convenient to suppose it *algebraically closed*.

A critical justification of our choice of the universal domain which, as ZARISKI says, is "*a reservoir of infinitely many transcendentals for the point coordinates in our geometry*" {ZARISKI [b], p. 78}, is given by A. WEIL: we remark, according to this author, that there is only one geometry for each value of the characteristic and independently of the choice of $\mathbf{K}$. In particular we have LEFSCHETZ's *principle*, stating the validity in characteristic zero of any result true over the complex field, whenever a finite number of geometrical objects is implied: so that topological and analytic methods may be legitimately used when **K** is of characteristic zero {see A. WEIL [a], p. 242}.

A k-variety V of ideal $\mathfrak{p}$ is then called *an absolute variety* when V is k'-irreducible over every extension k' of k, i. e. if the radical of the extended ideal $\mathfrak{p}k'[\xi]$ is prime for each field $k' \supset k$. This is certainly true if $\mathfrak{p}k'[\xi]$ is itself prime, which happens if, and only if, P being a generic point of V over k, $k(P)$ is a *regular* extension of k (that is to say: 1) k is algebraically closed in $k(P)$; 2) $k(P)$ is separably generated over k); we shall call such a field k *a field of definition* for V and write $k = \operatorname{def} V$. The necessary and sufficient condition for a variety V to be merely *absolutely irreducible* is, on the contrary, that every element of $R_k(V)$ should be either transcendental or p-radical over k, where p is the characteristic exponent of k {see SAMUEL [c], p. 32}: but then k may not be a definition field for V.

Every extension of a definition field k for V is still a definition field for V and the smallest of all these fields will be denoted by $k = \operatorname{def}^* V$ {see A. WEIL [a], p. 71}: if this coincides with the prime field of **K**, the variety V is called *universal* {this name has been proposed by ZARISKI: see [b], p. 79}.

Given an arbitrary but finite number of varieties, there is always a common definition field, over which **K** is of infinite degree of transcendence: then results related to one or more varieties may be enunciated, without explicitly stating the definition field.

Each k-variety is a finite union of absolute varieties over the algebraic closure $\bar{k}$ of k, which are associated to the minimal prime ideals of $\mathfrak{p}\bar{k}[\xi]$ in $\bar{k}[\xi]$. These so called *absolute components* are conjugate over k in the sense of the GALOIS theory. Henceforth we shall call an absolute variety simply a variety and denote by V/k or P/k respectively a variety with a definition field k or a generic point of a variety V over a definition field k for V. The affine or projective space over **K** will be simply denoted by $\mathbf{A}^n$ or $\mathbf{P}^n$ respectively.

The following terminology is sometimes useful: we shall say that "*almost all*" points of a variety V have a given property α if the points of V which do not have the property α belong to a proper algebraic subset of V. Clearly a property which is true for almost all the points of V is also true for any generic point P of V, but the converse is false as is proved by the property of being a generic point for V over a definition field k.

The "*almost all*" terminology has always been used by Italian geometers under the names of "*elemento generico*" and other analogous terms {see SEVERI [e], p.320, note (1)}: this was and may be allowable when there is no risk of confusion with the concept of a generic element in the abstract sense.

We give here a less trivial example, due to ZARISKI {[b], p. 80}, of a case where these two concepts, certainly never coincident, are also non equivalent: after having given, as in (I, 3), the pure algebraic notion of analytical irreducibility of a variety V

at a point, consider an algebraic curve W on V. Then V may be analytically reducible at almost any point of W and analytically irreducible at any generic point of the curve: over the complex field this may happen if V decomposes into several branches at almost any point of W, provided these are transitively permutable.

A first important application is related to the definition of the *order* of a variety V: let us call *linear* variety in $\mathbf{P}^n$, any variety which is represented by a system of linear equations. Then it is known that the totality of $(n-r)$-dimensional linear varieties has an algebraic GRASSMANN structure which is absolutely irreducible {see e. g. SEVERI [15], and HODGE-PEDOE [a], p. 309}; and one can show that almost all the $(n-r)$-dimensional linear varieties in $\mathbf{P}^n$ intersect an r-dimensional variety V/k of this space in a finite constant number of points which are algebraic and conjugate over k: this number is called the *order* of V {see SAMUEL [c], p. 38}.

The first half of this result, which is classical, also furnishes a means of defining dimension {see ZARISKI [a], p. 2}.

3. The local Rings

The Noetherian subring $\mathfrak{n}_k(P; V)$ of $R_k(V)$ composed of the fractions a/b, where b is not null at a point P of a variety V, is called *the local ring of V at P*. This is a local ring in the sense of KRULL, having only one maximal ideal $\mathfrak{m}_k(P; V)$ consisting of the fractions vanishing at P. If P is a variable point possessing the locus W over k, we shall also denote these rings by $\mathfrak{n}_k(W; V)$ and $\mathfrak{m}_k(W; V)$ respectively.

For notions in local algebra the reader may consult NORTHCOTT [a] and SAMUEL [b]. This chapter of modern algebra has been founded by KRULL [2] as a research in abstract algebra: its later history is intimately connected with algebraic geometry. We recall from SAMUEL [b] some basic definitions.

An $\mathfrak{m}$-adic ring is a topological Noetherian ring A, having as a fundamental system of neighbourhoods of 0 the powers of an ideal $\mathfrak{m}$, provided that this topology gives a HAUSDORFF space (this is equivalent to the condition that $1+\mathfrak{m}$ is free from zero-divisors {see SAMUEL [b], p. 5}). An $\mathfrak{m}$-adic ring A is called a ring of ZARISKI if in addition any ideal of A is closed {ZARISKI [12], SAMUEL [b], p. 6}. A semi-local ring, i. e. a Noetherian ring with only a finite number of maximal ideals $\mathfrak{p}_1, \mathfrak{p}_2, \ldots, \mathfrak{p}_h$ is a ring of ZARISKI, with $\mathfrak{m} = \mathfrak{p}_1 \cdot \mathfrak{p}_2 \cdots \mathfrak{p}_h$ {SAMUEL [b], p. 7}: these rings were introduced by CHEVALLEY [1]. A semi-local ring with a single maximal ideal is simply a local ring in the sense of KRULL [2].

The last type of ring is useful for the study of a variety in the neighbourhood of a point, the semi-local type, instead, presents itself in a natural manner when we need to study a variety in the neighbourhood of a finite number of points: finally the $\mathfrak{m}$-adic rings are capable of describing the behaviour of a variety along a subvariety.

The principal operations on these rings are: the topological completion, the construction of factor rings and quotient rings, and the finite extensions {SAMUEL [b], Ch. I}.

If the ideal $\mathfrak{n}$ defines the topology of a semi-local ring A — this is then called a defining ideal of A — SAMUEL has proved that the length of $\mathfrak{n}^n$, for large n, is a polynomial $P(n)$ in n: this generalises a well known theorem of HILBERT {see SAMUEL [b], p. 24}. The degree d of $P(n)$ is called, according to SAMUEL, the dimension of A. This is also the number of the smallest system which generates a defining ideal of A (a so-called system of parameters for A): this was the definition for dimension given by CHEVALLEY in [1]. Moreover if A is a local ring, d coincides with the greatest length of a chain of prime ideals in A, which gives the original definition of dimension for a local ring due to KRULL [2]. Further if the leading term of $P(n)$ is written in the form $e(d!)^{-1} n^d$, then e is an integer, called the multiplicity of $\mathfrak{n}$ in A.

The local rings which usually occur in algebraic geometry (geometric local rings) admit a rather formal characterisation due to CHEVALLEY [3], which may be seen in SAMUEL [b], Ch. III. ZARISKI has proved [15, 19] the fundamental result that any geometric ring A free from zero-divisor and integrally closed has a like ring for its completion: hence follow immediately the geometrical results at the end of (I, 5).

Finally we mention that the theory of complete local rings, due chiefly to COHEN [1], is known only for regular rings, i. e. the rings whose maximal ideal is generated by as many elements as their dimension, and with some additional conditions {see SAMUEL [b], Ch. V}.

If P' is a specialisation of P over k, the local ring at P' is a subring of the local ring at P, coinciding with it if and only if P is also a specialisation of P' over k (this specialisation is called by A. WEIL {see A. WEIL [a], p. 27} *a generic specialisation*). It follows that if $W' \subset W \subset V$ are all k-varieties, then $\mathfrak{n}_k(W; V) \subset \mathfrak{n}_k(W; V)$.

If V is a projective variety, the local ring is defined in an analogous way and is isomorphic to the affine local ring whenever P is not at infinity. The local ring $\mathfrak{n}(W;V)$ referred to $\mathbf{K}$ is called *absolute*. The local ring $\mathfrak{n}_k(W;V)$ can be completed in a local complete ring $\hat{\mathfrak{n}}_k(W;V)$ with respect to the topology defined by the powers of the maximal ideal. A k-variety V is called, following ZARISKI {see ZARISKI [15]}, *analytically irreducible* along a k-subvariety W if the completion $\hat{\mathfrak{n}}_k(W;V)$ is without zero-divisors.

4. Algebraic Product

If V and V' are affine varieties with $k = \operatorname{def}(V, V')$, respectively in $\mathbf{A}^n$ and $\mathbf{A}^{n'}$ (affine spaces over $\mathbf{K}$) we may consider the subset $V \times V'$ of the affine space $\mathbf{A}^{n+n'}$. There are always a generic point P/k of V and a generic point P'/k of V' independent over k (this is equivalent to linear disjointedness of $k(P)$ and $k(P')$ {see A. WEIL [a], p. 4 and p. 78}.

Then the ideals of V and V' generate a prime ideal $\mathfrak{J}$ in $k[X, X']$ and, $k(P, P')$ being regular over k, this is the prime ideal of a variety called *the algebraic product* $V \times V'$. Analogous definitions can be given for the projective case {see SAMUEL [c], p. 20}.

Clearly we have $\dim V \times V' = \dim V + \dim V'$. If, in the projective case, U is a subvariety of $V \times V'$ provided $k = \operatorname{def}(V, V', U)$, and Q/k be a generic point of U, one can always write $Q = P \times P'$ where $P \in V$, $P' \in V'$. Moreover P has a locus over k, which is a subvariety U^* of V: the points of U^* are all those points $\bar{P}$ of V for which there is some specialisation of Q of the form $\bar{P} \times \bar{P}'$ with $\bar{P}' \in V'$. This variety U^* is said *the set-theoretic projection* of U on V and is denoted by $U^* = \operatorname{pr}_V U$.

This operation preserves the notion of generic point. We always have $\dim U^* \leqq \dim U$ and, if $\mathfrak{J}$ is the ideal of U in $\mathbf{A}^{n+n'}$, U^* is defined by the intersection ideal $\mathfrak{J} \cap k[X]$ {see A. WEIL [a], pp. 80—81}.

Let us now consider three projective spaces $\mathbf{P}^n$, $\mathbf{P}^{n'}$, and $\mathbf{P}^N$ respectively with homogeneous coordinates (x), (x') and (y). The application $f: y_i = f_i(x, x')$ $(i = 0, \ldots, N)$, where the $f_i(x, x')$ are bihomogeneous polynomials of equal degrees a and a' respectively in (x) and (x'), is then a *rational* application of $\mathbf{P}^n \times \mathbf{P}^{n'}$ in $\mathbf{P}^N$, preserving the notions of k-variety, of generic point, and so on. If the $f_i(x, x')$ are general and $W = f(\mathbf{P}^n \times \mathbf{P}^{n'})$, the variety W, which is in birational correspondence without exceptions with $\mathbf{P}^n \times \mathbf{P}^{n'}$, is called a projective model of $\mathbf{P}^n \times \mathbf{P}^{n'}$: it is in particular the well known SEGRE variety {see, e. g., HODGE-PEDOE [a], p. 95} if the $f_i(x, x')$ are bilinear. If $\mathbf{P}^n = \mathbf{P}^{n'}$, the subset of $\mathbf{P}^n \times \mathbf{P}^n$ filled with the points of type (P, P) is called the *diagonal variety* Δ of $\mathbf{P}^n \times \mathbf{P}^n$: it is obviously birationally equivalent to $\mathbf{P}^n$.

The transform of Δ into W is parametrised by the monomials of degree $a + a'$ in the coordinates (x): if we consider only the distinct monomials we obtain a projection of Δ on a subspace of $\mathbf{P}^n$. This is called an *r-ple model of* $\mathbf{P}^n$ where $r = a + a'$; and is determined by r to within a birational biregular transformation. The transform into Δ of a k-set H of P^n is called an r-ple model of H and is denoted by $H^{(r)}$: the associated application transforms the sections of $H^{(r)}$ with the hyperplanes of its space into the sections of H with the forms of degree r in $\mathbf{P}^n$ {see SAMUEL [c], p. 22}.

5. Normal Varieties

An affine (or projective) k-variety is called, with ZARISKI {see ZARISKI [2, 4, 8]}, *affinely (or projectively) normal over* k or simply *k-normal* if the ring of the affine (or homogeneous) coordinates of V is *integrally closed.*

A ring R is said to be integrally closed if each element of its quotient field which is integral over R belongs to R. The integral closure of R is the set of elements of its quotient field which are integral over R.

It is clear that a k-variety projectively normal over k is affinely k-normal for each choice of the hyperplane at infinity. If the k-variety V is not k-normal in $\mathbf{A}^n$, one can construct a birational model of V which is such, simply by adding to $k[x]$ the elements, certainly finite in number, necessary for the integral closure of the ring $k[x]$. Then, if $k[y]$ is the integral closure of $k[x]$, the locus of (y) or of (x, y) over k is a k-normal model V^*: in the second case V is a projection of V^*.

In the projective case the integral closure $k' = k[y]$ of $k[x]$ is still an integral domain over k and one can suppose that the elements $y_0, y_1, \ldots, y_n$ are homogeneous in the x's. Then we may show that there is a positive integral number h, such that, if $(\omega_0, \omega_1, \ldots, \omega_\mu)$ is a linear base over k for the module of the homogeneous elements of k' of degree h, each element of k' homogeneous of degree $\varrho\, h$ is necessarily a form of degree ϱ in $\omega_0, \omega_1, \ldots, \omega_\mu$. Therefore the locus of the point (ω) is a k-variety V' and the ring $k[\omega]$ differs from its integral closure only in respect of the elements of $k(\omega)$ which are algebraic over k without being in k. If k is algebraically closed in $k(x)$ (and this is always the case for an absolute variety), V' is projectively k-normal, and it is precisely the normal model of the h-ple of V: if $h > 1$, V has no normal model in a proper sense.

A k-variety V projectively k-normal is characterised by the fact that *neither V nor any h-ple model is the projection of a k-variety of the same dimension and of the same degree properly contained in a higher space* {see SAMUEL [c], p. 27}.

If V/k is a projectively k-normal variety it is so also over each definition field $k' \subset k$ and over each definition field $k' \supset k$, provided k' be a simple transcendental extension of k or an algebraic extension of k, finite and separable. A variety V is absolutely normal, i. e. k-normal over every field k of definition for V, if it is k'-normal over a perfect field k': V is then called affinely or projectively *normal*. The product of two or more k-normal varieties is still a k-normal variety.

A point P of a k-variety V is called k-normal if $\mathfrak{n}_k(P; V)$ is integrally closed. The affine or projective k-normal varieties have only k-normal points, while the converse is only true in the former case. Therefore a projective k-variety will be called *locally k-normal* when all its points are k-normal: for this it is necessary and sufficient that the integral closure of the ring of the homogeneous coordinates should have as conductor *an improper ideal*, i. e. one containing a power of the ideal $(x_0, x_1, \ldots, x_n)$.

The conductor of an integral ring R in its integral closure R' is the largest ideal of R which is also an ideal in R' {see SAMUEL [a], Ch. II}.

The following is an important result due to ZARISKI {see ZARISKI [15, 19]}: *every k-variety which is k-normal along a k-subvariety W, is analytically irreducible and analytically normal along W:* i. e. $\hat{\mathfrak{n}}_k(W; V)$ is an integral ring integrally closed {for the proofs, besides the original works of ZARISKI, see SAMUEL [c], p. 66}. We add that locally normal varieties over any definition field will be called simply *normal varieties* (ZARISKI: *absolutely normal*).

6. Birational Transformations

Let F be a subvariety of the product $V \times V'$ of the projective varieties V, V' and also $k = \operatorname{def}(V, V', F)$, $V = \operatorname{pr}_V F$ and $\dim F = \dim V$. We shall call *index* of the projection of F on V the degree of the algebraic extension $k(x, y)$ over $k(x)$, where (x, y) denotes a generic point of F

over k, and (x) a generic point of V over k. The index furnishes the number of the points of F (defined over $\overline{k(x)}$ and conjugate over $k(x)$) which project into the point (x).

If $k(x, y) = k(x)$, a point of V determines one and only one point of V' and, if $V' = \mathrm{pr}_{V'} F$, the variety F is called the graph of a ***rational correspondence*** of V on V'. The coordinates (y) of a point on V' can then be written in the form $t\, y_i = f_i(x)$, where the f_i are forms of equal degrees over k and where (x) is a convenient point on V, t being a non-zero factor of proportionality. It is clear that if one specialises $x \to \bar{x}$ with respect to k, the point (y) has a specialisation over k, which is univoquely determined provided the $f_i(\bar{x})$ do not all vanish: if this is the case the transformation is said to be ***regular*** in (x); it is seen that the condition of regularity at a point P coincides with the condition for the quotients of the $f_i(x)$ to be contained in the local ring $\mathfrak{n}_k(P; V)$: then $\mathfrak{n}_k(P'; V') \subseteq \mathfrak{n}_k(P; V)$ where P and P' are corresponding points on V and V' respectively. If, besides, $k(x, y) = k(y)$ and then $k(x) = k(y)$, F is called the graph of a ***birational*** correspondence T. T is biregular at a pair of corresponding points (P, P') if T is regular at P and T^{-1} at P': then $\mathfrak{n}_k(P; V) = \mathfrak{n}_k(P'; V')$.

Among the various properties of a variety V we shall distinguish the following: (a) *Absolute (birational) properties*, which are invariant for all B. T. and are substantially properties of the field of the rational functions on V; (b) *Relative (birational) properties*, which are invariant only for everywhere biregular B. T.; (c) *Local properties*, which are properties of the local ring $\mathfrak{n}(P; V)$, and, more particularly, "*local analytic properties*", which refer to $\hat{\mathfrak{n}}(P; V)$, and may be expressed as properties of rings of formal power series and of finite extensions of such rings {see A. WEIL [a], p. 252}.

In particular a variety V/k will be called k-unirational if it is a k-rational transform of a $\mathbf{P}^r$, and birational if it is birationally equivalent over k to a $\mathbf{P}^r$: these notions are certainly distinct, as ROTH has proved in [10].

This is a sufficient scheme for the study of birational transformations as long as one considers only biregular pairs of corresponding points: otherwise it is convenient to go deeper into the situation. Following ZARISKI {see ZARISKI [8]} one may call corresponding in T two subvarieties W and W', respectively of V and V', if and only if there is a valuation v of the common field of the rational functions over V and V' which has as centres respectively W and W': evidently this definition extends the previous one, coinciding with it if W and W' are biregularly associated.

We recall here briefly some general definitions in valuation theory, quoting for this theory ZARISKI [8], pp. 496—504, SAMUEL [a], Ch. I and HODGE-PEDOE [a], Ch. XVII: the first general treatment is by KRULL {see [1] and [a]}.

A valuation v of the field $R_k(V)$, where V is a k-variety, is a homomorphic mapping of the multiplicative group of $R_k(V)$, zero excluded, upon an ordered additive Abelian group G, such that: 1. $v(f\,g) = v(f) + v(g)$; 2. $v(f + g) \geq \min(v(f), v(g))$; 3. $v(f) = 0$, for some $f \in R_k(V)$; 4. $v(c) = 0$, for all $c \in k$ with $c \neq 0$. Finally we put $v(0) = +\infty$.

When V is a k-curve, every valuation arises from a branch of V, and then $v(f)$ is the order of the function f at the branch, and is an integer: we have $v(f) \gtrless 0$ if the order of the branch is a zero or a pole of f, while if $v(f) = 0$ then f is definite and non-vanishing at the centre of the branch.

If we define as valuation ring R_v of v the set of all functions in $R_k(V)$ which have non-negative maps in G by v, it is evident that, in this case, the function-theoretic values of the elements of $R_k(V)$ are the cosets of R_v mod $\mathfrak{p}$ where $\mathfrak{p}$ is the divisorless ideal consisting of all $f \in V$, such that $v(f) > 0$. Consequently $R_v/\mathfrak{p}$ is a field, which is called the residue field of v: the ring R_v is a local ring whose maximal prime ideal is $\mathfrak{p}$. This consideration is independent of the dimension of V: if $\dim V = 1$ this field is k, otherwise it may be a transcendental extension of k, whose degree is called the dimension of v. We have always $\dim v \leqq \dim V - 1$. If $\dim v = 0$, v is called an algebraic place of $R_k(V)$. If (x) is a generic non-homogeneous point of V in $\mathbf{P}^n$, we can assume $v(x_i) \geqq 0$ by a suitable choice of the hyperplane at infinity. Then the elements $f \in k[x]$ for which the residues in $R_v/\mathfrak{p}$ have value zero form a prime ideal $\mathfrak{J}$ in $k[x]$, and $W = W_k(\mathfrak{J})$ is a subvariety of V, which is called the centre of v on V. We have $\dim W \leqq \dim v$, and the centre of an algebraic place is a point of V. The rank of the group G is called the rank of v, and v is called discrete or not, Archimedean or not, corresponding to the analogous characters of G. If v is $(r-1)$-dimensional, it is necessarily discrete of rank one.

We point out that valuation theory supplies a fine tool for treating the various modes of approach to a point or to a subvariety of a variety: it replaces the usual classical methods based on continuity when we pass to abstract fields {see ZARISKI [b], p. 87}.

Observing that V and a locally k-normal model V^* are birationally related, having isomorphic function fields, one can simplify the situation by replacing V and V' with two such models. Naturally this implies a previous study of the birational correspondence between V and V^*: the simple answer is that V can always be considered as a general projection of V^* This projection p preserves the dimension and is everywhere regular on V: its inverse p^{-1} associates with any subvariety W of V a finite number $\nu \geqq 1$ of subvarieties of V which all project onto W. This number is greater than unity for (and only for) the subvarieties of V defined by the conductor of the coordinates ring of V with respect to its integral closure: these subvarieties are a subset of the multiple subset of V {for proofs, see SAMUEL [c], p. 65}. Therefore we can at once suppose V and V' to be locally k-normal.

Now let W/k be a subvariety of V. We shall then distinguish the following cases: a) W determines only one corresponding variety $W' \subset V'$ in T with $\dim W = \dim W'$. Moreover either $\mathfrak{n}_k(W; V) = \mathfrak{n}_k(W'; V')$ and then T is regular along W and W', or necessarily $\mathfrak{n}_k(W; V) \supset \mathfrak{n}_k(W'; V')$ and then T is called *irregular* along W; b) W determines several corresponding varieties in V' and hence an infinity because V is k-normal along W: W is said to be *fundamental* for T, and one finds that $\mathfrak{n}_k(W; V) \supsetneqq \mathfrak{n}_k(W': V')$. Moreover $\dim W \leqq \dim V - 2$, because otherwise there would be only one valuation having W as centre on V {see ZARISKI [8], p. 514}.

It is sometimes useful to denote by $F[W]$ the transform of W as a variety and $F\{W\}$ the transform of W as a set of points algebraic over k. A fundamental theorem of ZARISKI {see ZARISKI [8], p. 522 and [16]} on birational transformations is the following: *if F is a rational correspondence of V/k into V'/k and if P is a k-normal point of V such that $F\{P\}$ is zero dimensional, then $F\{P\}$ is a single point.* This theorem {see ZARISKI [16]} follows easily from ZARISKI's theorem on analytic irreducibility given in (I, 5) and it is also a particular case of a more general theorem on connectedness due to the same author {see (VI, 4)}.

7. Simple Points

Let W be a k-subvariety of the k-variety V. We shall say that W is k-simple over V if the local ring $\mathfrak{n}_k(W;V)$ is regular: i. e. if $\mathfrak{m}_k(W;V)$ is generated by as many elements as its dimension and therefore by $d = \dim V - \dim W$ elements: these are called the *uniformising parameters* of V along W. A n. a. s. c. for this is that $\mathfrak{m}/\mathfrak{m}^2$, which can be regarded as a vector space over $\mathfrak{n}/\mathfrak{m}$, has dimension equal to d: if W is a rational point over k this vector space is called *the tangent space* of ZARISKI of V at the point P {see SAMUEL [c], p. 68}. It is isomorphic to *the tangent cone* at P if and only if P is k-simple {see SAMUEL [c], p. 71}. This cone is the set of the limit secants of V through P. It may be defined by a graded ring as in IGUSA [5]: see also NORTHCOTT [7].

From the fact that every regular ring is integrally closed {see SAMUEL [b], p. 29} it follows that V is k-normal along W if this is k-simple on V: the converse is true only if $\dim W = \dim V - 1$. The biregular birational transformations over k induce an isomorphism between $\mathfrak{n}_k(W;V)$ and the transformed local ring $\mathfrak{n}_k(W';V')$ {see (I, 6)}, and they thereby transform k-simple points into k-simple points.

ZARISKI has been able to extend to k-simplicity the classical JACOBIAN criterion, arriving at a matrix related to a system of equations for V, which vanishes only at the points of V which are not k-simple. From the ZARISKI criterion the set of the non k-simple points on V is an algebraic subset $\mathbf{S}$ of V: if V is k-normal we find that $\dim \mathbf{S} \leqq \dim V - 2$.

A point P of a variety V is called *simple* if it is k-simple over any definition field for V: for this is sufficient the k-simplicity over a perfect field k. A k-simple point of a variety V remains so over any separable extension of k. We also recall that, if the points P, P' are simple points of the varieties V, V', the same is true of the point $P \times P'$ of $V \times V'$: this is not always so in the relative case {for the proofs see SAMUEL [c], Ch. II, 4}.

8. The Intersection Multiplicity

We shall restrict our definition to a particular but fundamental instance. Let V and W be two varieties in $\mathbf{A}^n$ defined over k. The algebraic set over k $V \cap W$, if not empty, has all the absolute components of dimension equal to or higher than $d = \dim V + \dim W - n$: moreover, in the projective case, it is never empty.

This is rather difficult to prove: classically it was known by analytical methods. The first algebraic proof is in VAN DER WAERDEN [5, I, XII]. It may be connected with the so-called *"going down theorem"* of COHEN-SEIDENBERG [1], as in SAMUEL [c], pp. 22—23. See also A. WEIL [a], p. 86.

A component C of the set $V \cap W$ is called *proper* or *improper* according as $\dim C = d$ or $\dim C > d$: the number $e = \dim C - d$ is the excess of C. Now let us consider a copy $\mathbf{A}'$ of $\mathbf{A}$ (i. e. a biregular birational model of $\mathbf{A}$) and a copy W' of W in $\mathbf{A}'$. Let Δ be the diagonal of $\mathbf{A} \times \mathbf{A}'$ and C^Δ the subvariety of Δ projecting on $\mathbf{A}$ into C, where C is a component of $V \cap W$: then C^Δ is a component of $\Delta \cap (V \times W')$. The field $R(V \times W')$ contains both the local ring $\mathfrak{n}$ of C^Δ on $V \times W'$ and the differences $x_i - x_i'$ induced by the equations of Δ: the latter generate in $\mathfrak{n}$ an ideal $\mathfrak{q}$ wich is primary for the maximal ideal $\mathfrak{m}$, because C^Δ is a component of $\Delta \cap (V \times W')$. The multiplicity $e(\mathfrak{q})$ of $\mathfrak{q}$, which is independent of k when we prefer to take $\mathfrak{n}$ and $\mathfrak{q}$ relatively to a field $k = \operatorname{def}(V, W, C)$, is denoted by $i(C; V; W)$ and is called, with CHEVALLEY-SAMUEL {see CHEVALLEY [3], SAMUEL [1] and SAMUEL [c], p. 77}, *the intersection multiplicity of V and W along C*: it is invariant for birational transformations which are biregular along C. If C is a proper component, we have $\dim \mathfrak{n} = n$ and the ideal $\mathfrak{q}$ is generated by the parameters $x_i - x_i'$: this is the case considered by CHEVALLEY.

It is possible to show that then $i(C; V \cdot W) = i(C^\Delta; (V \times W') \cdot \Delta)$: this is a formula which reduces the definition of the intersection multiplicity to the case where one of the two varieties is linear {see SAMUEL [c], p. 83}.

Now let P be a proper component of $V \cap L$, where $\dim V = r$, $\dim L = n - r$ and where L is a linear variety in $\mathbf{A}^n$. Let L' be another such variety, but generic over a definition field k for V. One can now show that:

1. $V \cap L'$ is a point set algebraic over k and, more precisely, a complete system of conjugate points over $k(L')$.

2. In every specialisation of $V \cap L'$ over the $L' \to L$ with respect to k, the point P appears $i(P; V \cdot L)$ times.

3. If $k = \operatorname{def} L$, then P is algebraic over k and $i(P; V \cdot L)$ is a multiple of the order of inseparability of P over k.

If one proves the property (2) directly, one can assume the number so defined as a definition of the intersection multiplicity {a fact which is equivalent to ZARISKI's main theorem on B. T. of (I, 6), as A. WEIL has proved: see A. WEIL [a], p. 248, SAMUEL [c], p. 87 and NORTHCOTT [4]}: this is the A. WEIL definition {see A. WEIL [a], p. 116}, which can afterwards be extended to the more general case considered above by taking intersections with convenient linear spaces and using the diagonal variety, at least in the case of a proper component.

There are some formal properties which, if they hold, characterise an intersection multiplicity definition {see A. WEIL [a], p. 279, SAMUEL [c], p. 89}. These are, besides the symbol $i(P; V \cdot L)$ to be defined in the case above considered:

1) *The so called formulae of associativity and projection* {see SAMUEL [c], pp. 81, 84}.

2) *The criterion of multiplicity unity for proper components* {see SAMUEL [c], p. 79}.

We recall that the latter says that $i(C; V \cdot W) = 1$ if and only if: 1) *The tangent spaces to V and W at the generic point of C intersect properly.* 2) *C is simple on V and W. If C is proper this derives from* 1.

The WEIL definition assumes in the classical case (i. e. over the complex field), a well known analytic meaning: i. e. *the number of points of $V \cdot L'$ which approach to P when L' is specialised to L;* and one can then replace, as is commonly done, specialisation by the limiting process, thus arriving at the classical definition of intersection multiplicity due to SEVERI {see SEVERI [21]; the relations between SEVERI's definition and WEIL's definition have been investigated by SEVERI in [33]; we remark here that some idea of WEIL's theory is already in B. SEGRE [8]}.

We shall not describe how one can now pass to the more general case where the intersection is considered in a non-linear ambient U: a problem which can be reduced to the preceding {see A. WEIL [a], Ch. VI and SAMUEL [c], p. 98}. We merely remark that one is still able to define a number, denoted by $i_U(C; V \cdot W)$, where C is a component, simple on U, of the set $V \cap W$, V and W being subvarieties of the variety U.

If U is a subvariety of V, immersed in **A** or **P**, we call, with SAMUEL {(see [c], p. 94)}, multiplicity of U on V and denote by $m(U; V)$ the multiplicity of the maximal ideal of the local ring $\mathfrak{n}(U; V)$ {see (I, 3)}. We have then the following two important results: 1. $m(U; V) = i(U; U_1 \cdot V) = i(U; U \cdot V)$, where U_1 is a generic cone projecting U and of dimension $n + \dim U - \dim V$; 2. If P is any point of U, then $m(P; V) \geqq m(U; V)$, and $m(P; V) = m(U; V)$ if P is almost any point on U.

The associative property remains true only if C is proper: moreover if C is simple on U, SAMUEL has proved {see SAMUEL [c], p. 100} that $i_U(C; V \cdot W) = i(C; V \cdot W)$, the latter symbol being considered in the ambient space of U.

Obviously, in the member on the right, C is an improper component. For such a component C SAMUEL has proved {see [c], p. 85} the following important geometrical characterisation which had been suggested by SEVERI: $i(C; V \cdot W) = i(C; V' \cdot W')$, where V' and W' are almost any cones projecting respectively V and W and of such dimensions that C is proper in $V' \cap W'$.

This is evidently a *local* theory and in fact leads to the greatest generality in the results {see the observations in A. WEIL [a], p. 197}:

in the applications, however, it is more useful to have a *global* intersection theory, which, under the severe restriction that the ambient be *non singular*, has been systematically expounded by A. WEIL under the name of calculus of cycles.

9. The Calculus of Cycles

{See for all this section A. WEIL [a], pp. 197—214.}

A *(homogeneous) r-cycle* X^r on a variety V, or V-cycle of dimension r, is a linear combination with integral coefficients of some subvarieties V_i of V, all of dimension r: i. e. $X^r = \Sigma_i n_i V_i$. The varieties V_i are called the *components* of X and their algebraic union-set is called the *support* $||X||$ of X.

A cycle X is *null* if and only if $n_i = 0$ for each V_i. If a_i is the degree of V_i, we shall take as *degree* of X the number $\Sigma_i n_i a_i$. Moreover the cycle X is called *positive* ($X > 0$) or *effective* if and only if there is some $n_i > 0$ and no n_i is negative. A cycle neither positive nor null is called *virtual*. By these definitions one gives the structure of an Abelian group $\mathbf{G}^r(V)$ to the set of all r-dimensional cycles on V.

By abandoning the homogeneity condition with respect to the dimension of the components, one obtains similarly an Abelian group $\Gamma(V)$ of all the cycles on V. This group $\Gamma(V)$ is graded by the formula $\Gamma(V) = \Sigma_r \mathbf{G}^r(V)$, $(r = 0, \ldots, \dim V)$.

We shall now define two important operations on cycles: these are the *product* and the *algebraic projection*. The former is immediately defined by taking $X \times Y = \Sigma_{i,j} n_i m_j (V_i \times W_j)$, where $X = \Sigma_i n_i V_i$ is a V-cycle and $Y = \Sigma_j m_j W_j$ a W-cycle, V and W being two varieties. The latter will be defined only for an r-cycle with simple components over the variety V. Let $U \times V$ be the product of two varieties U and V and let A be at first a simple subvariety of $U \times V$. Then $A' = \mathrm{pr}_U A$ is also simple over U and we shall define as the algebraic projection of the variety A over V the U-cycle $\mathrm{pr}^* A = [A : A'] \cdot A'$ or $\mathrm{pr}^*_U A = 0$ according to whether A' is or is not of the same dimension as A, where $[A : A']$ denotes the index of A over A' {see (I, 6)}. For an r-cycle $X = \Sigma_i n_i A_i$ on $U \times V$ we shall then take $\mathrm{pr}_U X = \Sigma_i n_i \mathrm{pr}_U A_i$ and $\mathrm{pr}^*_U X = \Sigma_i n_i \mathrm{pr}^*_U A_i$. There is obviously a homomorphism of $\mathbf{G}^r(U \times V)$ into $\mathbf{G}^r(U)$.

A V-cycle $X = \Sigma_i n_i V_i$ is called *rational* over a field k of definition for V, if: 1. Every V_i is algebraic over k; 2. X is invariant for every automorphism of $\bar{k}$ over k; 3. n_i is a multiple of the order of inseparability of V_i over k, i. e. of $k(V_i)$ over k. It is obvious that such cycles form a subgroup of $\Gamma(V)$ and that a cycle rational over k is still so over every field $k' \supset k$.

The V-cycle X will be called *rational prime*, or, more briefly, *prime* if it is of the type $X = \Sigma_i p^f \cdot V_i$, the sum being extended only to all the subvarieties of V conjugate to the subvariety V_1 simple on V and p^f being the order of inseparability of V_1 over k. These characters are invariant with respect to the product and algebraic projection. A subvariety A simple on V is clearly a rational cycle over k if and only if $k = \operatorname{def}(V, A)$ and then A is a prime cycle.

This strictly algebraic notion of "*cycle*" on V must be carefully distinguished from the topological notion of "*cycle*" on the RIEMANN variety of V {see (IX, 2)}, especially when these concepts happen to be considered together {see e. g. (X, 6)}: generally the right sense will appear clearly from the context.

We shall say that two V-cycles X^r and Y^s *intersect properly* on V if and only if the components of the two cycles have as proper intersection components all those which are simple on V. Then, if $X = \Sigma_i n_i A_i$ and $Y = \Sigma_j m_j B_j$, we shall write $X \cdot Y = \Sigma_{i,j} n_i m_j (A_i \cdot B_j)$, where $A_i \cdot B_j = \Sigma_C i_V(C; A_i \cdot B_j) \cdot C$, the sum being extended to all the components C of $A_i \cap B_j$ simple over V: this operation is called the product-intersection of X and Y. When defined, it is commutative and coherent to the partial ordering in $\Gamma(V)$, and the variety V is its unity. Moreover it is distributive with respect to the sum and is also associative.

The following two relations connect this operation with the others above defined. The former says that: $(X \times Y) \cdot (X' \times Y') = (X \cdot X') \times \times (Y \cdot Y')$, where all the symbols denote V-cycles and where it is supposed that the product-intersections which appear are well defined; for this it is sufficient that they be defined, for example, on the right. The latter says: $\operatorname{pr}^*_U(X \cdot (Y \times V)) = (\operatorname{pr}^*_U X) \cdot Y$; the hypotheses are that V be projective and not singular, X be an $(U \times V)$-cycle and Y a U-cycle and that $X \cdot (Y \times V)$ be defined on $U \times V$.

By means of SAMUEL's extended theory of intersection we can associate with two (or more) cycles A and B, also non-homogeneous, in **A** or **P**, a cycle $A \perp B$, defined as $A \cdot B$ in the text but by taking also into account the improper components: see SAMUEL [c], p. 92.

We shall end this short account by referring to an important theorem of A. WEIL, which is the background to the theory of the algebraic systems {see A. WEIL [a], p. 204}.

(i) *Let U and V be two varieties with $n = \dim U$ and $k = \operatorname{def}(V, U)$ and let P be a generic point of U over k. Then:*

1. *If Q is any simple point on V, X the prime cycle over k on $U \times V$ determined by the generic point $P \times Q$ with respect to k, and $X(P)$ the V-cycle prime rational over $k(P)$ with the generic point Q over $k(P)$, then the cycle $X \cdot (P \times V)$ is defined on $U \times V$, and we have $X \cdot (P \times V) = P \times X(P)$.*

2. *If X is any rational $(U \times V)$-cycle over k, of dimension not less than n, then the cycle $X \cdot (P \times V)$ is defined and is of the fvom $X \cdot (P \times V) = P \times X(P)$, where $X(P)$ is a rational V-cycle over $k(P)$; moreover the cycle defined by this relation is 0 if and only if the projections on U of all the components of X are of dimension less than n.*

3. *Conversely, if $X(P)$ is any rational V-cycle over $k(P)$, there is one and only one rational $(U \times V)$-cycle X over k, such that $X \cdot (P \times V) = P \times X(P)$ and that every component of X has on U the projection U; if, moreover, $X(P)$ is a positive cycle, then the cycle X determined by these conditions is positive.*

The cycle X which appears in this theorem will be denoted by $X = \mathfrak{L}_k(X(P) \times P)$.

We give here some historical notices on the content of this chapter.

Algebraic sets, in the realm of classical geometry, were called varieties and our varieties were then characterised as irreducible {see e. g. the use in Bertini [a] and Zariski [a]}: these concepts are explicitly stated already in Kronecker's work [1]. An algebraic k-set is the same as the interference variety over k of Severi [38] or the bunch of varieties normally algebraic over k of A. Weil [a]: such a term has been used in the absolute case $k = \mathbf{K}$ by Chevalley [3] and in our relative sense it seems to be due to Samuel [c]. For a comparison of different terminologies used in algebraic geometry see Samuel [c], p. 129.

The notions of specialisation and generic point constitute a characteristic tool of abstract algebraic geometry: they are due to earlier work of E. Noether [1] and van der Waerden [1]. Specialisation arguments partially replace continuity arguments of classical algebraic geometry, which may be totally replaced by the use of general valuation theory, whose definitions are due to Krull {see [1] and [a]} and whose application to algebraic geometry is chiefly due to Zariski.

Specialisation and continuity both succeed in defining the value of an algebraic function f, given at a generic point of a variety V, also where f is not explicitly defined. This is the fundamental significance of van der Waerden's theorem concerning the extension of specialisations {see Zariski [b], p. 81}. Interesting considerations of the methodological convenience or otherwise of restricting the concept of generic point, a notion which may seem exceedingly broad, are given by Zariski in [b], p. 79.

The concepts of universal domain, absolute variety and so on are characteristic features of Weil's work. Product-sets were early defined in algebraic geometry by C. Segre and systematically used by Severi in the theory of correspondences; according to Severi, "algebraic correspondence" between two varieties V and V' is any subset of their product: see e. g. Severi [15]. This viewpoint is at present generally accepted and has been followed in particular by A. Weil in [a].

Set-theoretic projection is, instead, a recent tool in algebraic geometry: it is the abstract generalisation of projection in affine or projective spaces, belonging to the classical geometry, and has been systematically expounded by Chevalley [3] and A. Weil [a] as a method of giving the abstract quotient space: it is especially useful for giving formal conciseness to the results of intersection theory. Rational applications, Segre varieties and r-ple models are typical of the Italian school: see Bertini [a].

Normal varieties and the normalisation process are due entirely to Zariski: his first aim was to remove singular forms from a variety, preliminary to the clearing of singularities {see our Ch. II}. Successively they revealed themselves as outstanding tools in B. T., in the study of analytical irreducibility and normality, all topics whose invention and investigation is due to Zariski, as we have already pointed out in the text. Other significant applications are the principle of degeneration in (VI, 4) and the theory of linear systems in Ch. III.

Intersection theory, as we have seen, has been satisfactorily expounded by the classical methods by Severi in [21] {see also [32] and [c]}. Abstract intersection theories are due to VAN DER WAERDEN in [2] and [5, XII], but are restricted to the case where all the components of the intersection-set are proper: such a theory, with some improvements, may be studied in HODGE-PEDOE [a_2], Ch. XII. The theorem on the multiplicity of specialisations in (I, 8), keystone of WEIL's theory, is a strong extension of a case early considered by VAN DER WAERDEN {see A. WEIL [a], p. 247}.

Later we have the theories of CHEVALLEY [3] and A. WEIL [a] {see also B. SEGRE [8]}: we remark that, apart from the proof of uniqueness alluded to in (I, 8), direct proofs of the coincidence of both these definitions have been given by IGUSA [1] and SAMUEL [1]. Successive work of SAMUEL in local algebra has furnished the extension of CHEVALLEY's definition to irregular components, with many other geometrical complements.

SAMUEL has also investigated the possibility of extending intersection theory to components which are not simple on the ambient variety. Moreover both CHEVALLEY's and SAMUEL's theories can be applied to intersection multiplicity for algebroid varieties, which are defined by prime ideals in the ring of formal power series over a field k: all these subjects may be found in SAMUEL [c]. Other refined theories of intersection have been developed by BARSOTTI in [1, 2, 3], who in the last quoted work has also given an abstract exposition of SEVERI's methods regarding intersection between equivalence classes, for which see (VI, 9).

Cycles are called "virtual varieties" by the Italian school, and with this name they have been introduced into algebraic geometry by SEVERI in [8]: SEVERI has also given a notion of intersection-multiplicity of forms at a point of a projective space and thus, by improving an older method of KRONECKER, arrives at the definition of the cycle of intersection of the given forms: on these subjects see SEVERI [30, 33] and PERRON [1]; the abstract viewpoint in elimination theory may be found in VAN DER WAERDEN [b], Ch. IV.

We recall that sometimes the concept of a "base variety" may be useful: such varieties are simply identified with the ideals of polynomials; for contributions see SEVERI [38] and GAETA [a]. For curves and surfaces this concept may, in its geometrical content, be made precise by the notion of infinitely near points, thus generalising previous studies of ENRIQUES: this has been accomplished by ZARISKI in [1] (over algebraically closed fields of characteristic zero), by MUHLY in [1] (over arbitrary fields of characteristic zero) and by SEIDENBERG in [1] (over arbitrary fields).

We shall not consider the interesting and difficult problem of the classification of varieties with respect to their representation by ideals, restricting ourselves to recalling the notion of variety of the first species due to DUBREIL [a] and its identification with the projectively normal varieties of ZARISKI {DUBREIL [a] and [1]}: for further informations we refer the reader to the complete account of GAETA [a], to the outstanding works [1, 2, 4] of this author on the subject, and to DUBREIL-JACOTIN [1].

Finally we recall that if X and Y are two cycles in $\mathbf{P}^n$, such that $X \cdot Y$ is defined, then degree $(X \cdot Y) =$ degree $X \cdot$ degree Y: this is the famous theorem of BÉZOUT. See for proofs and extensions: SEVERI [33], SAMUEL [c], p. 107, VAN DER WAERDEN [5, I, XIV], NORTHCOTT [5], IVASAWA [1].

II. The Resolution of Singularities

One of the most important problems in algebraic geometry is the search for a birational model, free from singular points, of a variety V: apart from the obvious intrinsic theoretical interest a complete solution with an affirmative answer would allow us to consider, without any restriction, birational properties on a non singular model. As is well known, we should then be in the most favourable situation, chiefly because the intersection theory in its global form is at present restricted to such models and it does not seem that it can be extended without some radical change in its character.

Unfortunately this problem is very difficult: today an affirmative answer can be given only for varieties of dimension not higher than three and in the case of characteristic zero. These results are classical for a curve or a surface over the complex field: for the rest they are entirely due to ZARISKI. This chapter is devoted to a short account of ZARISKI's researches in the three-dimensional case {see ZARISKI [9]}.

1. The Local Uniformisation Theorem

A useful method of investigating the nature of a singular point of a variety V/k is to transform the variety V by a birational transformation T/k having at P *an isolated fundamental k-point:* one can then study the transform $T\{P\}$, if one has previously assured oneself that *no* subvariety of $T\{P\}$ is fundamental for the inverse T^{-1}.

This is more satisfactory if $T\{P\}$ is *not singular* for some T: we do not know if this favourable situation is always realisable; nevertheless ZARISKI has proved the following partial result, which is the so called *local uniformisation theorem* {see ZARISKI [5]}; where, and in all this chapter, k is a field of *characteristic zero:*

(i) *Let V be a k-variety in a projective space and let $R_k(V)$ be the rational function field over k. If v is an s-dimensional valuation* {see (I, 8)} *of $R_k(V)$ having as centre on V the k-subvariety C, then there is a birational transform V' of V such that the centre of v on V' is an s-dimensional simple k-variety C' with* $\mathfrak{n}_k(C;V) \subsetneqq \mathfrak{n}_k(C';V')$.

The very complex proof of the theorem consists, after reduction to the case of a form by projection, of a direct and effective construction of a succession of Cremona transformations (i. e. transformations which are birational in all the ambient space of the variety).

This yields the desired result in the case of zero-dimensional valuations of rank one: the construction is realised by a sort of induction, whose stages are marked by the PERRON algorithm. After this most difficult part the proof is completed by induction first on the rank and then on the dimension of v: it is here that one uses the Cremonian character of the above transformations {see for this proof the exposition in HODGE-PEDOE [a_3], pp. 290—314}.

Taking the particular case in which k is the complex field, let v be a valuation of $R(V)$ with centre at an isolated singular point P of V and let V' be the model constructed above: then v has its centre at a simple point P' of V' and $\mathfrak{n}(P;V) \subsetneqq \mathfrak{n}(P';V')$. If $t_1, t_2, \ldots, t_r$ are uniformising parameters in P' (I, 7), the homogeneous coordinates of P are given by power series in $t_1, t_2, \ldots, t_r$. These series converge in some neighbourhood $U(v)$ of P' on V'.

Let us now suppose that there is a finite number $v_1, v_2, \ldots, v_h$ of valuations of $R(V)$ with centre at P, such that these and any other of the same type have a simple centre on some $U(v_i)$. The set of series associated with all these valuations is then obviously capable of representing the complete neighbourhood of the singular point P on V: moreover this is a holomorphic representation. This example suggests the following theorem as an essential complement to the local uniformisation theorem:

(ii) *Let V be a k-variety with function field $R_k(V)$: there exists a finite system of k-varieties, each birationally related to V, such that every valuation of $R_k(V)$ has a simple centre on some of them.*

Such a finite system of birational models is called by ZARISKI a *(finite) resolving system,* while the local uniformisation theorem merely secures the existence of an *infinite* resolving system. The proof of the last theorem is achieved by introducing, in the set $\mathfrak{M}$ of all zero-dimensional valuations of $R_k(V)$, the following topology, called *the* ZARISKI *topology:* let us take as base for the closed sets in $\mathfrak{M}$, the sets of elements in $\mathfrak{M}$ associated with the k-points of any k-subvariety of a projective model V_α of V, when V_α varies in an arbitrary set. Then $\mathfrak{M}$ becomes a *compact space* endowed with *a finite covering:* this topological space is called *the abstract* RIEMANN *variety* {see ZARISKI [11]} *of* V. The theorem can also be proved by purely algebraic considerations {see HODGE-PEDOE [a_3], p. 315}.

2. Monoidal Transformations

On these transformations, similar to but not coinciding with the classical ones of the same name, the resolving process almost exclusively depends: to this end they have been throughly investigated by ZARISKI. We shall give here their most salient features {for proofs see ZARISKI [8], p. 532 or HODGE-PEDOE [a_3], p. 244}.

Let V and V' be two projective locally k-normal k-varieties, and T a birational transformation over k between them, without fundamental points on V'. After a substitution (if necessary) for V' of the graph of T in the SEGRE variety $V \times V'$, the equations of T can be taken in the form: $\varrho\, y_{ij} = x_i\, \varphi_j(x)$, where (x) is a generic point of V with respect to k, the φ_j are forms of equal degree, (y) is a generic point of V' with respect to k and ϱ is a factor of proportionality. Then it is easily shown that the fundamental locus of T on V coincides with the base set of the linear system $\Sigma_j\, \lambda_j\, \varphi_j(x) = 0$ on V {see (III, 6)}, if we exclude the fixed $(r-1)$-dimensional components, where $r = \dim V$. Such a fundamental locus cannot have a dimension greater than $r - 2$. If its ideal (φ) is *prime,* to within an irrelevant primary component, T is called (by ZARISKI) *a monoidal transformation.*

The k-variety W defined by (φ) is then called the *centre* of T, and when W is a *k-point,* T is called *quadratic.* A monoidal transformation (shortly: M. T.) dilates the centre W into a *k-hypersurface* of V' changing *simple* k-points for V and W *into simple* k-points for W': this is the principal reason for using M. T. in a resolving process. In fact a M. T. generally simplifies a singular k-point of V on W *provided the same point be simple on* W*:* therefore, before applying a M. T., we must desingularize its centre.

The M. T. have been called *dilatations* by B. SEGRE {see B. SEGRE [11, 13]} a name which seems preferable since it removes any confusion with the classical case: the reader may note that, in the plane, the classical quadratic transformation for example, is not monoidal, having three fundamental points.

3. ZARISKI'S proof for Threefolds

A locally k-normal model of a k-curve gives at once a non-singular model {see (I, 7)}. For a surface we can likewise find such a model; and the reader may consult a clear exposition of the more comprehensive proof, by ZARISKI, in the treatise of HODGE-PEDOE {see ZARISKI [3, 7] and HODGE-PEDOE [a_3], p. 322}. Hence we shall limit ourselves to giving a description of ZARISKI's proof for a *threefold* V over a field k of *characteristic zero:* this will occupy the remaining subsections of this chapter. Here we shall frequently call variety, curve or point respectively a k'-variety, a k'-curve or a k'-point, where k' is a fixed field. We remark that, since k is of characteristic zero, k-simple sets are also absolutely simple: see (I, 7).

a) {See ZARISKI [9], I.} The first part deals with some lemmas on the effect of a quadratic transformation (Q. T.) or of a M. T. T on V over k applied with the centre at a multiple k-point or k-curve of a fixed simple k-surface F on V. If P is ν-ple for F and simple for V a Q. T. of centre P furnishes on $F' = T[F]$ an algebraic 1-set $T[P] \cap F'$, possibly k-reducible. Then, if $T[P] \cap F'$ has a k-component ν-ple for F', it is itself irreducible and is a rational non-singular k-curve.

Similarly a M. T. associates with the base k-curve C, ν-ple for F and simple for V, an F'-set $D' = T[C] \cap F'$ of dimension unity: if C' now has a k-component ν-ple for F', it is itself irreducible and is birationally and biregularly equivalent to C at each simple k-point P of C, provided $m(P; F) = m(C; F)$: i. e. C and C' have *the same residue field.*

The residue field of a subvariety W of V is the field $\mathfrak{n}_k(W; V)/\mathfrak{m}_k(W; V) = k(z)$ where (z) is a generic point of W with respect to $k = \text{def}(V, W)$ {see (I, 3) and SAMUEL [c], p. 59}.

Moreover if P is ν-ple on F and if a corresponding k-point P' by a Q. T. of centre P is still ν-ple for F', then either $T[P] \cap F'$ is a non-singular rational k-curve or it consists of some k-curves containing P', and in the latter case *the residue fields of P and P' coincide* {see ZARISKI [9], th. 4}.

We shall say that a Q. T. or a M. T. reduces *uniformly* the multiplicity of its centre $H \subset F$ if, for each two corresponding k-points P and P' with $m(P; H) = 1$ and $m(P; F) = m(H; F)$, we have $m(P'; F') < m(P; F)$. Evidently a reduction process is a partial failure whenever it does not uniformly reduce the centre: the above properties, however, show that, with one exception, this set-back is always compensated, to a certain degree, *by the invariance of the residue field of the k-point or of the k-curve* under examination. ZARISKI shows that in fact, if the residue field is invariant, T operates locally similarly to the case *where k is algebraically closed.*

This is applied to the study of the *normal crossing* of two k-curves C and D of F: a normal crossing is a k-point P ν-ple for F and simple for C and D, where the two k-curves, both ν-fold for F, intersect *with distinct tangents.* We can here select three convenient uniformising parameters of V at P, such that the first two form a base for the prime ideal of C in V and the last two similarly for D. We can thus show that a Q. T. or a M. T. with centre in C (or in D) *maintains a normal crossing* of C and D, and that two M. T. with the centre respectively on C or D are *permutable* to within a biregular transformation, because such a change preserves the local ring at P: here one uses the above mentioned idea of local reduction. All these facts are proved by ZARISKI for a field k of *any characteristic.*

b) After an elegant proof of the theorem, known in the classical case, that one can desingularize a projective k-curve by using *only* Cremonian *Q. T. (for any k)*, ZARISKI goes on to give *a local reduction theorem* {see ZARISKI [9], 11, 13} *for a k-surface F embedded in V.*

Let v be a zero-dimensional valuation of the rational function field of F over k: let P be the centre of v on V, where P is simple for V and ν-ple isolated for F or a normal crossing of the total ν-ple 1-set for F. We shall call *permissible* a Q. T. with centre P or a M. T. with centre a ν-ple k-curve of F through P. Let T be such a transformation and $\overline{T}$ that induced by T between F and $F' = T[F]$. Using the results in (a) it is shown that, if the centre P' of $\overline{T}$ on F' is still ν-ple, then P' is either still ν-ple, but isolated for F', or it is a simple k-point of the ν-ple 1-set of F' or a normal crossing of this set.

One can now apply to the variety $V' = T(V)$ another permissible T', thus obtaining a variety V'', a k-surface F'' and a k-point P'', and so on till the transform of P maintains itself ν-ple: the theorem of local reduction under consideration assures us that *such a sequence is neces-*

sarily finite; then there exists a permissible sequence which lowers the multiplicity of P.

It is clear that this theorem is *stronger* than the local uniformisation theorem given in (II, 1) because the reduction is here realised only *by permissible transformations* acting in a *non-linear* ambient. The proof is obtained by using valuation theory to the full and, since at certain points the field k must be assumed to have characteristic zero, the result is true, or at least is proved, *only* in this case.

c) Now let F be a simple k-surface of V and $S_1, S_2, \ldots, S_k$ be the k-components of its multiple 1-set. Using the first theorem in (b) it is possible to resolve the singularities of each S_i by Q. T. in the ambient space of V: this may introduce some new, non-singular, multiple curve. Moreover, by Q. T. with centre at the intersections of the components of the new multiple curve, one can reduce the question to the case where the new multiple 1-set is *non-singular, except for some normal crossings of two k-components.*

We now call *normal* a sequence of permissible transformations such that the Q. T. of the sequence have the centre *in an isolated multiple k-point of F*. The following is then a *global reduction theorem for a simple k-surface embedded in V and endowed with singular k-points and k-curves of maximal multiplicity* ν:

Every normal sequence of permissible transformations which does not lower the multiplicity of the centre, is necessarily finite {see ZARISKI [9], III, 17}.

In the classical case of a k-surface, embedded in projective space, this is the theorem of BEPPO LEVI. In the general case ZARISKI begins, after the singular locus has been prepared as above, by proving that if the theorem is true, then two maximal normal sequences of permissible transformations lead to the same k-surface: this is substantially an effect of the commutativity of the operations in the sequences, following partially from (b). One can thus limit oneself to using in the first place all M. T. with centres in the ν-ple k-curves of F: it is seen that, after a finite number of operations, these are removed.

One can therefore proceed by applying only Q. T. to the new k-surface, save for occasional M. T. whenever a Q. T. creates some new multiple k-curve. Now let $F, F_1, \ldots, F_i, \ldots$ be a sequence of any birational transforms of F and $P, P_1, \ldots, P_i, \ldots$ a sequence of related corresponding k-points with $P_i \in F_i$. If $F_{i+1} = T_i[F_i]$, the sequence $\{P_i\}$ is called a *normal sequence of ν-ple k-points* if the following conditions are satisfied: 1. Every P_i is ν-ple for F_i, and is either isolated on F_i, or it is simple on a ν-ple k-curve or is a normal crossing of two ν-ple k-curves; 2. $\{T_i\}$ is a normal sequence of permissible transformations.

Therefore the theorem enunciated above is proved if we can show that *every normal sequence of k-points is finite.* Let us suppose this to be false:

then we can find an infinite sequence of k-points which are all ν-ple. But this is certainly impossible if the sequence is not obliged to be normal, as follows from the local reduction theorem in (b): thus we must suppose that non-normal sequences are more powerful than normal sequences in lowering multiplicities. ZARISKI completes the proof showing that this cannot be the case: the extended BEPPO LEVI theorem is thus established.

d) Let V' and V^* be two birational k-models of V, corresponding to T/k and T^*/k respectively, and let us suppose:

1. *The inverse T^{*-1} to be without fundamental k-points on V^**; 2. *The birational transformation between V^* and V' to be without fundamental k-points at the k-points of V^* which correspond to simple k-points on V.*

This situation, which, as we shall see, is very favourable to our problem, will be described by saying that *T^* has eliminated the simple fundamental locus of T*. The major part of the work {see ZARISKI [9], IV} will be devoted to showing that such an elimination can be accomplished by using *only* Q. T. or M. T. *which transform simple k-points into simple k-points:* the last condition is trivially true for the k-points lying *outside* the fundamental locus of T because we shall *only* use M. T. or Q. T. with the centre on that locus or on one of its transforms. The use of M. T. will also secure that the inverse of the definitive transformation will *not* have some fundamental point on the final model of V.

ZARISKI proceeds to transform the problem by introducing *the linear system of the hyperplane sections* of V' in its ambient space. The transform of this system in T^{-1} will be a linear system L on V: let F be a generic surface of L with respect to k. If V is locally k-normal along a point or a curve H, then H is fundamental for T if and only if H belongs to the base locus of the system L, i. e. if H lies on F. We know that the condition is certainly satisfied if H is k-simple on V (I, 7) and thus *the simple fundamental locus of T coincides with the base locus of L, outside k-curves or base k-points singular on V.*

The above theorem now becomes a particular case of the following, referred to an arbitrary linear system on V:

Given a linear system L/k, without fixed components, on V/k, it is possible, by using only M. T. to transform V into a variety V^* in such a manner that: 1. *If L^* is the corresponding system, free from any fixed components, every base point of L^* arises from a singular point of V;* 2. *Simple k-points of V go into simple k-points of V^*.*

Let us first suppose that L is *without singular base k-points outside the singular locus of V*. Apart from the normal crossings one can begin by eliminating by Q. T. every singularity of the simple base k-curve of L. Next ZARISKI shows that, by using only Q. T., *it is possible to eliminate all the base k-points of a linear system of k-curves without fixed components on a surface.*

Now this theorem can be applied to our case with the following typical procedure: let C be a simple base curve of L. If $R_k(V)$ is the field of the rational functions on V, we can extend the field k by adding an element ξ of $R_k(V)$ whose C-residue is transcendental over k: then $V/k(\xi)$ is a *surface*; the linear system $L/k(\xi)$ is *a system of curves* over this surface and $C/k(\xi)$ is *a base point* of this system.

A M. T. T/k with centre in C, which transforms V and L in V' and L', induces over $V/k(\xi)$ a Q. T. which has its centre on $C/k(\xi)$ and transforms $L/k(\xi)$ into $L'/k(\xi)$. Therefore, by the above theorem, we can eliminate all the simple base k-curves of L/k, except in the case where, on the transform of C by T, there is some k-curve arising from a simple point of C: such a k-curve can in effect evade the preceding reduction because ξ may be algebraic over it: ZARISKI proves that this is not the case at least when, as we have supposed, L is free from singular base k-points.

Thus *we can delete all simple base k-curves.* These operations *do not change* the nature of the base k-points of L which are *outside* the singular locus S of any successive model, but they can introduce some new *isolated* base k-point for the transform of L. We can now suppose that L possesses *only simple isolated base k-points.* The proof proceeds with a direct analysis by sequences of Q. T. acting on a single base k-point, with an application of the above process every time one of these gives new simple base k-curves, with the complication that such operations may again yield new base k-points.

At any rate the finiteness of the whole process is assured: the theoretical bases are: *the strict ascendancy of the local rings of the base k-point sequences and the fact that these points remain simple in every transformation.*

Let us now suppose that the system L has *some singular k-points outside the singular locus of V.* We shall consider the system L as a single surface F^* over $k^* = k(F)$, where F is a generic surface of L over k.

The theorem of BERTINI (in ZARISKI's form) on singular points {see (III, 7)} assures that *the singular set of F^* consists of: 1. A subset arising from the singular locus of V/k; 2. A subset arising from a subset of the base locus of L,* which, excluding some singular point arising from a simple base curve of L, along which F has a variable singular point, consists entirely of singular points or curves arising *from singular base sets of L of the same dimension.* Moreover M. T. on V/k with centre in some $W/k \subset V/k$, are reflected in M. T. of F^*/k^* with the centre in the subvariety of F^* associated to W/k^*. The theorem is now immediately achieved by applying the generalised theorem of BEPPO LEVI to the surface F/k^*.

e) By the theorems in (II, 2) there is certainly a finite resolving system for any zero-dimensional valuation set of the field $R_k(V)$ of a variety V: it is obvious that if we succeed *in reducing to one the number of models in a resolving system,* then the variety V possesses a model without singularities. This reduction can certainly be effected step by step if one proves the theorem:

If there exists a resolving finite system consisting of two models V and V' for an arbitrary set $\mathfrak{N}$ of zero-dimensional valuations of $R_k(V)$, then there exists also a resolving model for $\mathfrak{N}$.

The means for the proof are simply the theorem in (d) and the following lemma:

If P and P' are corresponding simple k-points on two birationally related surfaces F and F' and if $\mathfrak{n}_k(P;F) \subseteqq \mathfrak{n}_k(P';F')$, then either $\mathfrak{n}_k(\mathbf{P};F) = \mathfrak{n}_k(P';F')$ or P' can be obtained from P by successive Q. T.

Now let be T/k the transformation from V to V' and let T^*/k be the transformation which eliminates the simple fundamental locus of T on V {see (d)}: we put $V^* = T^*(V)$. Hence also V^* and V' are a resolving system for $\mathfrak{N}$, because simple points of V remain simple on V, and (as one can prove) the same is true of the pair V_1 and V', where V_1 denotes the graph of the correspondence between V^* and V' in the product $V^* \times V'$.

Therefore we find two resolving models for $\mathfrak{N}$ related to each other and without fundamental points on V; we can suppose that this is the original situation for the pair V and V' with respect to V'. Thus if H and H' are any two k-subsets of V and V' corresponding in T, we get $\mathfrak{n}_k(H;V) \subseteqq \mathfrak{n}_k(H';V')$, and moreover we can again use T^* to eliminate the simple fundamental locus on V: V^* and V' are again a resolving pair for $\mathfrak{N}$.

Now let P, P' and P^* be the centres of a valuation v of $\mathfrak{N}$ over the models V, V' and let V^* and let $\mathfrak{F}$ be the simple fundamental locus of T on V including the singular k-points (if any) of V lying on simple base k-curves. It easily follows that whether P lies on $\mathfrak{F}$ or not, one of the k-points P' and P is necessarily simple; its local ring contains that of the other, except in the case where P is singular for $\mathfrak{F}$, and P', which is consequently simple, is fundamental in the correspondence between V' and V, and then either isolated or on a fundamental k-curve C': in the latter case, by using the above lemma, one proves that $T(C') = P$.

But such elements, if appearing in $\mathfrak{N}$, must be simple on V' and by the theorem in (d) they can be eliminated between V' and V. If $\overline{V}'$ is the model of V' after this elimination, one shows that the centre of every valuation in $\mathfrak{N}$ is simple and not fundamental on one at least of the models V and $\overline{V}'$. But this is just the condition for the product of those varieties to be non-singular: so this is obviously one resolving model for $\mathfrak{N}$.

In the classical case and in the domain of purely algebraic methods the resolution of singularities for surfaces was attempted by the combined efforts of BEPPO LEVI, C. SEGRE, SEVERI, ALBANESE and CHISINI. For an account of these proofs see ZARISKI [a], p. 17. The most satisfactory proof was that of WALKER [1], based on function-theorethic methods: afterwards classical suggestions of SEVERI and ALBANESE were combined in a new attempt at a classical proof by DU VAL in [2]. New complements and viewpoints on the general problem were contributed by DERWIDUÉ [1, 2, 3] and by B. SEGRE [11, 13]. Finally we quote the interesting program of ZARISKI in [22].

III. Linear Systems

1. Divisors

{See SAMUEL [c], p. 96; A. WEIL [a], Ch. VIII}.

A *divisor* of $\mathbf{A}^n$(or $\mathbf{P}^n$) is an $(n-1)$-dimensional cycle $X = \Sigma_i n_i V_i$ of this space: it can be *uniquely* determined by the rational function $\pi_i F_i(\xi)^{n_i}$ where $F_i(\xi) = 0$ is the equation of the form V_i in $\mathbf{A}^n$ (or $\mathbf{P}^n$). Conversely X determines such a function to within an arbitrary constant factor, and the degree of X coincides with the degree of the associated function.

Now let V be a projective variety in $\mathbf{P}^n$ and f a homogeneous rational function over $V(f \neq 0, \infty)$. If (x) is a generic point with respect to $k = \operatorname{def}(V, f)$, then we can write $f(x) = F(x)/G(x)$, where $F(\xi)$ and $G(\xi)$ are homogeneous polynomials of the same degree over k.

Let X be the divisor of $\mathbf{P}^n$ of equation $F(\xi)/G(\xi)$. Then the intersection-product $Y = X \cdot V$ is *defined*, because V lies neither in $F(\xi) = 0$ nor in $G(\xi) = 0$: thus Y is an $(r-1)$-dimensional cycle ($r = \dim V$) and the coefficient of any Y-component W is given by $i(W; X \cdot V)$.

Moreover $\mathfrak{n}(W; V)$ is of dimension 1 and admits the function $f(x)$ as parameter, so that the number $i(W; X \cdot V)$ is also the multiplicity of the ideal generated by $f(x)$ in $\mathfrak{n}(W; V)$: hence it depends only upon f. The cycle Y is called the *divisor* of the function f on V and denoted by (f). The positive and negative parts of Y are respectively denoted by $(f)_0$ and $(f)_\infty$: on the components of these cycles f induces either the value zero or infinity, and $(f) = (f)_0 - (f)_\infty$.

We recall here the following definition of a place of the field $R_k(V)$:

A place $\mathfrak{p}$ of $R_k(V)$ over k is a proper homomorphism over k $f \to f\mathfrak{p}$ of $R_k(V)$ onto (Δ, ∞) where Δ is a subfield of $\mathbf{K}$, *which is called the residue field of $\mathfrak{p}$.*

The elements $f \in R_k(V)$ such that $f\mathfrak{p} \neq \infty$ form a valuation ring $R_\mathfrak{p}$ whose prime maximal ideal is the kernel of the homomorphism of $R_\mathfrak{p}$ onto Δ. Places which have isomorphic residue fields (this is always the case when they have the same valuation ring) are considered *equivalent* and may be identified. Referring to (I, 8), we remark that there is a (1, 1) correspondence between the places and the valuations of $R_k(V)$, where the associated elements have the same valuation ring.

The degree of transcendence of Δ over k is called the dimension d of $\mathfrak{p}$, and we have $0 \leqq d \leqq r-1$ ($r = \dim V$). If $(x_1, x_2, \ldots, x_n)$ is a non-homogeneous generic point of V with respect to k, we may suppose, by making, if necessary, a projective transformation, that $x_i \mathfrak{p} \neq \infty$: the point $(x\mathfrak{p})$ which belongs to V is then called the *centre* of the place $\mathfrak{p}$. Obviously we have: $0 \leqq \dim(x\mathfrak{p}) \leqq d$. Moreover if h is an integer such that $0 \leqq h \leqq d$, then there is some projective model V' of $R_k(V)$ on which the centre of $\mathfrak{p}$ has dimension h. The point $(x\mathfrak{p})$ is the generic point over k of the centre of the valuation v associated with $\mathfrak{p}$.

If $d = r - 1$ the place p is called a k-prime divisor which is called of the first or of the second kind according as $\dim(x\,p) = r - 1$ or $\dim(x\,p) < r - 1$. The number of the divisors of the first kind associated with an $(r - 1)$-dimensional k-subvariety W of V is finite in number and is precisely 1 whenever V is locally k-normal (in particular simple) along W: in such a case the valuation ring of the place is the local ring of W on V.

If, instead, $s = \dim W < r - 1$, there are infinitely many valuations whose centres are W, and the local ring of W on V is then the intersection of the rings which belong to the s-dimensional valuations whose centre is W. Consequently there is a (1, 1) correspondence between the $(r - 1)$-subvarieties of a normal variety V and the prime divisors of the first kind of the function field on V, so that, in this instance, an $(r - 1)$-cycle may be called a divisor, as is sometimes done.

Supposing W simple on V it defines a prime divisor of the first kind of $R(V)$, and moreover $\mathfrak{n}(W; V)$ is the ring of a discrete valuation v_W and if v_W is normalised, we find: $i(W; X \cdot V) = v_W(f(x))$ {see SAMUEL [c], p. 83}.

We recall that if k is an algebraically closed field of definition for W, V and f, where W is a locally normal hypersurface on V and f any function which does not vanish everywhere on V, then $v_W(f) = v_{W,k}(f)$ {see ZARISKI [21], p. 556}.

Thus *if V is normal*, $(f) = \Sigma\, v_W(f(x)) \cdot W$, with the sum extended over all the components W of $X \cdot V$. Let us now call G the graph of f, i. e. the set of the points $(x, f(x))$, in the product of V for the projective line $\mathbf{P}^1$, and ϑ the divisor $(0) - (\infty)$ of $\mathbf{P}^1$: then one can easily establish the relation $(f) = \mathrm{pr}_V(G \cdot (V \times \vartheta))$ by passing to the affine space and using the biregularity of the projection of $G \cdot (V \times \vartheta)$ on V along each component, which requires the invariance of the local rings and therefore that of the coefficients of $G \cdot (V \times \vartheta)$.

A function-divisor over a normal variety is *never positive if it is not null*, as one sees by recalling that an element belonging to each affine coordinate ring of V belongs to the base field. From the relations $(f\,g) = (f) + (g)$, $(1/f) = -(f)$, it follows that if $(f) = (g)$, then f/g is a constant over V.

2. The Definition of Linear System

Let V/k be a projective variety in $\mathbf{P}^n$ and $X(\lambda)$ the positive divisor of $\mathbf{P}^n$ of equation $\sum_{i=0}^{n} \lambda_i F_i(\xi) = 0$, with the λ's transcendental over k and the $F_i(\xi)$ forms of the same degree over k. Supposing that no $X(\lambda)$ contains V (i. e. that the quantities $F_i(x)$ are *linearly independent over k*, x being a generic point of V/k) the intersection-product $Y(\lambda) = X(\lambda) \cdot V$ is defined, so giving an $(r - 1)$-dimensional cycle rational over $k(\lambda)$: it is the divisor of the function $\sum_{i=0}^{n} \lambda_i F_i(x)$ on V {see (III, 1)}.

The cycle $Y(\lambda)$ is the generic cycle over k of an algebraic system on V {see (VI, 1, 2)}: this is called *a linear system*, since it is parametrised by the linear variety Λ whose generic point over k is (λ).

If $Y(\lambda)$ contains some algebraic component over k, it is easily seen that, $k(\lambda)$ being a pure transcendental extension, there is a rational cycle over k, say Y_0, such that $Y(\lambda) = Z(\lambda) + Y_0$, where now $Z(\lambda)$ is totally transcendental over k: Y_0 is called *the fixed cycle* of the linear system. The system L whose generic cycle over k is $Z(\lambda) + Z_0$, where Z_0 is an arbitrary fixed $(r-1)$-cycle, of V, is still called a linear system: certainly $Z(\lambda) \geqq 0$, and generally the notion of linear system *is restricted so that* $Z(\lambda) + Z_0 \geqq 0$ i. e. one generally considers only linear system having every member positive or null.

If $(\bar{\lambda})$ is a point of Λ and $\bar{P}$ a point of $Z(\bar{\lambda})$, not lying in each $F_i(X) = 0$, then $((\bar{\lambda}), \bar{P})$ is a specialisation {see VAN DER WAERDEN [b], p. 182} of $((\lambda), P)$ over k, P being a generic point of $Z(\lambda)$ over $k(\lambda)$: therefore the cycle Z in $V \times \Lambda$ associated with the cycle $Z(\lambda)$ by the relations $Z(\lambda) \times \times (\lambda) = (V \times (\lambda)) \cdot Z$ and $\mathrm{pr}_V Z = V$ {see (I, 9)}, is *a multiple of a k-variety* and then so also is $Z(\lambda)$. Moreover, if P is a generic point of V over k, putting: $P \times \Lambda(P) = Z \cdot (P \times \Lambda)$, the cycle $\Lambda(P) = \varphi^{-1}(P)$ {for the straightforward definition of φ^{-1} see (VI, 3)} must be a linear $(m-1)$-dimensional subvariety of Λ {see (I, 9) and use the fact that $\mathrm{pr}_L Z = L$}: it follows that m generic points $P_1, P_2, \ldots, P_m$ on V, independent over k, belong *to one and only one* cycle of L rational over $k(P_1, \ldots, P_m)$.

The previous definition, substantially resting on that of divisor given in (III, 1) and on the notion of intersection multiplicity, is obviously related to the particular *projective model* V of the function field $R_k(V)$ of V itself.

This is essentially due to the fact that V may possess some $(r-1)$-dimensional *multiple* subvarieties, and in this case the geometrical definition of divisor as $(r-1)$-cycle is not birationally uniquely determined: each such subvariety is really the centre of more than one discrete valuation of rank 1 of $R_k(V)$ {see ZARISKI [8], p. 511}.

This fact has some unpleasant consequences, especially if one wishes to establish, as it is necessary to do, *the invariance*, with respect to birational transformations, of the concept of linear system: nevertheless the difficulty can be overcome by replacing the model V by a *normal* model V' of V. Then, on V', the geometrical definition of divisor coincides with the definition by means of the *valuation* theory or, which is the same thing, with that obtained by using the fact that now the homogeneous coordinate ring of V' is *a normal ring* (and, moreover, a Noetherian ring integrally closed in its quotient field) for which there is a *one[to-one* correspondence between $(r-1)$-dimensional fractional ideals in the ring and the $(r-1)$-cycles on V': the latter is the so called *"quasi gleichheit"* method of VAN DER WAERDEN-ARTIN {see VAN DER WAERDEN [a_2], p. 93; HODGE-PEDOE [a_3], pp. 56—70; NORTHCOTT [a], pp. 76—83; see also the application in (III, 6)}.

We can then conveniently transfer each result obtained on V' onto V, by the birational correspondence between V' and V, which is a projection {see (I, 6)}, and by considering every multiple subvariety of $(r-1)$-dimensions as *a finite set of varieties distinguished by the respective valuations:* i. e. with respect to the simple subvarieties of V' which project over one of them. This is substantially the classical procedure, which is here clarified by the concept of normality.

Last, we observe that obviously the preceding process is no longer necessary if one considers linear systems free from fixed multiple components and with some generic element simple on V: in particular this is always the case if V itself is non-singular. The birational invariance of L is now a trivial consequence of the definition by a normal model which is invariantively associated with V.

3. Linear Equivalence

We shall now outline, with ZARISKI {see ZARISKI [21], pp. 555—565}, the theory of linear systems on a normal variety V, which is, as we have now seen, the fundamental case. We begin with the definition of linear equivalence in $\mathbf{G}^{r-1}(V)$ (shortly $\mathbf{G}(V)$):

Two divisors X_1, X_2 are said to be linearly equivalent ($X_1 \equiv X_2$) if $X_1 - X_2$ is the divisor of a function on V.

The subgroup of the divisors X, *with* $X \equiv 0$, will be denoted by $\mathbf{G}_l(V)$ and the factor group $\mathbf{A}(V) = \mathbf{G}(V)/\mathbf{G}_l(V)$ will be called *the group of linear equivalence on V.*

Now let $\mathfrak{L}$ be any finite module in $\mathbf{K}(V)$: i. e. a linear space in $\mathbf{K}(V)$ over $\mathbf{K}$. If f is any function of $\mathfrak{L}$, there is certainly a divisor X of V such that $(f) + X \geqq 0$: such an f is called a multiple of X. The definition of linear system given in (III, 2) can now be restated in the following terms:

A linear system L on the normal variety V is the set of all the divisors (f) such that $(f) + X \geqq 0$ $(f \neq 0)$.

Then $\mathfrak{L}$ is called a *definition module* for the linear system L. It is clear that any two divisors of L are linearly equivalent, consequently belonging to one and the same equivalence class. A divisor $X_0 \geqq 0$ is *a fixed component* for L, if for every divisor $X \in L$, $X \geqq X_0$, and X_0 may be eliminated or not, leaving, in any case, a linear system defined over the same module.

Moreover the modular structure of $\mathfrak{L}$ determines a *projective structure* in L, which, to within a fixed component, is invariant for transformations on $\mathfrak{L}$ of the form gf where $f \in \mathfrak{L}$, $g \in \mathbf{K}(V)$ and $g \neq 0$; therefore the projective structure of L *does not depend* upon $\mathfrak{L}$. If $X_0 = (f_0) + D$ is an arbitrary cycle of L, let us operate on $\mathfrak{L}$ the transformation: $f' = (1/f_0)\, f$. Then every divisor $X = (f) + D$ of L can be written in the form: $X = (f') + X_0$, and the new module $\mathfrak{L}'$ contains $\mathbf{K}$ since $\mathfrak{L}'$ consists, besides the zero function, of all the functions f' on V with $(f') = X - X_0$, $X \in L$. Such a module, uniquely defined by L and the divisor X_0, will be denoted by $\mathfrak{L}(X_0)$.

4. Complete Systems

Consider now two linear systems L and M belonging to the *same* linear equivalence class, a divisor $X_0 \in L$, the module $\mathfrak{L}(X_0)$ and the module $\mathfrak{M}$ determined by the zero function and the functions g for which $(g) = Y - X_0$ with $Y \in M$.

It is clear that $\mathfrak{M}$ is a definition module for M and that the sum module $\mathfrak{N}$ of $\mathfrak{L}$ and $\mathfrak{M}$ is of *finite* dimension. Then the linear system related to the module $\mathfrak{N}$ is called the minimal sum of L and M: it is obviously the least linear system on V containing both L and M.

If, on the contrary, L and M *do not belong* to the same equivalence class, it is obvious that there does not exist a minimal sum in the strict preceding sense. Take now the divisors $X_i + Y_j$, where the cycles X_i and Y_j describe a base for linear equivalence respectively in L and M: it is easily proved that, by applying the above result to these cycles, considered as systems all belonging to the same equivalence class, we obtain a system which contains all the divisors of the form $X + Y$ with $X \in L$, $Y \in M$: this system is still called *the minimal linear sum of L and M*.

At any rate the sum system admits as a definition module the $\mathbf{K}$-module which has for base the products of the bases of the defining modules for L and M. So one can easily go on to define the minimal linear sum of *any number* of linear systems, and, in particular, *the linear multiple mL* of L, where m is a positive integer.

This simple situation certainly arises if the two systems have *at least one common divisor:* if we take into account the fact that *the dimension of the linear systems containing a given divisor $X \geqq 0$ is finite,* we see at once that *the set of all the divisors $X \geqq 0$ linearly equivalent to a given divisor $X \geqq 0$, is a linear system,* consequently associated with a finite module. The assertion may be easily proved by using the fact that equivalent divisors have *equal degree* and that they are parametrised by *a subset* of the Chow set related to that degree, as we shall see in (VI, 1).

This comes to saying that the set of *the functions in $\mathbf{K}(V)$ multiple of $X \geqq 0$ and the zero, is a finite module:* this is also true for an analytic variety {see (IX, 1)} as A. Weil has proved {see A. Weil [4], p. 885}. Later in (III, 5) we shall have an elementary proof of the above assertion.

Calling the linear system so generated the *complete linear system* $|X|$ containing X, we can enunciate the following theorem of existence and unicity:

(i) *Every V-divisor* $X \geqq 0$ *belongs to one and only one complete linear system* $|X|$.

We have also readily the famous *residue theorem:*

(ii) *If* $X_0 \in L$, *the divisors* $Y \geqq 0$ *such that* $Y + X_0 \in L$ (which are called the *residues* of X_0 with respect to L) *form a complete linear system* $L - X_0$, *if itself is complete.*

We conclude this section by giving a theorem of A. WEIL {see A. WEIL [a], p. 240; ZARISKI [21], p. 560} which is useful in the abstract theory of linear systems:

(iii) *If* $X \equiv 0$ *is a V-divisor rational over an algebraically closed field* $k = \operatorname{def} V$, *then there is a function* $f \in k(V)$ *with* $(f) = X$ *and, conversely, the function divisors of* $k(V)$ *are rational over* k.

A system L is said to be *defined over* k if it is associated *with at least one module in* $k(V)$: from the unicity theorem it follows that this is the case if, and only if, k being algebraically closed with $k = \operatorname{def} V$, *there exists a maximal system* $Y_0, \ldots, Y_m$ *of linearly independent divisors in* L *such that the cycles* $Y_i - X_0$ *are rational over* k, where X_0 is the fixed component of L.

5. The Multiples of a Linear System

Let R_m, $m \geqq 1$, be the set of the *forms of degree* m in $R = K[x]$, where (x) is a generic point of V/k, and let $\mathfrak{L}_m(f_0) \supset K$ be the finite module of the functions $f(y)/f_0(y)$ with $f(y)$ and $f_0(y) \in R_m$. The linear system of the positive cycles $(g) + (f_0)_\infty$, with $g \in \mathfrak{L}_m(f_0)$ and $g \neq 0$ does not depend upon $f_0(y)$ in R_m but only upon m, and obviously these cycles are the zero divisors of the functions g.

This system is *without fixed components*, since $\mathfrak{L}_m(f_0)$ contains the functions y_i^m/f_0, and, by (III, 1), it is the linear system cut on V by the forms of degree m of its ambient space. Moreover $R_{m+m'}$ is a $\mathbf{K}$-module described by the products ff' with $f \in R_m$ and $f' \in R_{m'}$, and thus the system $L_{m+m'}$ is the minimal linear sum $L_m + L_{m'}$: in particular L_m is the minimal m-ple of the system L_1 of the hyperplane sections of V.

One can show directly that the linear system L_m is contained in a complete linear system by proving that every non-negative divisor linearly equivalent to one of L_m is of the form: $(g/f_0) + (f_0)_\infty$, where $g \neq 0$ is any element in the integral closure I_m of degree m of $\mathbf{K}[x]$ in $\mathbf{K}(x)$. Hence, recalling that there always exists an h-ple model of V, for a convenient h, which is projectively normal and for which $\mathbf{K}[x]$ is integrally closed in $\mathbf{K}(x)$ {see (I, 5)}, we deduce the important theorem:

(i) *The set of all forms of degree m of the ambient space of V cuts on V a complete linear system* L_m, *provided that m is sufficiently large* {see ZARISKI [21], p. 563}.

The number $\dim|L| - \dim L$, where $|L|$ is the complete system containing L, is called *the deficiency* of the system L and denoted by def$L$: then, if m is large, def$L_m = \dim I_m - \dim R_m = 0$. From all this and (I, 5) follows that:

(ii) *N. a. s. c. that a variety V be projectively normal is that* def$L_m = 0$ *for every m* {see Muhly [1]}.

We can now prove the theorem (i) in (III, 4): there are certainly forms F in $\mathbf{P}^n$ of a sufficiently high order containing a fixed but arbitrary divisor X of V (i. e. $V \cdot F \geqq X$) without containing V. Let then $X_0 = V \cdot F - X$. Therefore, by (i) and the residue theorem (ii) of (III, 3), the system $L_m - X_0$ is complete, if m is large, and moreover it contains X: this system is cut on V by the forms of order m containing the residue cycle X_0 of X. This, together with the observations in (III, 4), yields the proof of our theorem and, besides, the construction of the complete system $|X|$.

We observe that *an $(r-1)$-cycle rational over an algebraically closed field of definition for V determines a complete system $|X|$ defined over k* {see (III, 4)}, and that this is certainly the case for $|L_m|$. On the contrary, a system L, with fixed components, can naturally be defined over such a field k, *without* being necessarily so for the complete system $|L|$.

6. Ample Linear Systems

Let L/k be a linear system without fixed components on a variety V of $\mathbf{P}^n$, which here need not be normal, and let V' be a normal model of V, with $k = \operatorname{def} V$. Let (x) be a generic point of V with respect to k, and let $f_0(x), \ldots, f_m(x)$ be $m+1$ elements of $k(x)$ linearly independent over the algebraic closure of k in $k(x)$.

The polynomials $f_i(x)$ determine *uniquely* a set of $m+1$ integral divisors X_i of the field $k(x)$ with the following properties: (1) *Each X_i is a divisor of the first kind with respect to V*; (2) *$f_i/f_0 = X_i/X_0$ and the X_i are relatively prime.*

Each divisor X_i determines on V (and on V' is *conversely* determined by) an $(r-1)$-cycle X_i. The components of X_i correspond to the prime factors of X_i, but, if V is not locally k-normal, two distinct factors of X_i may very well correspond to one and the same component of X_i: while this is never the case on V'. If $\lambda_0, \lambda_1, \ldots, \lambda_m$ are arbitrary elements of k, then there exists a unique divisor $X(\lambda)$, such that $\sum_{i=0}^{m} \lambda_i f_i(x)/f_0(x) = X(\lambda)/X_0$. Let $X(\lambda)$ be also the $(r-1)$-cycle on V defined by the divisor $X(\lambda)$: it is then obvious that as the λ's vary in k the cycle $X(\lambda)$ varies in a *linear system L* associated with the module in $k(V)$ $(1, f_1/f_0, \ldots, f_m/f_0)$.

It is now easily proved, by the above representation, that the k-set of the points on V which lie in each $\|X(\lambda)\|$, called *the base set* of L, is the set associated with the radical of the ideal $(\xi_0, \xi_1, \ldots, \xi_m)$: moreover *if a point P is singular for the sets $\|X_i\|$, $i = 0, 1, \ldots, m$, it is so also for the set $\|X(\lambda)\|$* {for all that precedes see Zariski [8], p. 528}.

Let us now consider the graph W on $\mathbf{P}^n \times \mathbf{P}^m$ of the *rational* correspondence $T: \varrho\, y_i = f_i(x)$, $(i = 0, \ldots, m)$; one can show by the above representation, that *the fundamental locus* $\mathfrak{F}$ *of* T *on* V *is contained in the base set* $\mathfrak{B}$ *of* L: i. e. if $P \notin \mathfrak{B}$ and $P \in V$, then the intersection-product $(P \times \mathbf{P}^m) \cdot W$ is defined. Conversely, any base set along which V is normal belongs to $\mathfrak{F}$ {see ZARISKI [8], p. 528.} Moreover $\mathrm{pr}_{\mathbf{P}^m} W$ is a variety U with $k = \operatorname{def} U$ because (x), and hence (y), is regular over k.

If T is *birational* (and then $\dim U = \dim V$) the system L is called a *simple* system, otherwise L is said to be *compounded of the algebraic involutory system on* V {for the definition of such a system see (VI, 6)}, whose generic member over k, if Q is generic on U with respect to k, is $T^{-1}(Q)$: this is certainly defined since W projects onto V. *If* $\mathfrak{B}$ *is empty* (and there are no fixed components) the system L is called *ample* {this name is due to KODAIRA: see [5], p. 89}: it is clear that then T is *biregular* and conversely.

7. BERTINI's Theorems

The following are the well-known theorems of BERTINI in the *classical* case.

(i) *A generic element of a linear system* L, *without fixed components and not compounded of a pencil, is a variety.*

(ii) *A generic element of a linear system* L *has no singular points outside the base locus of* L *and the singular locus of* V.

These theorems have undergone a deep analysis on the part of ZARISKI with important contributions of MATSUSAKA and AKIZUKI {see ZARISKI [6, 10]; MATSUSAKA [2]; AKIZUKI [1]}. We shall give here a short description of these new results.

To begin with a linear system L on a variety V is said to be *compounded of a pencil*, i. e. a 1-dimensional, possibly irrational, algebraic system of index one on V {see (VI, 1, 6)} if the generic element of L has all its components on some element of that pencil: this may happen from the algebraic standpoint in two ways: (1) Either $\dim L = 1$, and then the field $\mathbf{K}(L)$, generated over $\mathbf{K}$ by a definition module for L containing $\mathbf{K}$, is *not* algebraically closed in $\mathbf{K}(V)$; (2) Or $\dim L > 1$, and then $\mathbf{K}(L)$ has *degree of transcendence one* over $\mathbf{K}$.

It is clear that both cases correspond to the geometrical case where the linear system L is *a linear series* of order greater than 1 and of dimension s on a pencil with respectively $s = 1$ or $s > 1$. *In the former case* one shows that, if L/k is *not* compounded of a pencil and if the extension $k(V)$ is *separable* over $k(L)$, then the generic element of L is *absolutely irreducible* {see ZARISKI [6], p. 64; MATSUSAKA [2]}: this demonstrates the first theorem of BERTINI for a linear pencil and then it can easily be proved for any linear system satisfying the same hypotheses.

For the latter theorem, in the abstract case, we begin {see ZARISKI [10], p. 130} by taking $k' = k(\lambda)$, where the λ's are $m + 1$ indeterminates, algebraically independent over $k(V)$: then $k' = \operatorname{def} V$. Each subvariety W/k of V/k defines a subvariety W/k', which is its *extension* to k' and has as generic point P over k' a generic point P of W with respect to k: the λ's are still algebraically independent over $k(P)$ and, if this is true, the reasoning can also be inverted.

This correspondence preserves the dimension, and *the singular locus of V/k' extends that of V/k.* If W'/k' is an arbitrary subvariety of V/k' with a generic point P' over k', there exists one and only one subvariety W/k of V/k with a generic point P such that $k(P')$ and $k(P)$ correspond in an isomorphism relating P' to P: such a W/k will be called the *contraction* of W'/k'. Obviously $\dim W'/k' \geqq \dim W/k$.

Next we consider the system L, supposed *free* from fixed components and *not* compounded of a pencil, as an $(r-1)$-dimensional variety V' over k', defined by a generic point (z) such that: (1) $k(x) \cong k(z)$; (2) $\sum_{i=0}^{m} \lambda_i f_i(z) = 0$, where $\sum_{i=0}^{m} \lambda_i f_i(z)$ is related to the system as in (III, 6).

Then, by the latter condition, V'/k' contracts into V/k; one can show that if W/k' extends W/k, then W/k' lies in V' provided W/k lies in the base set of L, and after that W'/k' is singular for V'/k' if and only if W/k is a singular base variety for L {see ZARISKI [10], p. 135}. From this follows the theorem:

(iii) *If W'/k' is a singular k'-subvariety of V'/k', then the contracted k-subvariety W/k on V/k is either singular or belongs to the base set of L.*

In the case where k has *characteristic zero*, the second theorem of BERTINI is an immediate consequence of this theorem and of the Jacobian criterion for simplicity: one has merely to prove that *there cannot be variable singular points on loci which extend over k' into simple subvarieties of V'* {see ZARISKI [10], p. 138}.

This *may*, instead, be the case if the characteristic is *not* zero, as is proved by the well known example $x^p + y^2 - 2\lambda y = 0$, taken over *a perfect field k of characteristic p* {see ZARISKI [10], p. 140}: in fact, in this case, there is the variable double point $x = (\lambda^2)^{1/p}$, $y = \lambda$.

AKIZUKI has continued the analysis with the search for *a n. a. s. c. such that the theorem be valid for every characteristic.*

Following the classical procedure more closely than ZARISKI, he proves that an absolutely simple point P' for V, not lying in the base set of L (f. i. with $f_0(P') \neq 0$), is singular for an element of L if and only if $\sum_{i=1}^{m} \lambda_i D_j(y_i) = 0$ at P', where $y_i = f_i(x)/f_0(x)$ and for *every* derivation D_j over $k(x)$. Let now C be an absolute component of the set, algebraic over $\bar{k}$, filled by these points (the so called *critical set*) and let $P' = (x')$ be a generic point of C over $\bar{k}$ with $f(x') = 0$ and $y'_i = f_i(x')/f_0(x')$: now one can show that, if $\overline{k(x')}$ is *separably* generated over $\bar{k}(y')$, then the set of the elements of L which contain a point of C is *a proper algebraic subset* of the linear variety, parametrising L, with generic point (λ). If the same hypothesis is satisfied for each component of the critical set and for each affine model of V, one can affirm that the second theorem of BERTINI *is still true in the classical formulation* {see AKIZUKI [1], p. 177}.

We can conclude that the theorems of BERTINI are unconditionally true *only* in fields of characteristic zero: otherwise we require the additional hypotheses, which have been mentioned above.

Sometimes, in applications, more particular criteria, such as the following, are useful:

(iv) *Let X be a V-divisor on the non-singular variety V, such that the system $|X - C_1|$ exists, and such that $|X|$ is without fixed points. Then almost all divisors of $|X + C_1|$ are non-singular varieties* {see MATSUSAKA [5], pp. 116—120}.

The demonstration is obtained by observing that: (1) $|X|$ contains C_1 and therefore it is not a pencil and moreover $k(L) \not\subset \{k(V)\}^p$; (2) The birational correspondence T associated with $|X|$ {see (III, 6)} is free from fundamental elements and then the graph W of $(P, F(P))$ is birationally and biregularly related to V; (3) A generic hyperplane section of W, certainly non-singular, projects on V into a non-singular element of the system $|X + C_1|$, which is then endowed with some non-singular element: the theorem now follows at once from the fact that the property of having some singularity is certainly algebraic for the element of a linear system and therefore such that it is not possessed by the generic element of the system, if some specialisation of that generic element is not endowed with it.

We have here used the property that *the generic section of a non-singular variety is non-singular:* this obvious consequence of BERTINI's theorem is a particular case of the following noteworthy fact:

(v) *Almost all elements of L_m/k are normal varieties having simple points at simple points of V.*

This has been proved, in the relative case, by A. SEIDENBERG {see SEIDENBERG [2]}.

We remark that SEIDENBERG's proof needs a great deal of algebraic work, so that it has been considered a "difficult proof". The same result, for the generic hyperplane section, is, instead, more elementary: in effect it has been proved by ZARISKI in [21], p. 565, by using only BERTINI's theorems, the classical Jacobian criterion for simple points and WEIL's characterisation of absolutely normal varieties {see A. WEIL [a], p. 270}. A direct elementary proof of the non-singularity of the generic hyperplane section has been given by NAKAI in [1].

Consequently we observe that *almost all* the elements of any ample linear system are *not singular*.

We conclude by recalling the following useful criterion for ampleness:

(vi) *The minimal linear sum of an ample system and any other system, without fixed components and base points, is ample. Moreover the system* $|C_m - X|$, $(X > 0)$, *is certainly ample if m is greater than the degree of the positive cycle X* {see KODAIRA [5], p. 91}.

The former assertion follows immediately from the definitions, and the latter by considering (with SEVERI) the cones of the ambient space of V which project X and cut on V elements of L: we see at once that on our hypotheses $|C_m - X|$ has no fixed components or base points. The proof is then achieved by adding $|C_1|$ and applying the former result, taking into account that $|C_1|$ is, by definition {see (III, 6)}, an ample system.

As we have already mentioned, this chapter is chiefly modelled on ZARISKI's work [21].

We recall that the theory of linear systems has been the fundamental tool in the researches of the Italian school, so that almost every concept of this chapter is originally due to Italian geometers: see e. g. the exposition of linear systems in Ch. X of BERTINI [a]. VAN DER WAERDEN has given an elegant treatment of our theory founded chiefly on the principle of counting constants {see (VI, 3)} and on specialisation theory: see [b], Ch. VII. ZARISKI, after obtaining many results scattered throughout his work {see e. g. [6, 8, 10]}, where he used prevalently the quasigleichheit method, arrived at the quite simple treatment [21]: we recall also the work [1] of FERNANDEZ BIARGE.

We add that other proofs of BERTINI's theorems have been given by VAN DER WAERDEN in [5, X] and by B. SEGRE in [9]. Finally the classical notion of a linear system on a surface which is complete relatively to a set of points, either ordinary or infinitely near {see ZARISKI [a], p. 29}, has been extended to higher varieties by VAN DER WAERDEN in [10] by means of valuation theory, in such a manner as to satisfy the fundamental condition of being a birationally invariant notion.

IV. The Geometric Genus

1. The Adjoint Forms

We now give an account of the classical theory of the geometric genus for a variety defined *over the complex field*, following a recent exposition by SEVERI {see SEVERI [35]}: later on we shall give a treatment of the same subject, over any field k, from a different standpoint {see (IV, 4)}.

Let V be an r-dimensional non-singular variety defined in $\mathbf{P}^n$ over the complex field. We recall that, projecting V from almost any $[n-r-2]$ {the symbol $[h]$ denotes an h-dimensional subspace of $\mathbf{P}^n$} onto a dual space $\mathbf{P}^{r+1}$, both taken in $\mathbf{P}^n$, we obtain in $\mathbf{P}^{r+1}$ a hypersurface V', endowed with a double $(r-1)$-dimensional locus S, having almost anywhere a bihyperplanar tangent cone, and with singular subsets of S of higher multiplicity: such singularities will be called *ordinary singularities.*

The forms of degree m in $\mathbf{P}^{r+1}$, which contain S, are called *adjoint forms* of V': they certainly exist *at least if m is sufficiently large.* One can prove, either directly by induction on m and by use of BÉZOUT's theorem, or by appealing to the results of (III, 5) that:

(i) *For large m, the adjoint forms cut on V, outside S, a complete linear system L_m.*

This can easily be generalised to the case where the adjoint forms must contain a fixed subvariety of V', and one can also prove the following two important complements {see SEVERI [35], p. 7}:

(ii) *If C_m is the general member of L_m, then, for large m, $i(F \cdot V'; C) = 1$.*

Here C_m is, by BERTINI's theorem, a variety, and F is the adjoint form cutting on V' the member in question:

(iii) *The system L_m has no simple base points on V' if m is large.*

This may be proved by observing that it is obviously so for the rational system of cones projecting S from the points of $\mathbf{P}^{r+1}$.

Consequently the system L_m cannot be compounded and (by BERTINI's theorem) almost all members of this system are *varieties.*

These theorems are also obviously true, if m is large, for the system residual to L_m with respect to *a fixed positive cycle* of V' {see (III, 4, 5)}.

2. The Canonical System

We now derive, with ENRIQUES {the original exposition of ENRIQUES was restricted to surfaces: see e. g. ENRIQUES [a], Ch. III; the one here given extends the method of ENRIQUES to higher varieties and is due to SEVERI [35]}, *the canonical system on a variety V.*

a) We begin with the theorem:

(i) *Let L be a simple r-dimensional linear system on the hypersurface V': then the locus of singular points belonging to the cycles of L is the support of an $(r-1)$-dimensional cycle.*

For the proof let $F(\xi) = 0$ be the equation of V' in $\mathbf{P}^{r+1}$, and $\sum_{i=0}^{r} \lambda_i F_i(\xi) = 0$ be the equation of a linear system of adjoint forms Φ of order m, and such that: $\Phi \cdot V' - S - D = C$, where C is any member of L and D a convenient fixed cycle on V'. The existence of such a system of adjoints follows at once from the completeness theorem and from the residue theorem {see (III, 4) and (IV, 1)}. Moreover we can suppose D to be prime and $i(F \cdot V'; D) = 1$.

Now the n. a. s. c. for a point P of V' to be singular for the algebraic set $C + D$ is that the Jacobian $J(X) = \partial(F_0, \ldots, F_r, F)/\partial(\xi_0, \ldots, \xi_{r+1})$ be null at P, observing that it is not identically null over V because the system L has been supposed simple: i. e. the functions $F_0, \ldots, F_r, f$ are functionally independent over V'. Then the cycle determined by the system $J(\xi) = 0$, $f(\xi) = 0$ is defined: it contains the cycles D and S; the residue, after these have been substracted, is still an $(r - 1)$-dimensional cycle. This cycle is called the *Jacobian cycle* of L and is denoted by C_j if C is a member of L.

b) Here we prove:

(ii) *The Jacobian cycles of all the r-dimensional linear systems* $\bar{L}$ *contained in a simple linear system L on V', belong to one and the same linear equivalence class of V'.*

The proof follows at once by observing that the degree of the form $J(\xi) = 0$ does not depend upon the particular subsystem $\bar{L}$ considered in L and that the adjoint forms of that degree cut on V' a linear system of cycles outside S. The complete system containing all the above cycles is called the *Jacobian system* L_j of L on V'.

c) Now let E be *a fixed* $(r - 1)$*-dimensional prime cycle* on V', simple on V'. We prove that:

(iii) *If L is an r-dimensional linear system with a defined Jacobian cycle, then* $(C + E)_j = (r + 1)\, E + C_j$, *where C is a member of L.*

For the proof we suppose that the system L is represented as in (a) and that $g(\xi) = 0$ is a form of degree l cutting on V' the cycle E, residually to a prime cycle D' simple on V'. Then the system $L + E$ is cut on V', outside the cycle $S + D + D'$, by the linear system $\sum_{i=0}^{r} \lambda_i\, g(\xi)\, F_i(\xi) = 0$ of forms in $\mathbf{P}^{r+1}$ of degree $l + m$, and the Jacobian cycle $(L+E)_j$ will be cut on V' by a certain Jacobian form $\bar{J}(\xi) = 0$.

One can now show, by a simple calculation {see SEVERI [35], pp. 10 to 13}, that $\bar{J}(\xi) = (1 + \lambda/l)\, g(\xi)^{r+1} \cdot J(\xi)$ on V'. But the form $g(\xi) = 0$ has with V', along E and D', intersection multiplicity unity, and, therefore, $\bar{J}(\xi) = 0$ cuts on V' a cycle which, residually to the components S, D and D', is the sum of the two cycles $(r + 1)\, E$ and C_j.

We observe that the cycle E might possibly present itself more than $(r + 1)$ times in $(C + E)_j$: this will obviously happen if and only if E is already a component of C_j.

d) The following is *the* ENRIQUES *adjunction theorem:*

(iv) *Let* $|C_1|$ *and* $|C_2|$ *be two simple linear systems, at least r-dimensional, on V'. Then there exists on V' the complete system* $|(C_1 + C_2)_j|$ *and we have the linear equivalences:* $(C_1 + C_2)_j \equiv C_{1j} + (r + 1)\, C_2 \equiv C_{2j} + (r + 1)\, C_1$.

To prove this we begin by observing that the complete system $|C_1 + C_2|$ is also, on our hypotheses, at least r-dimensional and simple: hence the system $|(C_1 + C_2)_j|$ is defined by (a). Denoting by L a generic r-dimensional linear system extracted from $|C_1 + C_2|$ and by $L^{(1)}$ any r-dimensional linear subsystem of $|C_1|$, the linear system $L^{(1)} + \bar{C}_2$ is a specialisation over the complex field of the generic system L, where $\bar{C}_2$

is a prime cycle of $|C_2|$. The theorem is proved by taking into account c), the definitions and the obvious fact that the notion of Jacobian system is covariant with respect to the specialisations.

The linear equivalence: $C_{1j} - (r+1)\,C_1 \equiv C_{2j} - (r+1)\,C_2$, now follows: this is *the adjunction formula,* and *the linear system* $K = |C_{1j} - (r+1)\,C_1| = |C_{2j} - (r+1)\,C_2|$ *is called the impure canonical system on* V'.

This system is consequently defined on V' *independently* of the system $C_i (i = 1, 2)$, provided it has some *positive or null* cycle: otherwise we have *a canonical linear equivalence class* represented by virtual cycles; sometimes, in this case, we say that the canonical system is *virtual.* The canonical system *on* V is defined as the corresponding system on V, by the inverse of the projection relating V to V'.

e) The canonical system admits the following projective generation. The polar form of a point M of the space $\mathbf{P}^{r+1}$ with respect to V' is an adjoint form for V' of order $\mu - 1$ if μ is the order of V': it cuts on V', outside S, the Jacobian cycle of the r-dimensional linear system cut on V' by the hyperplanes through M. Hence follows that:

(v) *The adjoint forms of order* $\mu - r - 2$ *cut on* V' *the canonical system, provided they, and hence the canonical system, exist.*

3. The Canonical System as a Birational Invariant

We next examine the behaviour of the impure canonical system under birational transformation.

a) Let T be a birational transformation between the non-singular varieties V and $\overline{V}$, both defined *over the complex field.* The subsets of $\overline{V}$ corresponding to the fundamental points of F on V are irregular in the sense of (I, 6) and are here more particularly called *exceptional:* the components of these subsets are obviously finite in number but they can have any dimension from 1 to $r - 1$.

In our analysis we shall leave out of consideration the exceptional components of dimension *less than* $r - 1$, since these have no effect on the transformation law for the linear systems of $(r - 1)$-dimensional cycles on V.

Now let P and $\overline{P}$ be any two corresponding points by T, and let us take $(u_1, \ldots, u_r)$ and $(v_1, \ldots, v_r)$ respectively as uniformising parameters on V and $\overline{V}$ at P and $\overline{P}$. The correspondence T can be locally represented by the equations: $v_i = f_i(u)$ and the inverse T^{-1} by $u_i = g_i(v)$, $(i = 1, \ldots, r)$. The functions f_i and g_i are meromorphic functions of the arguments in some neighbourhood of P or of $\overline{P}$ respectively: more precisely they are holomorphic, provided T be biregular at the pair $(P, \overline{P})$. This is certainly the case if P is not fundamental, V and $\overline{V}$ being here non-singular.

Suppose now that P be fundamental for T and $\bar{P}$ non-fundamental for T^{-1}: then the determinant $J(v) = \partial(g)/\partial(v)$ is itself holomorphic locally at $\bar{P}$, vanishing at $\bar{P}$, because otherwise T would be biregular at $(P, \bar{P})$. Since no point in a convenient neighbourhood of $\bar{P}$ is fundamental for T, we conclude that $J(v)$ *is necessarily null on each exceptional hypersurface of* $\bar{V}$ *through* $\bar{P}$.

b) We shall now prove:

(i) *If* L *is an* r*-dimensional linear system on* V*, with a defined Jacobian cycle* L_j *and without base points, then the Jacobian cycle* $\bar{L}_j$ *of the transformed system* $\bar{L}$ *of* L *by* T *is given by:* $\bar{L}_j = T[L_j] + \bar{E}$*, where* $\bar{E}$ *is an* $(r-1)$*-cycle on* $\bar{V}$*, consisting of exceptional components on* $\bar{V}$*.*

In the first place almost any cycle C of L is clearly neither fundamental nor exceptional for F, and one can suppose the same to hold for the Jacobian cycle L_j, with respect to all the components, by embedding, if necessary, the system L in a more ample linear system and then replacing L by almost any one r-dimensional linear subsystem of this larger system; in effect this does not change the linear equivalence class of L_j.

If P is a fixed point of V, one can represent the system L as that cut on V by a linear system of forms in $\mathbf{P}^n$, of a sufficiently large order, outside a subvariety D not containing P. Then the system L can be represented, locally at P, by the equation: $\sum_{i=0}^{r} \lambda_i\, \varphi_i(u) = 0$, where $\varphi_i(u)$ are holomorphic locally at P and the Jacobian variety L_j is consequently given by the equation: $J(u) = 0$, where $J(u) = ||\varphi_{ih}||$, $(i, h = 0, \ldots, r)$, $\varphi_{i0} = \varphi_i$, $\varphi_{ih} = \partial\, \varphi_i/\partial u_h (h \neq 0)$.

Naturally, on our hypotheses, $J(u)$ is always a holomorphic function in the neighbourhood of P. Since L is without base points we take for transform $\bar{L}$ of L the linear system consisting of all the transformed cycles of L by F. $\bar{L}$ will evidently be simple, and hence it follows that the Jacobian cycle $\bar{L}_j$ will be defined. Moreover the system $\bar{L}$, in a neighbourhood of $\bar{P} = T[P]$, will admit the equation $\sum_{i=0}^{r} \lambda_i\, \varphi_i(g(v)) = 0$, and the determinant $\bar{J}(v)$, analogous to $J(u)$, associated with this system will be meromorphic locally at $\bar{P}$ and related to $J(u)$ by the usual transformation law: $\bar{J}(v) = J(u) \cdot \partial(v)/\partial(u)$, which holds for each biregular pair of points for F, in the neighbourhoods of P and $\bar{P}$.

Recalling the result in a), we deduce at once from this relation that each analytic branch of an exceptional hypersurface of $\bar{V}$ through $\bar{P}$, is contained in the analytical set $\bar{J}(v) = 0$, i. e. in some component of the Jacobian cycle $\bar{L}_j$ of $\bar{L}$: in conclusion, this cycle contains all the exceptional $(r-1)$-dimensional hypersurfaces of $\bar{V}$ for T, and with a multiplicity completely determined by the order of $\partial(v)/\partial(u)$ at $\bar{P}$ and therefore depending only on T and not on the system L: thus the Jacobian cycle $\overline{L}_j$ is well defined.

c) Suppose now that the system L is contained in a larger linear system L^* satisfying the same hypotheses as in (b): by using the fact that, if L varies in L^*, then the related cycle $\overline{E}$ does not vary, we obtain the following corollary to (i):

(ii) *The (complete) Jacobian linear system of the transformed system on $\overline{V}$ of a linear system L^* on V, at least r-dimensional, simple and without base points, exists and is precisely that complete linear system containing the cycle $\overline{L}_j = T[L_j] + \overline{E}$, where L is related to any r-dimensional subsystem L of L^*.*

d) Now let C be a generic cycle of L on V and $\overline{C} = T[C]$ the transformed cycle on $\overline{V}$. The impure canonical systems of V and $\overline{V}$ are then respectively given by:

$$K = |C_j - (r+1)\,C|, \qquad \overline{K}^* = |\overline{C}_j - (r+1)\,\overline{C}|;$$

since, by b), $(\overline{C}_j) = (\overline{C})_j + \overline{E}$, putting $\overline{K} = T[K]$, we have:

$$\overline{K}^* = \overline{K} + \overline{E}\,. \tag{1}$$

This is the transformation law of the impure canonical system under an arbitrary birational transformation T between V and $\overline{V}$.

If $\overline{K} + \overline{E} \geqq 0$ then obviously: $\dim|\overline{K} + \overline{E}| \geqq \dim|\overline{K}|$ and $\dim|\overline{K}| \geqq \dim|K|$, where the systems $|K|$ and $|\overline{K}|$ are considered virtually free from base points {for the definition of this usual expression see e. g, ZARISKI [a], p. 30; ENRIQUES [a], p. 33}. Defining the integer $p_g(V) = \dim|K| + 1$ as the *geometric genus* of a variety V, endowed with some non-negative canonical cycle, we derive in this case the relation $p_g(\overline{V}) \geqq p_g(V)$, and, by symmetry, we have $p_g(V) \geqq p_g(\overline{V})$: thus the geometric genus is an absolute invariant under birational transformation; this can at once be extended to the case where $|K|$ is virtual, in which case $p_g(V) = 0$. Summing up we have the theorem:

(iii) *If V and $\overline{V}$ are two non-singular birationally related varieties, any canonical impure cycle $\overline{K}^*$ on V consists of the transform $\overline{K}$ of the analogous cycle on V, augmented by a certain cycle $\overline{E}$ having as components the $(r-1)$-dimensional exceptional subvarieties for T on $\overline{V}$. The geometric genus of V is an absolute invariant in the class of the non singular models of V.*

e) The theorem (iii) has some important consequences. First, the two systems $|K^*|$ and $|K|$, being either both effective or both virtual {a linear system will be called effective or virtual if it is endowed or not with some non-negative cycle} have, in the former case, the same dimension: therefore the cycle $\overline{E}$ is necessarily a fixed part of $|\overline{K}^*|$, and since an $(r-1)$-dimensional subvariety is called exceptional if it is so for some birational transformation T, we have the following corollary:

(iv) *The impure canonical system* $|K|$ *on a non-singular variety* V, *if effective (so that* $p_g(V) \geqq 0$*), contains all the exceptional varieties of* V *of dimension* $r-1$, *which, consequently, are finite in number.*

This fact is no longer true if $p_g(V)=0$: in this case it is well known, from the theory of surfaces, that one can have a variety with *an infinity of exceptional* $(r-1)$*-dimensional subvarieties*, as for instance the rational and scrollar surfaces: in fact, these are the *only* surfaces endowed with an infinity of exceptional curves {this is a famous theorem of CASTELNUOVO-ENRIQUES in [1]: see also ENRIQUES [a], p. 127; ZARISKI [a], p. 71}: it is not known whether a similar characterisation is possible in the case $r > 2$.

If $p_g(V) \geqq 0$, the linear system, obtained from the impure canonical system of V by removing all the fixed exceptional components, is called *the pure canonical system:* this is therefore an absolute invariant of V. We may summarise the properties of this system in the following theorem:

(v) *The pure canonical system on a non-singular variety* V, *of geometric genus* $p_g(V) > 0$, *is an absolute invariant of* V *and its dimension is* $p_g(V)-1$. *It can, nevertheless, lose or acquire base varieties without changing dimension, even if one considers these subvarieties as virtually non-existent. Moreover it can still have some non-exceptional fixed component.*

The second part of the theorem follows by observing that $\dim|\overline{K}^*| = \dim|K|$ since $p_g(V) > 0$, $|\overline{K}^*| = |\overline{K}+\overline{E}|$ and $p_g(V) = p_g(\overline{V})$ by (iv), where the two systems are considered *virtually free* from base points. The last assertion is proved by means of examples: an obvious one is given by a variety with $p_g(V)=1$ and a non null canonical hypersurface.

We remark that, while for $r=2$ all the exceptional components of the impure canonical system can be removed, if they are finite in number, by birational transformation, a similar result is not known for the case $r > 2$. Moreover we do not know, when $r > 2$, whether the exceptional hypersurfaces of V are necessarily of the *first species*, i. e. if each birational transformation associated with any of them cannot have a fundamental point on them: as is the case for $r=2$ {for a formal characterisation of these hypersurfaces see (VI, 5)}.

f) We have seen that, if the impure canonical system is effective, so also is the pure canonical system $|H|$: therefore, when $p_g(V) > 0$ the dimension, increased by unity, of the system $|i\,K|$, which is called the i-canonical system on V, is also an absolute invariant of V.

If $p_g(V)=0$ and the system $|i\,K| (i > 1)$ is effective, from the relation: $|i\,\overline{K}^*| = |i\,\overline{K} + i\,\overline{E}|$ which is a consequence of (1) and by repeating the proof given in (f), we again deduce *the absolute invariance of the integer* $p_g^i(V) = \dim|i\,K| + 1$, which is called *the i-genus of* V.

It follows also that the system $|i\,K|$ contains as fixed i-ple components all the exceptional hypersurfaces of V, which are consequently still finite in number: incidentally we find that a variety endowed with an infinity of such hypersurfaces must have *all the plurigenera zero.*

The complete system $|-K|$, when it is effective, is called, by SEVERI {see SEVERI [35], p. 38 and the historical note}, *the (impure) anticanonical system* and its dimension, increased by unity, is *the antigenus* of the variety V.

An unsolved question is the possibility or otherwise of deducing from the anticanonical system, which is evidently a relative invariant of V, an absolute invariant: this seems an important question because its solution might furnish a means for the birational characterisation of the unirational varieties, i. e. those varieties which can be mapped on an involution of a projective space {see (I, 6)}.

In effect both for these and for the birational varieties, the plurigenera are all zero, and it may be that there is an effective anticanonical system. When the system $|K|$ belongs to the null class of the linear equivalence, it is so also for the system $|-K|$, but if $|K|$ is reduced to the fixed exceptional components, then obviously $|-K|$ is virtual.

4. The Arithmetic Definition of the Canonical System

We shall now give, with ZARISKI {see ZARISKI [21], p. 585}, an alternative definition of the canonical system on a variety V, which will be supposed to be *normal* over a definition field in the universal domain of characteristic p. For the classical case this is simply the well known transcendental definition which goes back to CLEBSCH and M. NOETHER {see M. NOETHER [1]}.

If $x_1, x_2, \ldots, x_r$ are functions of $\mathbf{K}(V)$ forming a separating transcendence base of $\mathbf{K}(V)/\mathbf{K}$ {this signifies that $\mathbf{K}(V)$ is separably algebraic over $\mathbf{K}(x_1, x_2, \ldots, x_r)$; such a base certainly exists since the field $\mathbf{K}(V)/\mathbf{K}$ is separably generated over $\mathbf{K}$: see MACLANE [1], p. 384}, an *r-fold differential* $f\,d(x)$ on V is the differential $f\,d\,x_1\,d\,x_2\ldots d\,x_r$, where $f\in\mathbf{K}(V)$ and $r=\dim V$. Such differentials transform according to the usual Jacobian law: $f\,d(x)=f\,J\,d(x')$, where $J=\partial(x)/\partial(x')$ and (x') is the new separating base.

Let A be a prime cycle on V and ω an r-fold differential on V. Since V is normal, A is simple on V, and hence there exist uniformising coordinates $x_1, x_2, \ldots, x_r$ of A such that $\mathbf{K}(A)$ is a separable algebraic extension of the field generated over $\mathbf{K}$ by the functions induced on A by the x_i: this situation is briefly indicated by saying that the functions $x_1, x_2, \ldots, x_r$ form a set of *separating uniformising coordinates of* A. Then we can show that (x) is necessarily a separating transcendence base for $\mathbf{K}(V)/\mathbf{K}$, so that we can write ω in the form $f\,d(x)$.

Hence it easily follows that, if a function $g\in\mathbf{K}(V)$ induces on A a finite function $(g)_A$, then the induced functions $(\partial g/\partial x_i)_A$ are also finite: consequently if $\omega=f'\,d(x')$ with respect to another set (x') such as (x), the Jacobian $J=\partial(x)/\partial(x')$ and J^{-1} both induce finite functions on A

and therefore $v_A(J) = 0$ and $v_A(f) = v_A(f')$. The integer $v_A(f)$ can then be called the *order* of ω at A and denoted by $v_A(\omega)$. The cycle A is a *null* cycle or a *polar* cycle of ω according as $v_A(\omega)$ is positive or negative.

Now let $x_1, x_2, \ldots, x_r$ be the non-homogeneous coordinates of the generic point of V respect to $\mathbf{K}$: we may assume that the first r of these coordinates form a separating base of $\mathbf{K}(V)/\mathbf{K}$ and then, as one can prove, a set of separating uniformising coordinates on every prime cycle of V, outside a finite number of such cycles. Therefore, excluding these cycles, we have $v_A(\omega) = v_A(f)$ where $\omega \neq 0$ is the r-fold differential $f\,d(x)$ on V.

The cycle $\Sigma\, v_A(\omega) \cdot A$, where the sum is consequently finite, is called *the cycle of the differential* and denoted by (ω).

Since the r-fold differentials of $\mathbf{K}(V)$ form a vector space of dimension one over $\mathbf{K}$, *the cycles of any two such differentials are linearly equivalent, and any $(r - 1)$-cycle on V belonging to this equivalence class, is itself the cycle of an r-fold differential* determined to within an arbitrary non-zero constant factor: such an equivalence class is called *the canonical class of V*, and *the canonical system $|K|$ of V is defined as the complete linear system consisting of the non-negative canonical cycles.*

An r-fold differential ω is of *the first kind* on V if $\omega = 0$ or if $\omega \neq 0$ and $(\omega) \geqq 0$. The number of linearly independent r-fold differentials of the first kind over $\mathbf{K}$ will be called *the virtual geometric genus of V* and will be denoted by $p_g(V)$: obviously $p_g(V) \geqq 0$. If $p_g(V) > 0$ $|K|$ is effective and then: $\dim |K| = p_g(V) - 1$.

If V' is a normal variety birationally equivalent to V and V' is without fundamental points, an r-fold differential of the first kind on V' corresponds to an analogous differential on V: hence $p_g(V) \geqq p_g(V')$.

Since $p_g(V)$ is a non-negative integer, it follows that $p_g(V')$ attains a *minimum* in the birational class of V, which is called *the geometric genus of the field* $\mathbf{K}(V)$ or the effective genus of V, and is denoted by p_g. If, in particular, the variety V is non-singular, then one can prove that *the virtual geometric genus coincides with the effective genus* {see KOIZUMI [1]}.

If the variety V is defined over the complex field and is projected into a primal with ordinary singularities (as may be effected by almost any projection) represented by the equation $F(\xi_1, \xi_2, \ldots, \xi_{r+1}) = 0$, then {see e. g. HODGE [2]} every r-fold differential of the first kind on the primal is of the form $f\,dx_1\,dx_2 \ldots dx_r$ where $(x_1, x_2, \ldots, x_r)$ is a generic point of the primal with respect to K and $f = Q/F'$, Q being an adjoint polynomial to $F = 0$ of degree $n - r - 2$ and $F' = [\partial F(\xi)/\partial \xi_{r+1}]_{(x)}$: this gives immediately the result that the new definition of canonical system coincides with the preceding definition of ENRIQUES {see (IV, 3)} for non-singular varieties defined over the complex field.

5. Relations between Canonical and adjoint Systems

Let ω be an r-fold differential on V, with the same hypotheses as in (IV, 4) and let C_m be a generic cycle of L_m/k, where k is an algebraically closed definition field for V. Let t be a uniformising parameter of C_m, i. e. a function on V of order one on C_m. We can then find $r - 1$ other elements in $\mathbf{K}(V)$ which, together with t, form a separating set of uniformising coordinates of C_m. Let $\omega = f\, dt\, dz_1\, dz_2 \ldots dz_{r-1}$ and let us assume that C_m is neither a polar cycle nor a null cycle of ω.

If $\bar{f}$ and $\bar{z}_i$ are the functions induced on C_m respectively by f and z_i, it is easily proved by using (IV, 4) that the expression $\bar{f}d(\bar{z})$ is an $(r-1)$-fold differential on C_m, different from zero and depending only upon ω and t: this is called, with ZARISKI, the *C_m-trace* of ω relative to t and denoted by $\mathrm{tr}^t_{C_m}\,\omega$.

We shall sketch ZARISKI's proof {see ZARISKI [21], p. 587} of the following important formula:

$$(\mathrm{Tr}^t_{C_m}\,\omega) = [(\omega) + C_m - (t)] \cdot C_m\,, \tag{2}$$

where obviously the cycle on the right is a well defined $(r-2)$-cycle on C_m.

Let D be any prime $(r - 2)$-cycle on C_m: then, since D is simple on V, one can find a function u on V, such that no prime component of $(u) - C_m$ passes through D. Moreover, since D is also simple on C_m, one can choose $r - 1$ functions z_i on V such that the first is a uniformising parameter of D on C_m and the others form a set of separating uniformising coordinates of D on C_m.

It is now easily seen that ω can be written in the form: $\omega = f\,d\,t\,d(z) = g\,d\,u\,d(z)$ so that $\mathrm{Tr}^t_{C_m}\,\omega = \bar{f}d(\bar{z})$. Therefore the coefficient of D in the cycle $\left(\mathrm{Tr}^t_{C_m}\,\omega\right)$ is equal to $\bar{v}_D(\bar{f})$, where $\bar{v}_D$ is the valuation of $\mathbf{K}(C_m)/\mathbf{K}$ determined by the prime cycle D. To prove our relation, it is thus sufficient to show that $\bar{v}_D(\bar{f})$ is also the coefficient of D in the cycle on the right, which, to this end, may be replaced by the cycle $[(\omega) + (u/t)] \cdot C_m$.

After having proved that the cycle (ω) coincides locally at D with the cycle (g), ZARISKI reduces the proof to showing that:

$$\bar{v}_D\left(\mathrm{Tr}_{C_m}\, g u/t\right) = \bar{v}_D(\bar{f})\,.$$

Now, from one of the above forms of ω, one obtains: $g = f\,\partial t/\partial u$, where the partial derivatives are taken with respect to the r independent variables u, z_1, $z_2, \ldots, z_{r-1}$. Setting $t/u = \xi$, the function ξ has a finite trace on C_m and one finds: $\partial t/\partial u = \xi + u \cdot \partial \xi/\partial u$, whence: $gu/t = f + (u^2/t) \cdot \partial\xi/\partial u$. Since $\partial\xi/\partial u$ induces a finite function on C_m and u^2/t a null function, the formula is demonstrated.

The above formula has the following important geometrical significance. The system $L_m \cdot C_m$ cut on C_m by L_m is called *the characteristic system of L_m on C_m*. Let G be a characteristic cycle on C_m, so that $G = C_m \cdot \bar{C}_m$, where $\bar{C}_m \in L_m$. We take $C_m - \bar{C}_m = (t)$, and then t is a function on V. Therefore with this particular choice, the left hand side of (1) is a canonical cycle on C_m and the right hand side is of the form $(K + \bar{C}_m) \cdot C_m$, where $K = (\omega)$. We conclude with the following result:

(i) *If K is a canonical cycle on V such that C_m is not a component of K and if G is a characteristic cycle on C_m, then the cycle $G + K \cdot C_m$ is a canonical cycle on C_m.*

This fact may be stated in another form by introducing the notion of *adjoint system* $|X'|$ to a complete linear system $|X|$, defined as the system $|X'| = |X + K|$; we can then say that $|C'_m|$ *cuts on C_m canonical cycles of C_m.*

This also holds for any complete linear system $|X|$ which serves to define a birational biregular transformation of V. In particular, if A is any $(r-1)$-cycle on the non-singular variety V, the theorem is certainly true for the system $|X| = |A + C_m|$, for large m, as one deduces from (III, 7). Later we shall see that one can extend the theorem to more general cases.

Almost all results described in this chapter are classical, at least for surfaces: see e. g. ENRIQUES [a], Ch. III; ZARISKI [a], Ch. III.

The anticanonical system defined in (III, 3) has been introduced by SEVERI in [19] in order to establish an arithmetical property of the so called varieties of SEVERI-BRAUER, which are the varieties of linear order 1 in the sense of SEVERI {see [a], p. 20}: i. e. which can be mapped birationally without exceptions onto a linear space over a suitable algebraically closed extension of their definition field {see B. SEGRE [b]}. It appears incidentally in ENRIQUES and in CASTELNUOVO {see [a], p. 454}.

We remark that the anticanonical system of $\mathbf{P}^n$ is, as follows from the definitions, effective and given by the linear system of hypersurfaces of order $n + 1$: then the antigenus of $\mathbf{P}^n$ is $\binom{2n+1}{n}$ {see SEVERI [49], p. 132}.

The conjectures on the importance of the anticanonical system in the classification of birational or unirational varieties may be confronted with the interesting works [8, 9, 14] of ROTH, where there are also examples of classes of higher varieties endowed with effective anticanonical system: see ROTH [b], p. 79.

SEVERI in [40], p. 140 has proved that the absolute antigenus, defined as the maximum of the antigenera of the projective models of a given function field, is always zero for a surface, except for the rational surfaces, for surfaces with zero canonical curve and possibly for the surfaces which are birationally equivalent to elliptic scrollar surfaces: for the first class its value is 10, for the second class 1 and for the last class 2 or less. Hence SEVERI deduces a new characterisation of the rational surfaces as the surfaces having 10 as value of the absolute antigenus {see SEVERI [40], p. 140}. This theorem has, in relation to the classical characterisation of CASTELNUOVO {see e. g. ZARISKI [a], p. 75}, the advantage of using a single characterising invariant.

For the relations of the plurigenera or antigenera respectively with KÄHLER's covariant or contravariant rational tensors see KÄHLER [2, 4].

We recall that if a surface contains a suitable linear systems such that the process of taking the successive adjoint systems terminates after a finite number of steps, then, by a classical theorem of CASTELNUOVO-ENRIQUES in [1] {see ZARISKI [a], p. 74}, the surface is birationally equivalent to a ruled surface: this theorem has been extended partially to threefolds by ROTH in the researches [9, 12]: for a description of all these results see ROTH [b], Ch. V, p. 76. We finally observe that all results quoted in this note refer to varieties defined over the complex field.

V. The Arithmetic Genus

1. The Definition

This chapter is devoted to the algebro-geometric theory of the arithmetic genus of a variety, which has been recently reviewed by SEVERI {see SEVERI [35], pp. 42—81} and ZARISKI {see ZARISKI [21], Ch. III}, with a complement due to MUHLY-ZARISKI {see MUHLY-ZARISKI [1]}: the following exposition is chiefly modelled on ZARISKI's which is the most comprehensive, being applicable to *a normal variety over any field k*; nevertheless different and sometimes more far-reaching viewpoints of SEVERI are also referred to in the case where V is defined over *the complex field.*

We begin with ZARISKI's definition of the virtual arithmetic genus of an $(r-1)$-cycle on a variety.

a) Let W be a pure r-dimensional algebraic set in $\mathbf{P}^n$, i. e. one whose components are all r-dimensional. The maximum number of linearly independent forms in $\xi_0, \xi_1, \ldots, \xi_n$ with coefficients in $\mathbf{K}$ and of degree m, linearly independent also on W, is given, for m sufficiently large, by a polynomial in m, $\varphi(W, m)$, which is known as *the* HILBERT {see HILBERT [a], p. 233} *characteristic function of W.*

We recall that if K is an infinite field and R is the ring $K[\xi_0, \xi_1, \ldots, \xi_n]$, then the set of all forms of degree m of R is a K-module of rang $\binom{m+n}{n}$. If $\mathfrak{A}$ is a homogeneous ideal of R, the set of forms of degree m belonging to $\mathfrak{A}$ is a K-module, whose rank is denoted by $\chi(\mathfrak{A}, m)$. The difference $\varphi(\mathfrak{A}, m) = \binom{m+n}{n} - \chi(\mathfrak{A}, m)$ is the characteristic HILBERT function.

This function, as has been proved by HILBERT {see [a], p. 233}, is, for large m, a polynomial

$$a_0\binom{m}{r} + a_1\binom{m}{r-1} + \cdots + a_r.$$

The degree r of this polynominal is the dimension of $\mathfrak{A}$ and the coefficient a_0 is the degree of $\mathfrak{A}$. If $\mathfrak{A} = \mathfrak{J}(V)$, where V is a variety, $\varphi(\mathfrak{A}, m)$ gives the number of independent conditions in order that a hypersurface of degree m contain V, which is the so-called postulation of V: then the dimension and the degree of $\mathfrak{A}$ coincide with the usual characters for V {see e. g. GRÖBNER [a], p. 154}.

The coefficients $a_0, a_1, \ldots, a_r$ are all projective characters of $\mathfrak{A}$, and, if $\mathfrak{A} = \mathfrak{J}(V)$, they may be expressed in terms of the arithmetic genera of V {see below} and of its linear sections, as is proved by the formula (3) of the text which is due to SEVERI {see SEVERI [13]}.

SPERNER in [1] and DUBREIL in [2] have examined the interesting question of finding the conditions under which a sequence of given positive integers describes the values of HILBERT's function for any m. A related question is that of finding an expression for $\varphi(\mathfrak{A}, m)$ which would be valid for any m: this has been answered by GAETA in [2, 3]. This expression is a polynomial whose coefficients contain the orders of the forms of a base for $\mathfrak{A}$ and in addition certain characters related to the *syzygy chain* of HILBERT belonging to $\mathfrak{A}$ {see GRÖBNER [a], p. 185} and in which we regard as zero all combinatorial symbols of the type $\binom{a}{b}$ with $a < b$: this is in accord with a result of SEVERI in [2] for the case where the ideal $\mathfrak{A}$ is associated with a variety V which is a complete intersection of forms in its ambient space.

This gives, when in particular $\mathfrak{A} = \mathfrak{J}(V)$, a method of calculating the arithmetic genera of V and of its linear sections in terms of the characters considered above: this calculation has been developed by SEVERI in [2] for the particular case already quoted before. We recall that OSTROWSKI has replaced the chain of the first, second, third, ... *syzygies* of $\mathfrak{A}$ by forms in 2, 3, 4, ... series of variables {see [1] and GAETA [a]}. Denoting by $d_{1i}, d_{2i}, \ldots$, the degrees of these forms, the postulation formula is written by GAETA in the form

$$\varphi(\mathfrak{A}, m) = \binom{m+n}{n} - \sum_{i=1}^{s_1} \binom{m-d_{1i}+n}{n} + \sum_{i=1}^{s_2} \binom{m-d_{2i}+n}{n} - \cdots,$$

with the same convention as before for the combinatorial symbols {see GAETA [3] and [a]}.

Finally we recall that OSTROWSKI in [1] has demonstrated, by developing an earlier statement of MACULAY in [a], that the series $H(\xi) = \sum_{m=0}^{\infty} \varphi(\mathfrak{A}, m)\, \xi^m$, $(\varphi(\mathfrak{A}, 0) = 1)$, converges for $|\xi| < 1$ and has the sum $H(\xi) = g(\xi)/(1 - \xi)^{n+1}$ where

$$g(\xi) = 1 - \sum_{\alpha=1}^{h} (-1)^\alpha \sum_{i=1}^{s_\alpha} \xi^{d_{\alpha i}}:$$

h is the number of *syzygies* modules and $d_{\alpha i}$ the degree of OSTROWSKI's polynomials {see for other notheworthy facts GAETA [a]}. Effective postulation formulae with respect to 1-dimensional ideals in R with given projective characters have been calculated, generalising previous results of ROTH in [1, 2, 3], by TODD in [13] by means of a degeneration principle and by B. SEGRE in [7] by a direct calculation without assuming any unproved hypothesis.

Let $\mathfrak{A}$ be the homogeneous ideal of W in the polynomial ring $\mathbf{K}[\xi]$ and let R denote the residue class ring $\mathbf{K}[x]$ of $\mathfrak{A}$, where x_i is the $\mathfrak{A}$-residue of ξ_i. Then, for m sufficiently large, $\varphi(W, m)$ is equal to the dimension of the $\mathbf{K}$-module R_m of the homogeneous elements of degree m in R.

If, in particular, W is a normal variety, then the expression $\varphi(W, m)$ - 1 gives, for large m, *the dimension of the complete linear system* L_m cut on W by the forms of order m in $\mathbf{P}^n$ {see (III, 5, i)}.

We define as *virtual arithmetic genus of* W *the integer:*

$$p_a(W) = (-1)^r\,(\varphi(W, m) - 1), \tag{1}$$

which is independent of $\mathbf{K}$, since, if k is any algebraically closed field of definition for W, the ideal $\mathfrak{A}$ is the extension of the ideal which defines W in $k[\xi]$, and this does not affect either $\varphi(W, m)$ or $p_a(W)$.

b) Now let W^{r-i} be the intersection with a generic $[n-i]$ of $\mathbf{P}^n(k)$, where k is any algebraically closed field of definition for W: obviously W^{r-i} is a *pure* $(r-i)$-dimensional algebraic set.

The HILBERT function satisfies the equation:

$$\varphi(W^r, m) - \varphi(W^r, m-1) = \varphi(W^{r-1}, m)\,, \tag{2}$$

which, if we replace successively m by $m-1$, $m-2$, ..., 1 and use (1), gives, by induction on r, the following expression for $\varphi(W^r, m)$

in terms of the virtual arithmetic genera $p_a(W^i)$, $i = 0, \ldots, r$:

$$\varphi(W^r, m) = \sum_{i=0}^{r} (-1)^i \{p_a(W^{i-1}) + p_a(W^i)\} \binom{m+r-i}{r-i}, \quad (p_a(W^{-1}) = 1). \quad (3)$$

The induction proof is based on the reduction formula:

$$\sum_{i=s}^{h} \binom{i}{s} = \binom{h+1}{s+1}$$

and on the fact that (3) is true for $r = 0$, since W^0 is a finite set of points and $\varphi(W^0, m)$ is equal to the constant $1 + p_a(W^0)$, i. e. to the number of points in W^0.

c) We now generalise (3) by replacing the linear system of the hyperplanes in $\mathbf{P}^n(k)$, with its s-ple {see ZARISKI [21]}: one then obtains similarly the following formula, where $W_s^{(i)}$ denotes the intersection of W with $r - i$ generic and independent members of L_s:

$$\varphi(W^r, ms) = \sum_{i=0}^{r} (-1)^i \{p_a(W_s^{(i-1)}) + p_a(W_s^{(i)})\} \binom{m+r-i}{r-i}, \quad (p_a(W_s^{(-1)}) = 1), \quad (4)$$

where $W_s^{(r)} = W^r = W$.

If here we take $m = 1$ and replace s by m, we find *the new expression for* $\varphi(W, m)$:

$$\varphi(W, m) = r + 1 + \sum_{i=0}^{r} (-1)^i p_a(W_m^{(i)}), \quad (W_m^{(r)} = W). \quad (5)$$

Taking now $W = V$, where V is our normal variety, we have, for large m, $\varphi(W, m) = 1 + \dim L_m$ and, moreover, W_m^{r-1} is now a generic cycle of L_m/k. Denoting by $C_m^{[i]}$ the intersection of i generic independent cycles over k of L_m/k we derive the formula:

$$\dim L_m = r + \sum_{i=0}^{r} (-1)^i p_a(C_m^{[r-i]}), \quad (C_m^{[0]} = V), \quad (6)$$

where $C_m^{[i]}$ are $(r - i)$-dimensional *normal* varieties {see (III, 7)}, while $C_m^{[r]}$ is a set of gm^r distinct points {see ZARISKI [21], p. 579} g being the order of V, and therefore: $p_a(C_m^{[r]}) = gm^r - 1$.

2. The Modular Property of the Arithmetic Genus

a) We begin with a lemma:

(i) *Let Z and Z' be positive $(r-1)$-cycles on the normal variety V. If $Z \equiv Z'$, we have $\varphi(Z, m) = \varphi(Z', m)$ and hence $p_a(Z) = p_a(Z')$,*

where the same symbol denotes the cycles and their supports.

For the proof, we observe that, for large m, the system $L_m - Z$ exists and that it is cut on V by the hypersurfaces of order m which pass through Z.

Then:

$$\dim(L_m - Z) = \dim L_m - \varphi(Z, m) . \tag{7}$$

For large m, L_m and $L_m - Z$ are both complete and then $L_m - Z \equiv L_m - Z'$. Hence, for large m and consequently for any m, one finds $\varphi(Z, m) = \varphi(Z', m)$.

b) Now let k be an algebraically closed field of definition for V and for the components of the cycle Z taken as in (a). Then the genus $p_a(Z \cdot C_m)$ is well defined, because the $(r-2)$-cycle $Z \cdot C_m$ on C_m is positive and free from multiple components.

We shall prove the lemma:

(ii) *Let Z and Z' be positive $(r-1)$-cycles on the normal variety V, without multiple components, and let $k = \bar{k} = \operatorname{def}(V, Z)$. Then, if $Z' \equiv Z + C_m$, where C_m is a generic cycle of L_m/k, we have*

$$p_a(Z') = p_a(Z) + p_a(C_m) + p_a(Z \cdot C_m). \tag{8}$$

For the proof observe that the cycle $Z + C_m$, certainly positive, is free from multiple components because C_m is not a component of Z. By the lemma in (a) we have: $\varphi(Z', i) = \varphi(Z + C_m, i)$, for all i. If $\mathfrak{p}$ and $\mathfrak{p}'$ are the prime homogeneous ideals of Z and C_m in $\mathbf{K}[\xi]$, we have the well known relation:

$$\varphi(\mathfrak{p}, i) + \varphi(\mathfrak{p}', i) = \varphi(\mathfrak{p} + \mathfrak{p}', i) + \varphi(\mathfrak{p} \cap \mathfrak{p}', i) , \tag{9}$$

and, moreover, the relations: $\varphi(\mathfrak{p}, i) = \varphi(Z, i)$, $\varphi(\mathfrak{p}', i) = \varphi(C_m, i)$, $\varphi(\mathfrak{p} \cap \mathfrak{p}', i) = \varphi(Z + C_m, i)$, $\varphi(\mathfrak{p} + \mathfrak{p}', i) = \varphi(Z \cap C_m, i)$, where the last is obtained using the fact that $\mathfrak{p} + \mathfrak{p}'$ differs from the ideal of $Z \cap C_m$ in $\mathbf{K}[\xi]$ only by an irrelevant component.

Then (9) gives: $\varphi(Z', i) = \varphi(Z, i) + \varphi(C_m, i) - \varphi(Z \cdot C_m, i)$, which furnishes (8) by the definition of the virtual arithmetic genus in (V, 1).

3. The Definition of the Virtual Arithmetic Genus of a Cycle

Till now all we know is the definition of the virtual arithmetic genus for an $(r-1)$-cycle on V, free from multiple components. We shall now proceed with Zariski {see Zariski [21], p. 581}, to extend this definition to any $(r-1)$-cycle on V.

If $r = 1$ and $X = \Sigma a_i P_i$ is any zero-cycle on V, we set $p_a(X) = \Sigma a_i - 1$, according to the definition already given when the a_i are all equal to unity: in particular the virtual arithmetic genus of the null cycle on a curve is equal to -1. Assume now that *for any normal variety of dimension s, with $1 \leqq s < r$, we have a definition of $p_a(X)$ for any $(s-1)$-dimensional cycle X on V, satisfying the following conditions:*

I. If $X \equiv X'$, then $p_a(X) = p_a(X')$.

II. The virtual arithmetic genus of the null cycle of dimension $s-1$ is $(-1)^s$.

III. If $k = \bar{k} = \text{def}(V, X)$ and if C_m is a generic cycle of L_m/k, then we have *the modular property:*

$$p_a(X + C_m) = p_a(X) + p_a(C_m) + p_a(X \cdot C_m)\,, \tag{10}$$

where $X \cdot C_m$ is to be regarded as a cycle on C_m.

IV. $p_a(X)$ is invariant for any automorphism of the universal domain $\mathbf{K}$.

These conditions are obviously satisfied if $s = 1$, regarding $X \cdot C_m$ as a null cycle of dimension -1 and using formally II. Now, let V be of dimension r and let X be any $(r-1)$-cycle on V. If the complete system $|X|$ contains some positive cycle Z free from multiple components, we define $p_a(X) = p_a(Z)$: the definition is independent of the choice of Z by the lemma (V, 2, i). If this is not the case there exists a least non-negative integer $j(X)$ such that the system $|X + L_j|$ exists and contains some positive cycle free from multiple components.

We proceed by induction on $j(X)$. Assume that $p_a(X)$ has already be defined for all the $(r-1)$-cycles X such that $j(X)$ is less than a given positive integer h, and let X be any $(r-1)$-cycle on V such that $j(X) = h$. We now define $p_a(X)$ as follows: $p_a(X) = p_a(X + C_1) - p_a(C_1) - p_a(X \cdot C_1)$, where $p_a(X + C_1)$ is easily seen to be defined by the induction hypothesis and where $X \cdot C_1$ is to be considered on C_1.

Zariski afterwards shows, by a formal proof which we shall not reproduce here, that this definition is independent of the choice of the definition field k (which is actually used to define C_1) and that the conditions I—IV are satisfied. In the classical case, Zariski's definition coincides with the Severi's earlier definition {see Severi [13, 35]}.

4. The Birational Invariance of the Arithmetic Virtual Genus

We proceed to outline the Muhly-Zariski proof of the relative invariance of the character $p_a(V)$ for an r-dimensional variety V defined *over any algebraically closed field* k {see Muhly-Zariski [1]}.

Let V and $\bar{V}$ be models of a field Σ of degree of transcendence r over an algebraically closed field k and let (x) and (y) be generic homogeneous points respectively of V and $\bar{V}$. The ring $k[\xi, \eta]$ is then mapped onto the ring $k[x, y]$ by a homomorphism sending ξ_i into x_i and η_i into y_i, whose kernel is a doubly homogeneous prime ideal $\mathfrak{A}$ in $k[\xi, \eta]$, which obviously represents, in the projective bihomogeneous space $\mathbf{P}[\xi, \eta]$, the graph of the birational correspondence relating V and $\bar{V}$.

As a natural extension of the Hilbert function for a homogeneous ideal in a ring of coordinates, van der Waerden {see van der Waerden [3]} has called characteristic function or Hilbert function *the number of the bihomogeneous forms in* (ξ, η) *of degree* m *in* (ξ) *and* n *in* (η) *which are linearly independent over* k mod $\mathfrak{A}$: we shall denote this function by

$\varphi(m, n, \mathfrak{A})$. It is an enumerative function uniquely determined by the models V and $\overline{V}$ of the field Σ.

The main VAN DER WAERDEN result characterises the structure of this function in the form:

$$\varphi(m, n, \mathfrak{A}) = \Sigma_r\, a_{ij} \binom{m}{i} \binom{n}{j}, \quad \text{with} \quad m \geqq M,\ n \geqq N\,, \tag{11}$$

where the sum is extended over all integers i, j such that $i + j \leqq r$. The polynomial on the right side hand associated with the pair $V, \overline{V}$ will be denoted by $\varrho(m, n)$.

Now let L_m and $\overline{L}_n$ be the linear systems over k, cut respectively on V and $\overline{V}$ by the forms of order m and n. We can consider, on the SEGRE variety $V \times \overline{V}$, the graph W of the birational transformation T relating V to $\overline{V}$ and we can think of the system L_m and $\overline{L}_n$ transferred, without change of name, to W.

Let $L_m + \overline{L}_n$ be the minimal linear sum of these systems on W i. e. the least (not necessarily complete) linear system containing both L_m and $\overline{L}_n$ and let $|C_m + \overline{C}_n|$ be the complete sum. We define the two enumerative functions:

$$r(m, n) = 1 + \dim|C_m + \overline{C}_n|\,, \tag{12}$$

$$s(m, n) = 1 + \dim(L_m + \overline{L}_n)\,. \tag{13}$$

It follows from the definition of $\mathfrak{A}$ that: $\varphi(m, n, \mathfrak{A}) = s(m, n)$ and therefore the equation (11) can be written in the form:

$$s(m, n) = \varrho(m, n)\,, \quad \text{when} \quad m \geqq M,\ n \geqq N\,. \tag{14}$$

We shall show that, under suitable conditions for T, V and $\overline{V}$, the sum $L_m + \overline{L}_n$ is complete for all m and large n or for all n and for large m; thus for such m, n we shall have: $r(m, n) = s(m, n) = \varrho(m, n)$.

a) By a lemma of CASTELNUOVO {see CASTELNUOVO [a], p. 440 (Note **)} which ensures the completeness of the minimal sum of two linear series on an algebraic curve, when the first g_ν, complete and non-special, contains partially the second g_μ, free from fixed points, in such a manner that the residual series $g_\nu - g_\mu$ be non-special, we easily deduce the following lemma, which is the first stage of our next inductive proof:

(i) *If in the pair V, $\overline{V}$ the variety $\overline{V}$ is a normal curve, there is an integer n_0 such that $L_m + \overline{L}_n$ is complete for $n > n_0$ and any m. Moreover if μ and ν are respectively the orders of V and $\overline{V}$ and if π is the common genus of V and V, then:*

$$\varrho(m, n) = m\,\mu + n\,\nu - \pi + 1\,, \tag{15}$$

and $r(m, n) = \varrho(m, n)$ whenever $m\,\mu + n\,\nu > 2\,\pi - 2$.

The latter half of the lemma follows at once by the RIEMANN-ROCH theorem applied to the series $|C_m + \overline{C}_n|$.

b) Here are two new definitions.

One is that of a *sectionally normal variety:* beginning with a non-singular curve, we pass, by induction, to any variety whose hyperplane sections are sectionally normal; the other definition is that of *a proper birational transformation* of a variety V: this is a birational transformation which transforms almost all sections of V by the linear subspaces of its ambient space into normal subvarieties, the transform of V also being normal.

It is clear that all non-singular varieties are sectionally normal and that regular birational transformations defined over sectionally normal varieties are certainly proper.

We can now prove the theorem:

(ii) *If the transformation* $T: V \to \overline{V}$ *is proper, then* $L_m + \overline{L}_n$ *is complete when* $n \geqq n_0$, $m \geqq 0$.

The proof proceeds by induction with respect to the dimension r of Σ/k.

By the lemma in (V, 4a) this is true if $r = 1$: assume that this is also true if Σ is of dimension $r - 1$ over k. We shall then prove the theorem by induction with respect to m when $\dim \Sigma/k = r$. In fact $L_m + \overline{L}_n$ is certainly complete if $m = 0$ and n is large, because T is proper and V is normal. Let us assume the completeness of $L_m + \overline{L}_n$ for large n. Since T is proper, there is a cycle C_1 of L_1 such that $\overline{C}_1 = T(C_1)$ is a normal variety and the induced transformation $T': C_1 \to \overline{C}_1$ is also proper.

The systems L_m and $\overline{L}_n$ cut on C_1 and $\overline{C}_1$ the multiple systems $L_m^{(1)}$ and $\overline{L}_n^{(1)}$: then $L_m^{(1)} + \overline{L}_n^{(1)}$ is complete, by the inductive hypothesis, if $n \geqq \bar{n}_0$. We put: $\bar{r}(m, n) = 1 + \dim(L_m^{(1)} + \overline{L}_n^{(1)})$ and let $n_0 \geqq \bar{n}_0$ be such that $\overline{L}_n$ is complete if $n \geqq n_0$. Then both the systems $L_m + \overline{L}_n$ and $|C_m + \overline{C}_n|$ cut on C_1 the complete system $L_m^{(1)} + \overline{L}_n^{(1)}$ and the residual system of both those systems with respect to C_1 is the minimal sum $L_{m-1} + \overline{L}_n$, complete by the induction hypothesis.

We conclude that the systems $L_m + \overline{L}_n$ and $|C_m + \overline{C}_n|$ have equal dimension and therefore coincide: this proves the theorem. Moreover we find the relation:

$$r(m, n) = r(m - 1, n) + \bar{r}(m, n), \quad (n \geqq n_0). \tag{16}$$

c) It is now easy to prove that: $r(m, n) = \varrho(m, n)$, *if* $n \geqq n_0$ and m is arbitrary.

In fact let $\bar{\varrho}(m, n)$ be the ϱ-function associated with the pair $(C_1, \overline{C}_1)$. The assertion is true if $r = 1$ by the lemma in (V, 4a) and then we assume that it is still true if $\dim \Sigma/k = r - 1$. Then $\bar{\varrho}(m, n) = \bar{j}(m, n)$ if $n \geqq \bar{n}_0$ and $m \geqq 0$. Since the sum $L_m + \overline{L}_n$ is complete, $r(m, n) = s(m, n)$, so that, if m and n are both large, we have $r(m, n) = \varrho(m, n)$.

From (16) follows:

$$\varrho(m, n) = \varrho(m - 1, n) + \bar{\varrho}(m, n), \tag{17}$$

for large m and n, and hence for any m or n because the ϱ-functions are polynomials. The equations (16) and (17) together imply that if $r(m, n) = \varrho(m, n)$ then also $r(m-1, n) = \varrho(m-1, n)$.

d) We next have the theorem of the relative invariance of the arithmetic virtual genus:

(iii) *If V and $\bar{V}$ are normal models of Σ and if $T: V \to \bar{V}$ and $T^{-1}: \bar{V} \to V$ are both proper transformations, then $p_a(V) = p_a(\bar{V})$.*

As immediate corollary of this theorem we find:

If V and $\bar{V}$ are sectionally normal models (in particular non-singular varieties) *in regular birational correspondence, then $p_a(V) = p_a(\bar{V})$.*

The proof of the above theorem is simple. Let us put $\varrho_1(m) = \varphi(V, m) = \Sigma\, c_{1i}\binom{m}{i}$, $\varrho_2(n) = \varphi(\bar{V}, n) = \Sigma\, c_{2j}\binom{n}{j}$ and $r_1(m) = r(m, 0) = 1 + \dim|C_m|$, $r_2(n) = r(0, n) = 1 + \dim|\bar{C}_n|$. Then by (V, 4c), since T and T^{-1} are proper, we know that there are integers m_0 and n_0 such that: $r(m, 0) = \varrho(m, 0)$, if $m \geqq m_0$, and $r(0, n) = \varrho(0, n)$, if $n \geqq n_0$. It follows that $\varrho_1(m) = \varrho(m, 0)$, $\varrho_2(n) = \varrho(0, n)$, for all the values of m and n. If a_{00} is the constant term in $\varrho(m, n)$, then $c_{10} = a_{00} = c_{20}$: hence, by the definition of virtual arithmetic genus given in (V, 1), $p_a(V) = p_a(\bar{V})$.

5. The Absolute Invariance of $p_a(V)$ ($r \leqq 3$)

Assuming that $\dim \Sigma/k = 2, 3$, *that V and $\bar{V}$ are non-singular models of Σ and that k is of characteristic zero*, Muhly and Zariski have proved {see Muhly-Zariski [1], p. 83} that *$p_a(V)$ is then an absolute character of V.*

The proof is essentially based on the possibility of eliminating, by means of regular transformations, the fundamental elements of the birational transformation relating the models V and $\bar{V}$: unfortunately this has only been proved subject to the above hypotheses. We remark that this is the only partial proof we know for the absolute invariance of $p_a(V)$ over the complex field and therefore in cases where one can freely use transcendental or topological tools.

a) We begin with a definition. If $T: V \to \bar{V}$, we shall say that $V < \bar{V}$ ($\bar{V}$ dominates V) *if T^{-1} has no fundamental points on $\bar{V}$*. Then, if, as we shall suppose, V and $\bar{V}$ are non-singular, T is proper.

Now we shall prove the following lemma where, and in all this section, it is supposed that $\dim \Sigma/k = 2$ or 3:

(i) *If the transformation $T: V \to \bar{V}$ is quadratic with centre at a point P of V or monoidal with centre along a non-singular curve $D \subset V$ (in which case* $\dim V = 3$*), then T and T^{-1} are proper transformations.*

Meanwhile, in both cases, $V < \bar{V}$, so that T is proper.

Now let T be a Q. T. and let $\mathfrak{p}$ be the prime ideal of P in the ring $k(V)$. The system $T^{-1}(B_1)$, where B_1 is the system of the hyperplane sections of V, will be of the form $f(\lambda) = \sum_{i=0}^{k} \lambda_i f_i = 0$, $(\lambda_i \in k)$, where (f) is a basis of the forms in $\mathfrak{p}$ of order ν so large that the ideal $\mathfrak{A} = (f_0, f_1, \ldots, f_k)$ can differ from $\mathfrak{p}$ by at most an irrelevant component: then (f) is a generic point of V. By BERTINI's theorem, the generic member of that system, and hence almost all members, can be singular only at the base point P, which, as one easily sees, is not the case. Therefore the system $T^{-1}(B_1)$ has almost all its members normal, and the assertion is true if $r = 2$.

If $r = 3$ one must, moreover, prove that the characteristic curves of the system $T^{-1}(B_1)$ are almost all normal: this follows from the fact that T still induces a Q. T. on any subvariety of V through P and therefore on almost all members of $T^{-1}(B_1)$.

If T is a M. T. the proof proceeds similarly: by using the fact that every point of the curve centre D is simple both for D and V, one finds first that almost all members of $T^{-1}(B_1)$ are normal: we have now only to prove that the characteristic curves of the same system are almost all normal, and for this it is sufficient to show that they are almost all non-singular.

This is accomplished by observing that the characteristic system of $T^{-1}(B_1)$, on a generic member of the same system has no base points outside D, when the fixed component D is removed, since otherwise any hyperplane section of V would contain at least a curve of the pencil which is the transform of the points of D, and this is evidently absurd. The proof is completed by BERTINI's theorem.

b) By (V, 4c) and (V, 5a) we see that the virtual arithmetic genus of a non-singular model of a field Σ of dimension two or three over k is invariant by Q. and M. T.

ZARISKI has shown {see ZARISKI [9], Ch. IV} that, if V and $\bar{V}$ are non-singular models of a threedimensional field Σ, then there exist models V_1 and $\bar{V}_1$ in regular birational correspondence such that $V_1(\bar{V}_1)$ is obtained from $V(\bar{V})$ by a sequence of Q. and M. T. with a non-singular centre {see (II, 6)}. It follows that $p_a(V) = p_a(V_1) = p_a(\bar{V}_1) = p_a(\bar{V})$. From the similar statement for surfaces, we deduce the theorem:

(ii) *If Σ is a field of dimension two or three, then any two non-singular models of Σ have the same virtual arithmetic genus.*

ZARISKI and MUHLY prove also the very interesting complement, for $r = 2$, that *if V and $\bar{V}$ are normal models* of Σ *and* $V < \bar{V}$, *then* $p_a(V) \geq p_a(\bar{V})$ {see MUHLY-ZARISKI [1], p. 86}. It follows that any normal model of Σ has the virtual arithmetic genus equal to or greater than that of any singular model: the latter can also be considered as a character of the field and denoted by $p_a(\Sigma)$.

6. The Virtual Characters of a Cycle

SEVERI has obtained {see SEVERI [35], p. 53—65; ALBANESE [4]} a wide extension of the statements in (V, 2, 3) arriving at some important formal relations between the virtual arithmetic genus of a cycle A of $(r-1)$-dimensions on an r-dimensional variety V and the virtual arithmetic genus, suitably defined, of the characteristic cycles $A^{[i]}$.

We shall give an account of this work in the case where V is defined *over the complex field*, though all of it could be generalised to any algebraically closed field of definition: this may be found partially realised in ZARISKI's work.

a) We begin with a definition: a linear system L is called *totally irreducible* if almost all its members and almost all its characteristic cycles are varieties. SEVERI shows {see SEVERI [35], p. 22} that:

(i) *If L is at least r-dimensional and free from base points, then it is totally irreducible if and only if it is simple, except when it is compounded of an involutory system of point-sets.*

The proof can obviously be reduced to showing that almost any 1-dimensional characteristic cycle C is a curve. Suppose now that C is such and put $d = \dim L$. Then the linear subsystem $\bar{L}$ of L formed of all the members of L, whose support contains almost any point P of V, is $(d-1)$-dimensional and cannot have C as base curve: in fact, otherwise, L would have a dimension equal to $r - 2$. Conversely, if L is simple, apart for a zero-dimensional system, it is not compounded of a pencil, and then we can find a member A of L which (by BERTINI's theorem), is a variety: this argument can be repeated for the first characteristic system on A, and so on, thereby proving the theorem.

b) When L is totally irreducible it is easily verified that *almost any characteristic variety is non-singular:* it is sufficient for this to repeat the latter argument in (a) and use BERTINI's theorem on singular points.

Consequently the arithmetic genus of any one of these varieties is well defined by (V, 1) and, therefore, the arithmetic genus of a generic member of L and of every characteristic system $L^{[i]}$, $i = 2, \ldots, r$, over the complex field, are also well defined: by using the invariance with respect to the linear equivalence of the arithmetic genus of a divisor on a variety (V, 3) and an inductive argument with respect to the index i of the system $L^{[i]}$, we can prove that the value of this genus is independent of any isomorphism over the complex field.

Therefore that value can be called *the virtual arithmetic genus of index i of the linear system L* and will be denoted by $p_a(L^{[i]})$ {see also (X, 9)}.

The equality of the arithmetic genera of two independent generic members with respect to the complex field of any characteristic system can be also derived from the following stronger result, due to KODAIRA {see KODAIRA [5], p. 119 and (XI, 3)} and MATSUSAKA {see MATSUSAKA [5], p. 121 and (VII, 5)}: *The virtual arithmetic genera of any two algebraically equivalent divisors on a normal variety over any definition field k are equal.* We shall later sketch the proof of this important theorem in (VII, 5).

We remark that instead the virtual arithmetic genus of a cycle of dimension less than $r - 1$ cannot be an algebraic invariant on V, since such a cycle may very well belong to several maximal families, as for instance a multiple line in $\mathbf{P}^n$.

We shall also use the notion of *virtual degree of L*, that is the order of the generic member of the last characteristic system $L^{[r]}$.

c) Now let A, B and C be generic members respectively of the linear systems L, M and N such that $N = L + M$. Moreover we shall suppose that the intersection cycle of any two generic characteristic members of L and M respectively is defined and, further, that it is a non-singular subvariety of V.

We then have the following relations:

$$[C^{[r]}] = [(A + B)^{[r]}] \,, \tag{18}$$

$$p_a(C) = p_a(A) + p_a(B) + p_a(A \cdot B) \,, \tag{19}$$

where by $[A^{[r]}]$, and similar notations for B and C, we denote the degree of the system L.

The former is obvious, and the latter is obtained by SEVERI by using the relation (9) between the HILBERT functions associated with the ideals of A and B on V and the invariance of the virtual arithmetic genus under linear equivalence {see (V, 2a)}.

The formula (19) is afterwards generalised to the following:

$$p_a(C^{[i]}) = \pi((A + B + A \cdot B)^{[i]}) \,, \tag{20}$$

where π is *a linear operator* on the sum obtained after evaluating the symbolic power, *defined by:* $\pi(l\,X + m\,Y + \cdots) = l\,p_a(X) + m\,p_a(Y) + \cdots$: moreover the symbol $p_a(A^{[h]} \cdot B^{[k]})$, when $h + k > r$, is defined as equal to $(-1)^{\varepsilon - 1}$ where $\varepsilon = h + k - r$, in agreement with (V, 3) when $h + k = r + 1$.

The proof is by induction with respect to the index i based on the linear equivalence on $C^{[i-1]}$:

$$(C^{[i]}) \equiv (A \cdot C^{[i-1]}) + (B \cdot C^{[i-1]}) \,. \tag{21}$$

We observe that formula (20) can be written more explicitly in the form:

$$p_a(C^{[i]}) = \sum_{\alpha=0}^{i} \sum_{\beta=0}^{i-\alpha} \binom{i}{\alpha} \binom{i-\alpha}{\beta} p_a(A^{[\alpha+\beta]} \cdot B^{[i-\beta]}) \,, \quad (\alpha + \beta \leqq i) \,, \tag{22}$$

which, if $i = r$, is substantially the same as (18), as one could foresee from the definition of the virtual arithmetic genus of a set of points given in (V, 1).

The formula (22) does not contain terms without geometrical significance, if and only if, $2\,i \leqq d$. Otherwise it may be convenient to transform it into another in which all such terms are replaced by their numerical value: such a formula is the following, obtained from (22) by quite simple calculations

$$p_a(C^{[i]}) = \sum_{\alpha=0}^{\delta} \sum_{\beta=0}^{i-\alpha} \binom{i}{\alpha}\binom{i-\alpha}{\beta} p_a(A^{[\alpha+\beta]} \cdot B^{[i-\beta]}) + (-1)^{\delta-1} + (-1)^{\delta} \times \times \sum_{\alpha=0}^{\delta} (-1)^{\alpha} \binom{i}{\alpha} 2^{i-\alpha}, \tag{23}$$

where $i > \delta = r - i$, and where now all the terms have geometrical significance.

Severi generalises {see Severi [35], p. 58} the preceding relations to more than two systems, but we shall not reproduce these more general formulae which may be found by a straightforward induction with respect to the number of the systems.

d) With the aim of extending the above relations to *arbitrary cycles*, we prove the lemma:

(ii) *Every $(r - 1)$-cycle of V is linearly equivalent on V to a cycle of the form $C - B$, where C and B are non-singular hypersurfaces of V, such that the complete systems $|C|$ and $|B|$ are totally irreducible and free from base points.*

We begin by proving that if X is a non-negative cycle on V then the residual linear system $L_m - X$, is, for large m, simple and free from base points: the proof is easily accomplished by using Severi's so-called method of cones.

We consider the conical hypersurfaces Γ of the ambient space $\mathbf{P}^n$ of V, which project the cycle X, from a generic $[n - r - 1]$ of $\mathbf{P}^n$. These form a system which is contained in L_d, where d is the degree of X and the system of the residual cycles $\Gamma \cdot V - X$ is then contained in the residual system $L_d - X$.

But the former system is easily seen to be free from base points and simple, and, consequently, totally irreducible by (a): therefore so also is the latter. The same statement is then obviously true for any L_m if $m \geq d$.

Now, we can write the cycle A in the form $A = H - K$, where H and K are non negative cycles: the theorem follows immediately by applying the above fact to both the cycles H and K and using the linear equivalence: $H - K \equiv C - B$, where C and B are respectively two cycles of the systems $L_m - K$ and $L_m - H$, both satisfying the required conditions for large m.

e) Now let A be an arbitrary $(r - 1)$-cycle of V: by the lemma in d) it can be represented by the linear equivalence $A \equiv C - B$, where C and B are non-singular hypersurfaces of V satisfying the other conditions required by the lemma.

Suppose that we already know the definition of the virtual arithmetic genus of the cycles $X^{[i]}$ for any cycle X supported by any non-singular subvariety of V, which is permissible, since the definition is known when V is a curve. We shall attempt *to define the characters $p_a(A^{[i]})$.*

To this end we proceed by applying formally (20) as if A were a non-singular hypersurface. We then observe that one can solve, for $i=1, \ldots, r$, every one of the corresponding relations with respect to the term $p_a(A^{[i]})$, obtaining this character in terms of $p_a(C^{[i]})$ and other characters of the type $p_a(A^{[h]} \cdot B^{[k]})$, with $h+k>0$ and $k>0$: but all these are already known, and therefore we obtain a definition of $p_a(A^{[i]})$, provided that we establish *its independence of all the arbitrary elements employed:* this is accomplished by SEVERI by some rather delicate considerations which we omit {see SEVERI [35], p. 63}.

In conclusion, the symbols $p_a(A^{[i]})$ have a definite meaning for *any* $(r-1)$-dimensional cycle on V. Moreover SEVERI states, as a formal consequence of the definition, that *these new virtual characters still obey the same laws* (20), when A, B and C are arbitrary $(r-1)$-cycles with $C \equiv A+B$: in particular, the extended modular property (19) remains valid.

7. Virtual and Effective Dimensions

If we apply (19) to the case where the cycle B is linearly equivalent to $-A$, then $p_a(C) = (-1)^r$, and hence (19) yields the following expression for $p_a(-A)$:

$$p_a(-A) = (-1)^r - p_a(A) - p_a(-A^{[2]}). \tag{24}$$

Generally one can consider, on the same hypothesis for B, the linear equivalence $A \cdot C \equiv A \cdot A + A \cdot B$ formally on A as if this were a variety, and apply the modular property: this is in fact equivalent to the formula (20) which is valid also on our hypothesis. More simply one can formally apply (24) to $A^{[i-1]}$ instead of to V, obtaining:

$$p_a(-A^{[i]}) = (-1)^{r-i+1} - p_a(A^{[i]}) - p_a(-A^{[i-1]}), \tag{25}$$

where i takes the values $1, \ldots, r$. Multiplying (25) by $(-1)^{i-1}$ and adding for $i=1, \ldots, r$, we derive:

$$p_a(-A) = (-1)^r (r+1) + \sum_{i=1}^{r} (-1)^i p_a(A^{[i]}). \tag{26}$$

We shall now call *virtual dimension of the cycle A* the expression:

$$\delta(A) = \sum_{i=1}^{r} (-1)^{r-i} p_a(A^{[i]}) + (-1)^r p_a(V) + r, \tag{27}$$

which, by (26), can be written in the more compact form:

$$\delta(A) = (-1)^r \{p_a(V) + p_a(-A)\} - 1, \tag{28}$$

due to ZARISKI {see ZARISKI [21], p. 584} for *the case* $A = C_m$ and for a *normal* variety V over any algebraically closed field.

By (6), *when we consider this case* and suppose *m large, the virtual dimension coincides with the effective dimension of the complete system* L_m, which is then called a *regular* system by SEVERI {see SEVERI [35], p. 69}.

From (28), and by using *the extended modular property*, one immediately obtains SEVERI's formula {see SEVERI [35], p. 66}:

$$\delta(C) = \delta(A) + \delta_B(B \cdot C) + 1, \tag{29}$$

where $C \equiv A + B$, and $\delta_B(B \cdot C)$ is formally defined by:

$$\delta_B(B \cdot C) = (-1)^{r-1} \{p_a(B) + p_a(-B \cdot C)\} - 1\,, \tag{30}$$

A, B and C being arbitrary $(r-1)$-cycles.

Since, when A, B and C are generic members of three complete irreducible linear systems, with $C \equiv A + B$, the effective dimensions are related by the obvious formula:

$$\dim(B \cdot |C|) = \dim|C| - \dim|A| - 1\,, \tag{31}$$

one can say that the virtual dimensions satisfy the same relation as the effective dimensions.

We now prove the theorem:

(i) *If* X *is any* $(r-1)$*-cycle on* V *then, for large* m, *the system* $|X + C_m|$ *is regular.*

For the proof we take an integer s such that the linear system $|C_s - X|$ exists and is irreducible, and we fix a prime cycle E in this system. We have then $|X + C_m| = |C_{m+s} - E|$, and hence, in order to prove the result, we have only to show that: $\dim|C_m - E| = \delta(C_m - E)$. To this end, take $A = C_m - E$: then, by (29) we have: $\delta(C_m) = \delta(A) + \delta_E(E \cdot C_m) + 1$.

But, as we have seen above, $\dim L_m = \delta(C_m)$ and $\dim E \cdot L_m = \delta(E \cdot C_m)$, for large m, and consequently: $\dim A = \delta(A)$ for the same m; which proves the theorem.

The theorem is also true on a normal variety defined over an algebraically closed field, as ZARISKI has shown {see ZARISKI [21], p. 584}.

8. A Second Definition of the Arithmetic Genus

{See SEVERI [35], p. 76.}

Let us suppose that V, still defined *over the complex field and non-singular*, be embedded in $\mathbf{P}^n$ with $n > 2r$: this hypothesis is not restrictive from our viewpoint since the biregular birational correspondence related to the system L_m on V, for large m, certainly transforms V into a model of the required type.

It is now possible to ensure the existence of $n - r$ forms of orders $d_1, \ldots, d_{n-r}$ passing through V, which, if the orders are sufficiently large, are non-singular and intersect, outside V, in a non-singular variety W, such that $V \cdot W$ is a non-singular $(r-1)$-subvariety of V {see SEVERI [1], p. 74 or [e], p. 54}.

In this situation one can prove {see SEVERI [2] and (VIII, 8)} that *the forms of* $\mathbf{P}^n$ *of order* $N = \sum_{i=1}^{r} d_i - n - 1$, *passing through* W, *cut on* V, *outside* W, *the virtual impure canonical system* {see (VIII, 8e)}; *and that the forms of* $\mathbf{P}^n$ *of order* $m + N$ *cut on* V, *outside* W, *the system* $|K + C_m| = |C'_m|$, which, for large m, is regular {see (V, 7, i)}: hence

$$\delta(K + C_m) = \dim|K + C_m| = \dim|C'_m|\,, \quad (m \geqq m_0)\,. \tag{32}$$

On the other hand, the effective dimension of $|K + C_m|$ can be calculated, for large m, by projecting V into a hypersurface V' of $\mathbf{P}^n$, with ordinary singularities. The transform of $|K + C_m| = |C'_m|$ is then cut on V', outside the double singular set S, by the forms of order $m' = m + d - r - 2$ through S, where d is the order of V'.

This system has the dimension given, for large m, by:

$$\varrho_{m'} = \binom{m'+r+1}{r+1} - \binom{m'-m+r+1}{r+1} - \varphi(S, m') - 1\,. \tag{33}$$

Calling $r_{m'}$ the polynomial in m, which is the right hand side of (33), we derive, for large m, and hence, obviously, for any m, the equality:

$$r_{m'} = \delta(K + C_m)\,. \tag{34}$$

When $m = 0$, we find, in particular, by (33):

$$\begin{aligned} r_{d-r-2} &= \binom{d-1}{r+1} - \varphi(S, d-r-2) - \binom{-1}{r+1} - 1 \\ &= \binom{d-1}{r+1} - \varphi(S, d-r-2) + (-1)^r - 1\,. \end{aligned} \tag{35}$$

But $\delta(K + C_m) = \delta(K)$, as one immediately deduces from (29), applied to the case $A \equiv K + C_m$, for $m = 0$: therefore we have, by (28) and (34),

$$\delta(K) + 1 - (-1)^r = (-1)^r \{p_a(V) + p_a(-K)\} - 1\,. \tag{36}$$

If we define P_r^a as

$$P_r^a = \delta(K) + 1 - (-1)^r\,, \tag{37}$$

(36) gives the noteworthy formula:

$$P_r^a = (-1)^r \{p_a(V) + p_a(-K)\} - 1\,. \tag{38}$$

The integer P_r^a, whose definition generalises the classical definition of arithmetic genus for a curve and a surface, *coincides, at least in the classical case, with the arithmetic genus* $p_a(V)$, previously defined: later we shall expound {see (X, 8)} the KODAIRA-SPENCER transcendental proof of this fact. As we shall see in the next section, from the strict algebraic viewpoint very little is known concerning this problem.

9. The Virtual Characters of $|K|$

We shall define, with SEVERI {see SEVERI [35], p. 78}, *the virtual characters of the canonical impure system as the integers* $\omega_i = p_a(K^{[r-i]})$, $i = 0, \ldots, r-1$. These characters are connected with the arithmetic genera on both definitions by the relation of SEVERI:

$$\begin{aligned}\omega_0 - \omega_1 + \omega_2 - \cdots + (-1)^{r-1}\,\omega_{r-1} + r + (-1)^{r-1}\\ = P^a_r(V) + (-1)^r\, p_a(V)\,,\end{aligned}\tag{39}$$

which follows at once from (27) and (37). Hence, using the same property already established for the arithmetic genus p_a, we can conclude that $P^a_r(V)$ is a relative birational invariant of V.

If V is a non-singular curve of genus $p = p_a(V)$, then $-K$ is a divisor of degree $2 - 2p$ and hence $p_a(-K) = 1 - 2p$: consequently, by (37), $P^a_1(V) = p_a(V)$. If V is a non-singular surface, consider the complete linear system $|X| = |-K + C_m|$, with m so large that $|X|$ is irreducible and such as to define a biregular birational transformation of V. Then the generic member X of $|X|$ is a non-singular variety and the system $|C_m| \cdot X$ is formed by canonical cycles of the curve X. Therefore $p_a(-C_m \cdot X) = 1 - 2p_a(X)$: hence, since $X \equiv -K + C_m$, we easily obtain $p_a(-K) = 1$, and thus, by (37), $P^a_2(V) = p_a(V)$.

ZARISKI observes {the preceding consideration is also due to ZARISKI, see [21], p. 589} that, more generally, if one has proved that $P^a(V) = p_a(V)$ when r is odd, then $p_a(-K) = 1 - 2p_a(V)$, and one can prove by the same argument that consequently $p_a(-K) = 1$ if $\dim V = r + 1$. Therefore the algebraic proof of the equality of P^a and p_a *can be reduced to the case of odd dimension.*

On the other hand SEVERI has proved {see SEVERI [35], p. 79} *that the problem can be reduced to showing that the genus P^a enjoys the modular property*; in effect one could then proceed formally as in the case of $p_a(V)$, arriving at the relation:

$$\delta^*(K) = \Omega_0 - \Omega_1 + \cdots + (-1)^{r-1}\,\Omega_{r-1} + (-1)^r\, P^a(V) + r\,, \tag{40}$$

where $\delta^*(K)$ is the virtual dimension defined as in (28) with respect of $P^a(V)$, and where $\Omega_0 = [K^{[r]}] - 1$, $\Omega_i = P^a(K^{[r-i]})$, $i = 1, \ldots, r-1$.

This relation is analogous to the following, which is an immediate consequence of (27):

$$\delta(K) = \omega_0 - \omega_1 + \cdots + (-1)^{r-1}\,\omega_{r-1} + (-1)^r\, p_a(V) + r\,. \tag{41}$$

But obviously $\delta^*(K + C_m) = \dim|K + C_m|$, for large m, and hence one easily sees that $\delta(K) = \delta^*(K)$. Using now the above expressions (40) and (41), with induction with respect to r, we at once complete the proof.

SEVERI gives also a sketch of how one could proceed to prove the modular property (19) of P^a: there remain points which need a formal proof in order to achieve a complete algebraic demonstration of the equality of the two arithmetic genera. Nevertheless, by virtue of the transcendental proof, over the complex field and for a non-singular variety, we have the following relation

$$\Omega_0 - \Omega_1 + \cdots + (-1)^{r-1}\Omega_{r-1} + r + (-1)^{r-1} = (1 + (-1)^{r-1})\, p_a(V), \quad (41)$$

which had been foreseen by SEVERI {see SEVERI [13], p. 87} and is a first example of a relation connecting the virtual characters Ω_i with the arithmetic genus.

This has been generalised by ALBANESE {see ALBANESE [4]} and by MAXWELL-TODD {see MAXWELL-TODD [1] and TODD [18]}, who have obtained the following formulae:

$$\sum_{i=0}^{r-1} (-1)^i \Phi_i = 2(p_a(V) - 1), \quad (r \text{ odd}), \quad (42)$$

$$\sum_{i=0}^{r-2k} (-1)^i \left\{\binom{k+i-1}{i} + \binom{k+i}{i}\right\} \Phi_{r-2k-i} = 0,$$

$$(r \text{ odd}, k = 1, 2, \ldots, (r-1)/2), \quad (43)$$

$$\sum_{i=0}^{r-2k-1} (-1)^i \binom{k+i}{i} \Phi_{r-2k-i} = 0, \quad (r \text{ even}, k = 0, 1, \ldots, r/2 - 1), \quad (44)$$

where $\Phi_0 = \Omega_0$ and $\Phi_i = \Omega_i + (-1)^i$. We remark that (42) and (44) with $k = 0$ give SEVERI's formula (41): moreover they reduce to a previous formula of B. SEGRE {see B. SEGRE [3]} when k is maximum. They are proved by induction and by applying the formulae to the $(r-1)$-dimensional canonical system.

We give here some historical notes on the development of the theory of the arithmetic genus for higher varieties: for surfaces the reader may consult the classical accounts in ENRIQUES [a], Ch. IV and in ZARISKI [a], Ch. IV.

Our theory was founded by SEVERI in his memoir [13] (1909), which contains almost all the fundamental facts regarding algebraic geometry on higher dimensional varieties: in particular the definition of $p_a(V)$ by the postulation formula, that of $P^a(V)$ as in (V, 8) and the modular property of $p_a(V)$. The equality $p_a(V) = P^a(V)$ is proved for threefolds by means of the property of the completeness of the characteristic system of a complete continuous system {see (XI, 4)}: a result which is not invariably valid even on surfaces {see (XI, 3)}. Nevertheless SEVERI's proof actually requires only the existence of *some* complete continuous system with complete characteristic system: since this has been proved transcendentally by KODAIRA {see (XI, 4)}, SEVERI's proof thus remains valid.

The same remarks may be made concerning the similar proof of ALBANESE for the case $\dim V = 4$, in [2]. In the same work SEVERI reduces the proof of that equality for any dimension to the proof of the fact that "any pure algebraic set may be regarded a specialisation over the complex field of a non-singular variety": hence in effect it follows that $p_a(V)$ and $P^a(V)$ can be expressed as linear combinations with constant coefficients, independent of V, of a finite number of projective characters. The demonstration is then accomplished by SEVERI by proving in a particular case the identity of the coefficients in such expressions for $p_a(V)$ and $P^a(V)$.

Moreover SEVERI proves that the demonstration of the absolute invariance of the arithmetic genus $P^a(V)$ may be reduced to the proof of another still unproved

Conjecture, to the effect that the $(r-1)$-dimensional irregularity of the hypersurfaces on $V(\dim V = r)$ is bounded, where by i-dimensional irregularity of an i-dimensional variety we mean the difference (which may be positive, null or negative if $\dim V > 2$: see SEVERI [13]) between its geometrical genus and its genus P^a.

Successively TODD in [1], by using some enumerative formulae of ROTH in [1] concerning the projective characters for a threefold in $\mathbf{P}^n$, has expressed the genus $P^a(V^3)$ by the projective characters of V^3: moreover ROTH has given in [3, 4] the numerative expressions for the characters Ω_i (defined in (V, 9)) and for P^a with respect to any dimension of V. Finally TODD in [9] has given another proof of the equality of both genera and of their relative invariance: this proof, which follows the general line of SEVERI's work, is unfortunately based on an unproved Conjecture which is substantially equivalent to SEVERI's hypothesis.

Nevertheless this research establishes many interesting connections between the characters of a variety and SEVERI's first sketch is here greatly improved. Further contributions were made by ALBANESE in [3, 4]. In [4] this author gives the greater part of the statements in (V, 6) with, however, a lesser degree of precision, especially with respect to the case of an arbitrary cycle: here are also to be found the formulae of ALBANESE-TODD in (V, 9), which have been later and independently found by MAXWELL-TODD in [1]. The effective equivalence of these formulae is not evident: the proof of their equivalence has been given by TODD in [18].

Finally we have the works [35] of SEVERI, [21] of ZARISKI and [1] of MUHLY-ZARISKI which have been described in the text: apart from the extension to abstract reference fields these works close the question of the relative invariance of $p_a(V)$ and of the absolute invariance of this character if $\dim V \leqq 3$. The equality of both arithmetic genera over the complex field has been proved by KODAIRA-SPENCER in [1] by means of CARTAN's theory of the cohomology of stacks {see (X, 8)}: algebraic proofs of this fact and of the absolute invariance of the arithmetic genus, which all appear plausible, are not available. We point out that in Ch. X we shall expound KODAIRA's theory of the genera over the complex field yet, which may be regarded as a treatment quite independent of the above.

Lastly we mention that the concept of virtual arithmetic genus of a cycle, thoroughly scrutinised by SEVERI in [35] and by ZARISKI in [21] for divisors, is the object of the earlier work [1] of LEFSCHETZ. On this subject see also our Ch. X.

VI. Algebraic and Rational Equivalence

1. The Associated Variety

We proceed to give an account of the theory of algebraic systems and of the related notions of algebraic and rational equivalence: these are chiefly based on the theory of *the* CHOW *point*, whose definition, therefore, will be briefly recalled.

Let U be a variety of $\mathbf{P}^n$, with $\dim U = r$, defined over any field k of the universal domain $\mathbf{K}$, and let P/k be a generic point of U. Consider $(r+1)$ hyperplanes, generic and independent over $k(P)$ and passing through P. Such a set can be considered as a generic point, in the product of $r+1$ copies of the dual space of $\mathbf{P}^n$, for a variety which, as one can easily see by calculating its dimension by the principle of counting constants, is in fact a hypersurface of that product.

Such a hypersurface will be represented in the $(r+1)$-linear product space by an equation of the form $F(u^{(0)}, u^{(1)}, \ldots, u^{(r)}) = 0$, *homogeneous* and, as follows by symmetry, *of the same degree d* in each series $u^{(i)}$ of variables: one can easily prove that *d coincides with the order of U*. This form has been called by VAN DER WAERDEN and CHOW, *the associated form of U/k*.

One can show that this form, save for linear transformations over the prime field, has *the same coefficients* as the equation representing the projection U' of U from a generic subspace $[n-r-1]$ of $\mathbf{P}^n$ over k: since this generic projection U' determines a hypercone, generic with respect to k, containing U, one sees that U is definable as the intersection of such hypercones; hence we deduce that *there exists a one-to-one correspondence between the varieties of* $\mathbf{P}^n$ *and their associated forms:* the point whose coordinates, in a suitable projective space, are the coefficients of the associated form of U, is called the CHOW point of U. Consequently there is also an one-to-one correspondence between the varieties of $\mathbf{P}^n$ and their CHOW points.

The question of representing biunivocally r-dimensional varieties of a given order by forms has been solved by CHOW and VAN DER WAERDEN in [1], chiefly by reason of being able to give a rigorous definition of an algebraic system and of proving that the set of all varieties V of given dimension and degree embedded in $\mathbf{P}^n$, is an algebraic system. These are obviously quite natural questions, which arise more or less explicitly in the works of many mathemathicians (e. g. CAYLEY, GRASSMANN, BERTINI).

CAYLEY's form, applied only to conics in $\mathbf{P}^3$, is precisely the CHOW form. BERTINI used instead in [a] the hyperplanar envelope of V, obtaining, however, a representation with exceptional features corresponding to varieties which do not have a well determined envelope. This BERTINI form has been investigated by B. SEGRE in [10], who has shown how it can be used to solve the above problems. Moreover the associated form is derived explicitly, in its dual aspect, by SEVERI in [16]. For further notices see SEVERI [d].

The fullest account of the associated form is in HODGE-PEDOE [a_2], pp. 32—62, where it is called CAYLEY's form. See also SAMUEL [c], p. 44; DE BAER [1]; PEDOE [1]; SEVERI [37], which last must be compared with GODDARD [1].

This correspondence can be extended to *any positive r-cycle* of $\mathbf{P}^n$, assuming as associated form of the cycle X the product of the associated forms of the single components, taken with convenient powers equal to their coefficients in an expression of the cycle X. It is clear that, if X is a rational cycle over k, then the related CHOW point is defined over k: the converse is not always true if the characteristic of k is different from zero. If the cycle X is not positive, and one considers the decomposition $X = X_1 - X_2$, where X_1 and X_2 are non-negative (sometimes called the *canonical* decomposition), one can then assume as the CHOW point of X the pair of CHOW points associated with X_1 and X_2 in a biprojective space.

The fundamental result of CHOW {see CHOW-VAN DER WAERDEN [1]; see also SAMUEL [c], p. 48} is now the following:

(i) *The set of* CHOW *points related to the positive r-cycles of the same degree* μ *supported by a variety V is an algebraic set* $\mathbf{C}^{r,\mu}(V)$, *defined over the algebraic closure of* def V.

This establishes the existence of *an algebraic-geometric structure* in the subset $\mathbf{G}^r_+(V)$ of $\mathbf{G}^r(V)$ formed by the *positive* cycles. Hence forth we shall call *algebraic system of positive r-cycles of V any set of such cycles, whose* CHOW *points fill an algebraic subset of* $\mathbf{C}^{r,\mu}(V)$: in such a manner one can interpret all the geometry of a variety and the related terminology (irreducible algebraic systems, absolute or not; generic cycles with respect to a field of definition for the system; ...). An algebraic system will be called *maximal* if it is parametrised by an absolute component of $\mathbf{C}^{r,\mu}(V)$.

2. Specialisation of a Cycle and Algebraic Systems

An immediate application of the CHOW point is to the definition of *the specialisation cycle* X' of a cycle X with respect to a field k {this problem has been characterised axiomatically by A. WEIL in [a], p. 255 and it has been solved by MATSUSAKA in [1] and by SAMUEL in [1]; see also CHOW's work [3]}. When X is positive this is equivalent to saying that the CHOW point P' associated with X' is a specialisation of the CHOW point P associated with X with respect to k.

This definition satisfies all the formal properties, which are usually required for specialisations, such as transitivity, the constancy of rational cycles over k, compatibility with the sum of cycles, algebraic projection, product and intersection-product; moreover if X is an r-cycle supported by the variety V, then any cycle specialisation of X is a V-cycle which exists for any specialisation P' of P over k. Lastly, the theorem of extension of the specialisations of cycles is still true.

It is obvious that the support $||X||$ will consequently be specialised into the support $||X'||$, in the sense that, if a point $Q \in ||X||$ and if (Q', X') is a specialisation of (Q, X) then $Q' \in ||X'||$. The following uniqueness criterion is important:

(i) *If U/k and V/k are varieties and V is non-singular, let P/k be a generic point of U and let P' be a simple point of U, such that, if X is a $U \times V$-cycle rational over k, then the V-cycle $X(P') = \mathrm{pr}_V(X \cdot (P' \times V))$ is defined; then the analogous cycle $X(P)$ has the uniquely determined specialisation $X(P')$ over $P \to P'$ with respect to k.*

It is sometimes useful to extend the notion of algebraic system *to a set of cycles not necessarily positive:* for this, one can proceed, as we have already suggested, by using the CHOW points in biprojective spaces and constraining them to lie in some algebraic set of that space. The following representation, due to A. WEIL {see A. WEIL [a], p. 204, 208; see also SAMUEL [c], p. 107}, is generally more suitable.

Let us consider a projective variety U/k and fix a rational cycle X over k on $U \times V$, with all the components projecting onto U: we then shall call *virtual algebraic system of cycles on V the set of the specialisations over k of the cycle* $X(P) = \mathrm{pr}_V(X \cdot (P \times V))$, which is certainly defined by (I, 9, i). From the same theorem and the definition of specialisation it follows at once that the notion of algebraic system of positive cycles given in (VI, 1) coincides with the present one when $X > 0$.

We state the very useful theorem:

(ii) *If a generic cycle of an algebraic system is a non-singular variety, then so also are almost all cycles of the system. If the generic cycle of an algebraic system has a specialisation which is a non-singular variety, then the generic cycle is likewise non-singular.*

The proof follows {see NÉRON [1], p. 117; MATSUSAKA [5], p. 118} easily from the definition of specialisation and from the fact that the condition for a variety to decompose or to become singular is an algebraic condition i. e. one such that the CHOW points of the cycles which satisfy the condition have as locus an algebraic set.

3. Algebraic Correspondences

Let U/k and V/k be two projective varieties and let $\dim U = n$, $\dim V = m$. We shall call, following SEVERI {see e. g. SEVERI [a], p. 94}, *algebraic correspondence* T between U and V, any cycle X, possibly non-homogeneous, on the product $U \times V$ taken in this order.

We shall usually consider only the so called *pure* correspondences, i. e. those which have as graph on $U \times V$ a *homogeneous* cycle rational over k. The correspondence T is called *non-degenerate* if the projection of each component of X on U is U; it is called *irreducible over* k if X is *rational prime over* k (*absolutely* if X is defined over k).

By the WEIL theorem (I, 9, i), with any non-degenerate correspondence T is associated the algebraic system, whose generic cycle is the cycle, rational over $k(P)$, $X(P) = \mathrm{pr}_V(X \cdot (P \times V))$, where P is a generic point of U over k.

The locus of the support of $X(P)$ is an algebraic subset of V, coinciding with the projection of X on V, and this is called *the support of the algebraic system:* if it coincides with V, for each component of X, by the same theorem (I, 9, i), there exists the *inverse* correspondence T^{-1}, which is also non-degenerate and which associates with a generic point Q of V over k the U-cycle $Y(Q)$ rational prime over $k(Q)$ defined by the relation: $X \cdot (U \times Q) = Y(Q) \times Q$.

In such a case we have the following dimensional relation:

$$\dim T = \dim X = \dim U + \dim X(P) = \dim V + \dim Y(Q)\,. \quad (1)$$

This formula gives the famous principle *of counting constants* {this principle, which is due to SCHUBERT (see [a], p. 12), has been closely scrutinised by SEVERI in [15, 32] and by VAN DER WAERDEN in [5, VI]}: it is valid provided one knows that T and T^{-1} are non-degenerate with respect to a component of X, if this is a rational prime cycle, or, otherwise, with respect to all the components.

The reader who needs more information on the purely algebraic theory of correspondences may consult, besides the quoted treatise [a] of SEVERI, the following works: VAN DER WAERDEN [b], Ch. V; SAMUEL [c], p. 108; BARSOTTI [1, 2]; ABELLANAS [2, 3, 4] and IVASAWA [2].

We turn now to consider an algebraic system of positive cycles on a variety V: let $X(P)$ be a generic cycle of such a system, where P now denotes the generic point of the CHOW variety U which, as we shall suppose, parametrises the system. We shall suppose that $X(P)$ is rational prime over $k(P)$ and that V is the support of the system. Then the $U \times V$-cycle X defined by (I, 9, i) is rational prime over k and the correspondences T and T^{-1} are non-degenerate.

Therefore T is an algebraic correspondence which is k-irreducible and totally non-degenerate: it is called *the incidence correspondence* between U and V related to the algebraic system considered {see SAMUEL [c], p. 58}. The name comes from the fact that if (P', Q') is any pair of T, then $Q' \in ||X(P')||$, as one easily verifies.

We remark that for this particular algebraic correspondence the properties of the CHOW point secure that the cycle $X \cdot (P' \times V)$ is defined *for any* $P' \in U$, i. e. the incidence correspondence has *no fundamental points on* U, where one uses a similar terminology as for birational correspondences. Moreover the formula (1) is valid for this correspondence, where now $Y(Q)$ coincides with the CHOW variety parametrising the system of cycles whose support passes through Q.

4. The Degeneration Principle of ENRIQUES-ZARISKI

This principle which, in the classical case, is due to ENRIQUES, states that:

(i) *The continuous transform of a connected set is always connected.*

It is obviously a simple topological observation, which, however, is very useful in applications {see ZARISKI [a], p. 33}.

B. SEGRE has given in [6] a bolder principle of higher character, which states that: "A specialisation C' of a curve C on a surface V, over the complex field, may be reducible, with only one exception, if and only if C' has acquired at least $1 + p_g(V)$ double points." The exception concerns the case where the new double points present dependent conditions to the canonical curves of the surface. This theorem presumably admits many extensions.

The situation in the abstract case is quite different. The extension of the principle to any field k, by ZARISKI, has required the creation of a new theory: that of *the holomorphic functions on a projective variety* V/k {the fundamental work of ZARISKI on this subject is [20]; this is preceded by the algebraic work [12], where the so-called rings of ZARISKI are introduced: see (I, 3); see also the account ZARISKI [18] and the memoir ZARISKI [17]}.

We shall here limit ourselves to giving a descriptive account of the relations between these facts.

We recall that a k-set is k-connected if it is not the union of two non-empty and k-disjoint sets. We also recall that a subfield k' of a field k is said to be quasi-algebraically closed in k if every element of k which is separably algebraic over k' belongs to k'.

a) To begin with, a formulation of our principle for the abstract case may be given as follows:

(ii) *If the generic cycle of an irreducible algebraic system is a variety, then each cycle of the system is absolutely connected.*

Here by *absolute connectedness* we mean connectedness over the algebraic closure of any definition field.

The theorem can be translated into the language of algebraic correspondences by invoking the associated incidence correspondence T which relates the support V of the system to the CHOW variety U of the system: let X be the cycle which is the graph of T on $U \times V$. Now (i) states that, if $P' \in U$, then $X(P')$ is absolutely connected. Since this is not true for every correspondence, even if non-degenerate and k-irreducible, as one sees from elementary examples in the case of a birational correspondence, one must make explicit some other property implicit in the fact that T is the incidence correspondence.

To this purpose we normalise U into a model $\overline{U}$ and call F the graph-cycle of the correspondence between U and $\overline{U}$ on $U \times \overline{U}$. If we now interchange the second and the third factors in the product $U \times \overline{U} \times V \times V$, the subvariety $F \times \Delta$, where Δ is the diagonal of $V \times V$, goes biregularly into a subvariety Z of the new product, which is the graph of a birational correspondence between $U \times V$ and $\overline{U} \times V$. Moreover, Z being biregular along each component of $X, Z \cdot (X \times \overline{U} \times V)$ is defined: we put $\overline{X} = \mathrm{pr}_{\overline{U} \times V}(Z \cdot (X \times \overline{U} \times V))$. It follows that, if $(P, \overline{P})$ is a generic pair of F, then $X(P) = \overline{X}(\overline{P})$. One can also see that the same is true for any specialised pair $(P', \overline{P}')$ over $P \to P'$ with respect to k. Since $\overline{U}$ is normal, it is also, by the theorem in (I, 5), analytically irreducible and this suggests the following generalisation of (ii):

(iii) *Let U and V be two projective varieties, X a $U \times V$ cycle rational prime over a definition field k for U and V and P' a point of U, where U is analytically irreducible. Then:* (1) *If the cycle $X \cdot (P \times V)$ is defined, the support of the cycle $X(P') = \mathrm{pr}_V(X \cdot (P' \times V))$ is k-connected*; (2) *If the cycle $X \cdot (P' \times V)$ is not defined, the set of the specialisations of $X(P)$ over $P \to P'$ with respect to k (which form an algebraic subsystem of our system) fills a support which is an algebraic k-subset k-connected of V.*

After having proved this theorem, one can immediately consider the absolute connectedness by extending the base field up to the algebraic closure of the definition field of the cycle considered, i. e. up to $\overline{k(P')}$. We remark that the theorem (iii) is substantially more general than (ii), because it considers also the case where the correspondence of graph X has fundamental points on U: in this case, as ZARISKI remarks, the connectedness of the total transform is a non-trivial fact *even in the classical case.*

b) A first step in the proof of (ii) is the search for *an arithmetic criterion for connectedness* of a k-subset W of V/k. In the classical case it is known that W is analytically irreducible along W if and only if the ring of the holomorphic functions on V defined along W is an integral ring; moreover the following criterion for connectedness holds: "If W is not connected, V is analytically reducible along W. If W is connected and V is analytically irreducible at each point of W, then V is analytically irreducible along W."

By analogy we begin by giving the definition of holomorphic function in the abstract case, starting from the known notion of sequence $\{x_i\}$ of functions $x_i \in R_k(V)$ convergent at the point Q of V. With the introduction on V of the ZARISKI topology, where the algebraic subvarieties are the closed sets, V becomes a compact topological space.

Now if (G_α) is a finite open covering of the subspace W, ZARISKI defines *a k-function holomorphic on V and defined along W as a set of sequences (x_{i_α}) with $x_{i_\alpha} \in R_k(V)$ such that:* (1) *(x_{i_α}) converges uniformly on G_α.* (2) *If the point $Q \in G_\alpha \cap G_\beta$, then the sequences (x_{i_α}) and (x_{i_β}) are equivalent in the local ring at Q, i. e. they have the same limit.*

By this definition, suitably formalised, one can prove the above criterion over any definition field k, assuming as definition of analytic irreducibility of V along W the fact that the ring $\mathfrak{o}_k^*(W)$ filled by the holomorphic functions on V along W, *is an integral ring,* a definition which extends in a natural way that given in (I, 5) when W is a subvariety. We point out explicitly that the analytical irreducibility of V along W and the analytical irreducibility of V at a generic point P of W, if any, mean different things, and neither implies the other.

We now enunciate the fundamental *invariance theorem* of ZARISKI:

(iv) *Let T be an algebraic correspondence between two varieties V and V' defined over k with (P, P') as a generic pair over k. Let W be a k-subset of V and $W' = T(W)$ the total transform of W onto V'. Suppose that:* (1) *The field $k(P)$ is algebraically closed in $k(P, P')$;* (2) *V is normal at any point of W;* (3) *T^{-1} is rational, i. e. $k(P) \subset k(P')$;* (4) *T^{-1} is semiregular at each point of W', i. e. if (Q, Q') is a pair of T, with $Q \in W$ and $Q' \in W'$, then $\mathfrak{n}_k(Q', V') \geqq \mathfrak{n}_k(Q, V)$, which implies that Q' is the unique correspondent of Q. Then, with these hypotheses, the two rings $\mathfrak{o}_k^*(W)$ and $\mathfrak{o}_k^*(W')$ are k-isomorphic.*

The proof of this theorem {see ZARISKI [20], p. 59} is very interesting, but unfortunately it is too long to reproduce. It implies that the k-connectedness of W induces that of W': in fact, if W is k-connected, the hypothesis (2) and the above criterion show that $\mathfrak{o}_k^*(W)$ is an integral ring. Therefore, by (iv) $o_k^*(W')$ is also an integral ring, and hence, by the same criterion, the k-connectedness of W' follows.

Using the fact that an everywhere regular rational correspondence transforms a connected set into a connected one, we see that the theorem is still true if one *omits the hypotheses* (3) *and* (4). Lastly, when we are interested *only in connectedness*, we can also replace (1) and (2) by the following: (1') *$k(P)$ is quasi algebraically closed in $k(P, P')$, and then the set* $\mathrm{pr}_V(X \cdot (P' \times V))$ *is an absolute subvariety of V*; (2') *V is analytically irreducible at each point of W.* The theorems (ii) and (iii) follow at once.

5. Fundamental and Exceptional Varieties

The above researches have led ZARISKI to prove or to conjecture {see ZARISKI [18], p. 131} some interesting algebraic and geometric facts.

Let $o_k(W)$ be the subring of $R_k(W)$ filled by those functions which are holomorphic along W, hence $o_k(W) \subset o_k^*(W)$. It is clear that, if W is a k-subvariety of V, then every element of $o_k(W)$ induces on W a constant: therefore $o_k(W)$ has the maximal ideal filled by the elements which induce zero on W. If W is a pure connected algebraic k-set of V this is still true, while, if W is not connected, $o_k(W)$ will have a finite number of maximal ideal associated to the single connected components of W. Hence we could deduce that $o_k(W)$ is a local ring in the sense of KRULL {see KRULL [2]; see also note 15 to Ch. I}, or a semi-local ring {see (I, 3)} in the sense of CHEVALLEY {see CHEVALLEY [1]}, as soon as we knew that $o_k(W)$ is *a Noetherian ring*. This has not been generally established.

Nevertheless this is certainly the case *when W is a point*, and therefore one could attempt to extend this character at least when W is transformable into a point by some transformation which would not change the Noetherian character of the ring $o_k(W)$. From this one arrives at the consideration of the exceptional subvarieties or subsets of V, which are defined by ZARISKI in the following manner:

An algebraic subset W of a variety V is called exceptional if there exists a rational transformation T of V into another variety V', such that: (1) *T is semi-regular at each point of W*; (2) *$W_0' = T(W)$ is a zero-dimensional set*; (3) $T^{-1}(W_0') = W$.

If V is a normal variety one can also suppose T such that V' is still normal and $R_k(V')$ algebraically closed in $R_k(V)$. Then the valuation theory furnishes $o_k(W) = o_k(W_0')$, and hence the assertion that *the former ring is semi-local when W is exceptional.* Moreover, there may exist some points Q of V, but not of W, such that every rational k-function on V, holomorphic and null along W, is also such at Q.

Let $H(W) \supseteqq W$ be the set of these points and let $\overline{H(W)}$ be the least algebraic set over k containing $H(W)$. One can now show that, if

$H(W) = \overline{H(W)}$, this set is *exceptional*, and one can still prove the following characterisation of an exceptional connected subset of V: *The ring $o_k(W)$ must contain some non-constant functions and we must have $H(W) = W$.*

We shall now give a definition which formalises another classical concept: Let L/k be a linear system on V. A subset W of V, pure and k-connected, will be called *fundamental for L, when, if A_1 and A_2 are two members of L such that $W \subset A_1 \cap A_2$ and if x is a function of $R_k(V)$ with $(x) = A_1 - A_2$, then the function x induces a constant on W.*

Now let L be the linear system determined by a submodule of $o_k(W)$: it is then clear that *W is a fundamental set for L* and *that L is free from base points on W.* The converse is also true: this signifies that there is a member of L free from intersections with W and hence the relevant consequence that L and its n-ple L_n satisfy *both or neither* of the above conditions.

Taking into account that $o_k(W) = o_k(H(W))$, we see that the proof of the Noetherian character of $o_k(W)$ is reduced to showing *that* $H(W) = \overline{H(W)}$ by applying the above fact that $H(W)$ is then exceptional. This equality *may be transformed* in terms of linear systems.

First, if L_n fullfils the above conditions, every k-component Z of $\overline{H(W)}$ is fundamental for L: in fact, if L is defined by a base of functions (x_i) of $o_k(W)$ and if P is a generic point of Z over k, then $P \in H(W)$ and therefore the functions x_i induce constants on Z. Let us suppose that Z is not a component of $\overline{H(W)}$ and let Q be a point of Z and not of $H(W)$.

There will exist a function x belonging to all the maximal ideals of $o_k(W)$ and non-determined at Q: hence x determines a pencil of cycles of L which still satisfies the above condition and has a base point at Q. Further, Q is a base point for L_n since otherwise every element of L_n through Q would contain Z, which is fundamental for L_n: this is absurd since a member of L_n which is the sum of n members of L cannot contain Z.

Concluding this analysis, we find that, *if $H(W)$ is not an algebraic set, then there exists a linear system L, a subvariety Z of V and a point Q of Z such that Z is fundamental and Q a base point for L_n with any n.*

Here ZARISKI *conjectures* that, if Z is *fundamental* for *all* the multiples of a given system L, then the system L_n, for sufficiently large n, will be *free from base points on Z*, and, more boldly, that *if L is a complete linear system on a non-singular variety V, with* $\dim L \geqq 1$ *and without fixed components, then the complete system L_n will be free from base points for large n.*

Both conjectures, and especially the former, present difficult and interesting problems: at present all we know is that the latter holds for a surface, as we easily deduce from the RIEMANN-ROCH theorem in the case where n is large {see (X, 9)}.

6. A Property of Chow Varieties

Apart the fundamental defining property, the algebraic structure of the Chow varieties is little known: on the other hand one can easily see that any variety can be regarded as the Chow variety of a suitable algebraic system of positive cycles. Therefore the problem can be considered only with some additional conditions or from some particular viewpoint.

Chow has sought {see Chow [3]} for some property characterising *the incidence correspondence* T in relation to the possibility (or otherwise) of foreseeing when a point of a Chow variety will be *non-singular* by merely observing the behaviour of T and of the support V of the system.

More precisely, let P' be a point of U and $X(P')$ the associated V-cycle: we look for a local property of the cycle $P' \times X(P')$ on the graph of T, which will determine in a birationally and analytically invariant way the nature of P'. We recall that an algebraic system is called involutory if the inverse of the incidence correspondence is a rational application of V onto U: in such a case by (1) we have: $\dim V = \dim X(P) + \dim U$, and through a generic point of V passes only one cycle of the system.

Then Chow's result is as follows:

(i) *Let V be a variety supporting an involutory algebraic system and let U be the associated* Chow *variety: if P' is a U-point and Z is a simple component of the associated cycle $X(P')$, along which the inverse of the incidence correspondence is regular, then V and the product variety $U \times Z$ are analytically equivalent at Z and $P' \times Z$ respectively.*

Here analytic equivalence means *isomorphism* of the respective local rings.

The proof uses some structural relations of the local rings on the product varieties and the reduction to dimension zero. It rests chiefly on the following two algebraic facts, which are consequences of our hypotheses. If k is a definition field for all the varieties under examination and if (P', Q') is a pair of T, with $Q' \in Z$, then: (1) *The local ring $\mathfrak{n}_k(P'; U)$ is a subspace of the local ring $\mathfrak{n}_k(Q'; V)$, with respect to the natural topology in both rings*; (2) *The prime maximal ideal of $\mathfrak{n}_k(Q'; V)$ admits a finite base over $\mathfrak{n}_k(P'; U)$.*

An immediate corollary of Chow's theorem is *a condition for points or subvarieties to be simple:* $U \times Z$ is simple at $P' \times Z$ if and only if U and Z are simple at P' and Z respectively, i. e. *if and only if U is simple at P'*. Finally we see that *V is simple along Z if and only if P' is simple on U.*

7. Algebraic Equivalence

Let V be a non-singular variety and let $\mathbf{G}^r(V)$ be the group of the r-cycles, $r = 0, \ldots, \dim V$, on V. We introduce the following subgroup of $\mathbf{G}^r(V)$:

Def. a — *$\mathbf{G}_a^r(V)$ is the group of the r-cycles on V which are algebraically equivalent to zero on V: a cycle $A \in \mathbf{G}_a^r(V)$ if and only if there exist a cycle X on $V \times U$, where U is a non-singular curve, and a 0-cycle a of degree zero on U, such that:*

$$A = \mathrm{pr}_V(X \cdot (V \times a)).$$

The quotient group $\mathbf{G}^r(V)/\mathbf{G}_a^r(V)$ is called *the group of algebraic equivalence* for the r-cycles on V: two cycles A, A' of the same coset will be called *algebraically equivalent* and denoted by the relation $A \equiv A'$.

We now prove the theorem:

(i) *Any two r-cycles Y_1, Y_2 of the same irreducible algebraic system on V, are such that $Y_1 - Y_2 \in \mathbf{G}_a^r(V)$.*

Let $X(P) = \mathrm{pr}_V(X \cdot (P \times V))$ be the cycle of the given system related to the generic point P/k of the Chow (projective) variety U/k of the system. Then, if Y_1 and Y_2 are any two specialisations of $X(P)$ over k, we must prove that $Y_1 - Y_2 \in \mathbf{G}_a^r(V)$. We observe that there certainly exist two points P_1 and P_2 of U such that Y_i is a specialisation of $X(P)$ over $P \to P_i$ with respect to k, $(i = 1, 2)$.

Therefore the assertion is true if U is a non-singular curve, since P_1 and P_2 are then simple U-points and since the cycle $\mathrm{pr}_V[X \cdot (V \times (P_1 - P_2))]$ is defined, U being a Chow curve: taking into account the uniqueness of specialisations at simple points (VI, 2), this cycle must coincide with $Y_1 - Y_2$. If, instead, U is a singular curve we proceed by normalisation, falling back on the preceding case.

Suppose now $n = \dim U > 1$, and consider two generic points P_1 and P_2 of U with respect to k, denoting by $X(P_1)$ and $X(P_2)$ the associated cycles. We shall now fibre U by the sections with two hyperplanes $L^{(a)} (a = 0, 1)$ of the ambient space $\mathbf{P}^N$ of U, defined over the fields $k(u^{(a)})$, $a = 0, 1$, where the $u^{(a)}$ are independent variables over $k(P_1, P_2)$ and where we suppose, as we can without any restriction, that P_1 and P_2 are independent over k. Then a generic point of $L^{(a)} \cdot U$ over $k(u^{(a)})$ is generic for U/k and a generic point for $L^{(0)} \cdot L^{(1)} \cdot U$ over $k(u^{(0)}, u^{(1)})$ is generic for $L^{(a)} \cdot U$ over $k(u^{(a)})$, all the intersection products being varieties: the proof for this case is then easily accomplished by induction with respect to n.

If, finally, P_1 and P_2 are not generic, we observe that Y_i is still the unique specialisation of $X(P)$ over $P \to P_i$ with respect to k and then use Weil's result {see A. Weil [a], p. 276} according to which there exists a generic specialisation Q of P over k and a field $K \supset k$ such that Q has as locus over k a curve containing P_1 and P_2: which reduces the case to the above. This is Igusa's proof of (i) {see Igusa [3]; A. Weil [6]}.

The calculus of cycles, def a, and specialisation applied to the cycles immediately give the following properties:

(ii) *If V and U^m are two varieties and $A \in \mathbf{G}_a^r(V)$, then $A \times U^m \in \mathbf{G}^{r+m}(V \times U)$.*

(iii) *If U^m is a simple subvariety of V^n, $A \in \mathbf{G}_a^r(V)$ and $(A \cdot U)_V$ is defined, then $A \cdot U \in \mathbf{G}_a^{r+m-n}(V)$.*

(iv) *If V and U are two varieties, $A \in \mathbf{G}^{r+m}(V \times U)$* and *$W \subset V \times U$*, with $\dim V = \dim W$ and $\mathrm{pr}_V W = V$, then $\mathrm{pr}_V(A \cdot W) \in \mathbf{G}_a^r(V)$.

Denoting by U^m a generic section, over a transcendental extension of a definition field k for both V and U, by a linear variety of the space $\mathbf{P}^N$ which contains V and is of dimension $N - \dim A - 1$, we find, applying (iii) that A is of degree zero, and hence:

(v) *The cycles of* $\mathbf{G}_a^r(V)$ *have degree equal to zero, and if* $A - A' \in \mathbf{G}_a^r(V)$ *then* $\deg A = \deg A'$.

We now give another definition:

Def. *b*. *Two cycles* A, A' *are called chainwise equivalent on* V, *if there are* $(l+1)$ *r-cycles* $A = A_1, A_2, \ldots, A_l = A'$, l *non-singular curves* $C_1, C_2, \ldots, C_l$ *and a* $(C_i \times V)$*-cycle* X_i, *such that, for each* $i = 1, 2, \ldots, l$, $A_i = \mathrm{pr}_V(X_i \cdot (P_i \times V))$, $A_{i+1} = \mathrm{pr}_V(X_i \cdot (P' \times V))$, *where* $P_i \in C$, $P'_i \in C_i$.

This definition would seem to be more general than def. *a*: instead, we prove:

(vi) *In* $\mathbf{G}^r(V)$ *algebraic equivalence is the same as chainwise equivalence.*

Obviously it is sufficient to prove that, if A and A' are equivalent by def. *b*, then they are also equivalent by def. *a*. In order to show this let $C = C_1 \times C_2 \times \cdots \times C_l$, $P = P_1 \times P_2 \times \cdots \times P_l$, $P' = P'_1 \times P'_2 \times \times \cdots \times P'_l$ $(P \in C, P' \in C)$, $D_i = C_1 \times C_2 \times \cdots \times C_{i-1} \times C_{i+1} \times \cdots \times C_l$ $(i = 1, 2, \ldots, l)$ and last let $X = \sum_{i=0}^{l} X_i \times D_i$ which can be considered as a $C \times V$-cycle. Hence we readily derive the relations:

$$A + \sum_{i=2}^{l} A_i = \mathrm{pr}_V(X \cdot (P \times V)),$$

$$A' + \sum_{i=2}^{l} A_i = \mathrm{pr}_V(X \cdot (P' \times V)).$$

Then (i) gives immediately $A - A' \in \mathbf{G}_a^r(V)$.

Let us now consider two cycles A and A' of $\mathbf{G}_+^r(V)$: if they are algebraically equivalent, they have, by (v), the same degree μ: therefore they belong to the CHOW set $\mathbf{C}^{r,\mu}(V)$ and are here chainwise equivalent with the supplementary condition that in def. *b* one can take each X_i to be *positive*. Then the theorem (vi) at once ensures the existence of a cycle $B \in \mathbf{G}_+^r(V)$ such that $A + B$ and $A' + B$ belong to one and the same irreducible algebraic system of cycles of $\mathbf{G}_+^r(V)$. Hence follows the theorem:

(vii) *The group* $\mathbf{G}_a^r(V)$ *can also be defined as that generated by all the cycles of* $\mathbf{G}^r(V)$ *which are differences of cycles of* $\mathbf{G}_+^r(V)$ *belonging to the same algebraic irreducible system of* $\mathbf{G}_+^r(V)$.

From the fact that $\mathbf{C}^{r,\mu}(V)$ is an algebraic set it follows that any cycle of $\mathbf{G}_+^r(V)$ belongs to one and only one maximal system of algebraically equivalent positive cycles: this comes to saying that the intersection of $\mathbf{G}_+^r(V)$ with an element of $\mathbf{G}^r(V)/\mathbf{G}_a^r(V)$, which coset, as a set of cycles, is neither algebraic nor dimensional, is *an algebraic system* (absolutely connected or not; irreducible or not).

Algebraic equivalence has been chiefly considered by the Italian school in its work on continuous systems on surfaces {for history see ZARISKI [a], Ch. V} with outstanding contributions by ALBANESE [1] and SEVERI [10, 12, 14] and, besides, by POINCARÉ in [4, 5] and LEFSCHETZ in [a]. A close examination of the principles has been made by SEVERI in [31]. See also SEVERI [c] and, for an axiomatic characterisation, A. WEIL [a], p. 259. Further simplifications are due to SEVERI [40].

8. Rational Equivalence

We now introduce, in a manner analogous to def. *a* in (VI, 7), the following subgroup of $\mathbf{G}^r(V)$:

Def. *a*. $\mathbf{G}^r_l(V)$ *is the group of the r-cycles on V rationally equivalent to zero on V. A cycle* $A \in \mathbf{G}^r_l(V)$ *if and only if one can find a cycle X on* $V \times D$, *where D is a projective line, and a zero-cycle of degree zero a on D, such that:*

$$A = \mathrm{pr}_V(X \cdot (V \times a)) .$$

The quotient group $\mathbf{G}^r(V)/\mathbf{G}^r(V)$ is called *the group of rational equivalence* for the r-cycles on V: two cycles A, A' of the same coset will be called *rationally equivalent* ($A \equiv A'$).

It is hardly necessary to point out that, if $r = \dim V - 1$, i. e. if A is a V-divisor, then *rational equivalence is the same as linear equivalence* for divisors defined in (III, 3): this follows at once by considering the graph on $V \times D$ of a function of $k(V)$.

Now theorem (i) in (VI, 7) can be expressed in terms of rational equivalence, as soon as one supposes the algebraic system to be *rational:* we have only to pass to a birational model of this system which is a projective space and such that the representation is biregular at the Chow points of the cycles Y_1 and Y_2.

Moreover, all the theorems in (VI, 7) are valid *for rational equivalence, when to* (iv) *one adds the restrictive condition* $[W:V] = 1$. The analogue of def. *b* can also be stated by replacing the curves C_i by projective lines. Observing that the prime cycle C of the proof of (vi) is then associated with a support, which is a rational variety {see (I, 6)}, one finds at once *that* (vi) *holds also for rational equivalence.* By the obvious inclusion:

$$\mathbf{G}^r_l(V) \subset \mathbf{G}^r_a(V) \subset \mathbf{G}^r(V),$$

and by what has been said about algebraic equivalence in (VI, 7), we can assert that a rational equivalence class, i. e. an element of $\mathbf{G}^r(V)/\mathbf{G}^r_l(V)$, if intersected by $\mathbf{G}^r_+(V)$, determines an algebraic system of cycles. Hence we may deduce *the uniqueness of the total system of cycles of* $\mathbf{G}^r_+(V)$ *rationally equivalent to a given cycle of* $\mathbf{G}^r_+(V)$: naturally this system may well be reducible or not connected.

There is, however, a marked difference between rational and algebraic equivalence, since one cannot here be sure that *the absolute components of these systems are still rational:* this is substantially due to the fact that two rationally equivalent cycles, which belong to the same component of $\mathbf{G}^r_+(V)$, can always be embedded in a rational system of cycles of $\mathbf{G}^r(V)$, by def. *a*, but we do not know if they can also be embedded in a rational system belonging to $\mathbf{G}^r_+(V)$: this question has been proposed many times by Severi {see Severi [31]}, but is still unsolved even in the simplest case of curves of a projective space.

If, for instance, one considers the group $\mathbf{G}^{0,1}(V)$, i. e. the totality of V-points, one must then decide if $\mathbf{G}^{0,1}_+(V)$, i. e. V considered as a set of points, is or is not birational, when it is contained in a rational system of $\mathbf{G}^{0,1}(V)$: in the case $\dim V = 2$, SEVERI has shown that V is certainly of geometric genus null and free from torsion {see (VI, 10)}: so the above question turns on deciding whether such a surface is or is not rational, but, unfortunately, this problem also is unsolved.

It is interesting to observe that, if V contains an algebraic system of cycles of $\mathbf{G}^r_+(V)$ such that any two are rationally equivalent on an $(r-1)$-dimensional subvariety W of V, then, as one readily sees, the system is a rational system of $\mathbf{G}^r_+(V)$ and conversely: it follows that the above question is equivalent to that of deciding whether or not, for any rational system S of $\mathbf{G}^r(V)$, the above hypothesis is verified by the system $S \cap \mathbf{G}^r_+(V)$.

Lastly, we observe that, while, if $A - A' \in \mathbf{G}^r_l(V)$ and $U \subset V$, then $A - A' \in \mathbf{G}^r_l(V)$, the converse is *not* always true, as is proved by the trivial example of two simple points of an irrational plane curve.

Rational equivalence and the consequent important notion of a system of equivalence is due to SEVERI, who has considered it in various works {e. g. [18, 20, 23, 28]}, later summed up partially in [a]. See also SEVERI's criticism in [31, 41] and the historical account [a] of CONFORTO: further notices will be found below.

9. The Intersection-product for Equivalence Classes

Let A^r and B^s be two cycles on the non-singular variety V^n. It may be that the intersection-product $A \cdot B$ is not defined on V, because some component of the set $\|A\| \cap \|B\|$ has dimension higher than $r + s - n$. In this situation we have to decide the important question whether there is a cycle A' on V, with $A - A' \in \mathbf{G}^r_a(V)$, and such that $A' \cdot B$ is defined on V.

By linearity one can reduce the question to the case where A and B are varieties, and, then, if $V = \mathbf{P}^n$, one proves, without difficulty, that A, for example, can be included in an algebraic system of positive r-cycles, whose generic cycle has a defined intersection product with B: such is the transform of A by a homography having generic parameters over the field $k = \mathrm{def}(A, B)$, as SEVERI and VAN DER WAERDEN have proved {see SEVERI [22] and VAN DER WAERDEN [5, XIV]; a more correct proof is by HODGE in [3], reproduced in HODGE-PEDOE [a_2], p. 141}.

When V^n is not a projective space the subvariety A may very well not belong to an algebraic system of positive V-cycles. SEVERI has overcome {see SEVERI [21], and the fuller account in SEVERI [c]} the difficulty by using virtual algebraic systems: the principal tool of SEVERI's research is *the method of projecting cones* which is here applied in successive stages, as follows: we omit the proofs, which rest on the principle of counting constants and on specialisation theory {see HODGE-PEDOE [a_2], Ch. XII; VAN DER WAERDEN [5, XIV]}.

Let $\mathbf{P}^N$ be the ambient space of V^n and let A^r be a subvariety of V defined over a field k which is supposed to be also a definition field for V and for any other variety considered later. In the following theorems, we denote by W the cone projecting A from a subspace $[N-n-1]$ of $\mathbf{P}^N$, generic over k:

(i) *The cycle* $W \cdot V$ *is defined on* $\mathbf{P}^N$.

(ii) *The components of the cycle* $W \cdot V$, *among which there is* A, *are all simple.*

(iii) *A simple point of* A *does not lie on any other component* A_i *of* $W \cdot V$.

(iv) *If* B *is a subvariety of* V, *then each component of* $A_i \cap B$, *not contained in* $A \cap B$, *is proper and each component of* $A_i \cap B$ *contained in* $A \cap B$ *is properly embedded.*

Suppose now that $A \cap B$ has some component of maximal dimension $d > r + s - n$ and take $X = W \cdot V - A$ where by (iv) X is an r-cycle of V: then still by (iv) the set $B \cap \|X\|$ has only components of dimension less than d. On the other hand there is a $(N + r - n - 1)$-cycle of $\mathbf{P}^N$, let it be $\overline{W}$, such that W and $\overline{W}$ are algebraically equivalent and, moreover such that the cycle $\overline{W} \cdot B$ is defined on $\mathbf{P}^N$, as follows from the case $V = \mathbf{P}^N$ already considered.

Hence the intersection-product $\overline{W} \cdot V \cdot B$ is also defined by the formal properties of intersection and we obtain, by (VI, 7), $W \cdot V - \overline{W} \cdot V \in \mathbf{G}_a^r(V)$: it follows that $\overline{W} \cdot V - X$ is algebraically equivalent to A on V and that, besides, the set $||\overline{W} \cdot V - X|| \cap B$ has only components of dimension less than d. Hence, proceeding by induction with respect to the maximal dimension of the components of $A \cap B$, one finds a cycle $A' \equiv A$, such that $||A'|| \cap B$ is defined on V.

Moreover one can readily show that, at the cost of passing to an algebraic extension k' of k, if this is finite, one can suppose A' to be defined over k' when the vertex of the cone W is taken to be rational over k' instead of generic.

The cycle $A' \cap B$ belongs to an equivalence class with respect to $\mathbf{G}_a^r(V)$, which is uniquely defined by the equivalence classes of A and B on V, as follows easily from the properties of the specialisation of cycles, which preserves algebraic equivalence and intersection-products {see (VI, 2)}. Since, moreover, in the above construction, we have used only rational algebraic systems (for, as we shall see later in (VI, 10), W and $\overline{W}$ are certainly rationally equivalent cycles on $\mathbf{P}^N$ and hence so also are the cycles $W \cdot V$ and $\overline{W} \cdot V$ on V {see (VI, 8)}, we can extend the results for algebraic equivalence to rational equivalence on V.

In conclusion: *any two equivalence classes, algebraic or rational, determine uniquely a third equivalence class, algebraic or rational,* as the case may be, associated with the intersection cycle of any two cycles respectively representing the two given classes and such that their intersection-product is defined.

This third class will be called *the intersection-product class* of the other two. Such an operation is consequently defined *in all cases*. More generally, we can replace each formula of the calculus of cycles on a non-singular ambient variety, which for the cycles is usually valid only on some suitable hypothesis of regularity, by an analogous formula operating on the algebraic equivalence classes, which is then always valid without exception: this follows at once from the fact that the only operation with exceptions is the intersection-product itself. We thus obtain a calculus of equivalence classes similar to that for topological classes on a topological variety.

10. A Theorem of SEVERI and its Consequences

In order to go deeper into the study of rational equivalence we enunciate a theorem of SEVERI {see SEVERI [a], p. 62}, which, moreover, is a useful tool in other researches {see e. g. the use made by SAMUEL in intersection theory: [c], p. 102}:

(i) *Let A be an r-cycle of $\mathbf{P}^n$: then there are $n-r$-cycles $U_1, U_2, \ldots, U_{n-r}$ of $\mathbf{G}^{n-1}(\mathbf{P}^n)$ such that their intersection-product is defined, with*

$$A = U_1 \cdot U_2 \cdot \cdots \cdot U_{n-r}\,.$$

Omitting the proof, which is in SAMUEL's work {see SAMUEL [c], p. 98}, we observe that the cycles U_i, whose existence is assured by (i), are not generally positive, as is well known from elementary examples. The following is an immediate corollary of the proof of (i):

(ii) *The divisors of* (i) *can be supposed such that U_i is a cone and the other divisors of degree unity, i. e. the so-called virtual hyperplanes of* SEVERI. *Moreover all the U_i are defined over the perfect closure of a definition field for A.*

Writing each U_i as the difference of two positive divisors, we have the following corollary to (i):

(iii) *Any r-cycle of $\mathbf{P}^n$ can be expressed as a cycle whose components are intersections of positive divisors of $\mathbf{P}^n$.*

Let us now consider a non-singular variety V^n defined over k, and suppose that $\mathbf{P}^m$ is a projective space parametrising a rational system of V-cycles defined over k. More precisely, if X is a $(V \times \mathbf{P}^m)$-cycle rational over k and homogeneous of dimension d, we consider the system on V whose generic cycle is the cycle $X(P) = \mathrm{pr}_V(X \cdot (P \times V))$, where P is generic for $\mathbf{P}^m$ over k.

It may be that some component of X has on V a projection different from V: let $\bar{X}$ be the residue cycle after the removal of such components, which is still rational over k, and, as we suppose, non-null; moreover,

$\mathrm{pr}_{\mathbf{P}^m}||\overline{X}|| = \mathbf{P}^m$. Then, if Q is a generic point over k for V, the cycle $Y(Q) = \mathrm{pr}_{\mathbf{P}^m}(\overline{X} \cdot (Q \times \mathbf{P}^m))$ is defined and rational over $k(Q)$ {see (I, 9, i)}.

Applying theorem (i) to this cycle and supposing k perfect, which is permissible after introducing, if necessary, a suitable extension, we have $Y(Q) = U_1 \cdot U_2 \ldots U_{m-s}$ where, by (ii), U_i are divisors of $\mathbf{P}^m$ which are all rational over $k(Q)$ and where $s = r + m - n$, as we find by counting constants {see (VI, 3)}. Every $U_i(Q)$ determines on $V \times \mathbf{P}^m$ the locus of $Q \times U_i(Q)$ over k, which is a divisor rational over k and it is at once seen that the intersection-product $X_1 \cdot X_2, \ldots, X_{m-s} \cdot (Q \times \mathbf{P}^m)$ is defined on $V \times \mathbf{P}^m$, if Q is generic on V over k, and equal to $X \cdot (Q \times \mathbf{P}^m)$.

This ensures that $\overline{X}$ and $X_1 \cdot X_2, \ldots, X_{m-s}$ differ at most by a $(V \times \mathbf{P}^m)$-cycle whose projection on V is not V. Hence we have the theorem, which can be considered as an extension of the SEVERI theorem (i):

(iv) *If X is an $(m + r)$-cycle of $V \times \mathbf{P}^m$ rational over a perfect field k of definition for V and if* $\mathrm{pr}_{\mathbf{P}^m}||X|| = \mathbf{P}^m$, *there are $n - r$ divisors of $V \times \mathbf{P}^m$, rational over k, such that*

$$X = X_1 \cdot X_2 \ldots X_{n-r} + Z\,,$$

where Z is an $(m + r)$-cycle of $V \times \mathbf{P}^m$ with $\mathrm{pr}_V||Z|| \neq V$.

An algebraic system of the type $\mathrm{pr}_V((X_1 \cdot X_2, \ldots, X_{n-r} + Z) \cdot (P \times V))$, with P generic in a projective space and $\mathrm{pr}_V||Z|| \neq V$, is called, by SEVERI, who first considered such systems, *a partial system of intersection.*

Observing that each divisor X_i determines on V the virtual linear system $\mathrm{pr}_V(X_i \cdot (P \times V))$ and calling *semifixed* any component of the generic cycle of an algebraic system when its locus on V is a proper subset of V, we have the following important theorem of SEVERI {see SEVERI [a], p. 112}:

(v) *The generic cycle of a rational system, apart from the semifixed components, is obtained by intersecting the generic cycles of some suitable linear systems on V. Consequently it is a linear combination of the generic cycles belonging to some suitable effective linear systems, apart from the semifixed components.*

The preceding theorem has been taken by SEVERI as the definition of the so-called systems of rational equivalence, which, therefore, coincide with our rational systems: this identification of the rational systems with the systems of partial or complete intersection is no longer true if we remain in $\mathbf{G}_+^r(V)$, as is proved by examples: the simplest is afforded by the system of all pairs of lines in a projective space; this is obviously rational but cannot, as one readily shows, be regarded as an intersection system in $\mathbf{G}_+^r(\mathbf{P}^n)$.

In particular, when $V = \mathbf{P}^n$ by (i) and (iv) we have $Z = 0$: therefore any rational system of $\mathbf{P}^n$ is a system of complete intersection: hence the totality of r-cycles of degree d on $\mathbf{P}^n$ is a coset of $\mathbf{G}^r(\mathbf{P}^n)/\mathbf{G}^r_l(\mathbf{P}^n)$. Such a coset contains the cycles of the form $d \cdot \mathbf{P}^r$, where $\mathbf{P}^r$ is a subspace of $\mathbf{P}^n$: hence any rational equivalence class of degree d on $\mathbf{P}^n$ is the d-ple of the class of degree unity. This fact, having as corollary the existence, in any rational equivalence class, of cycles whose only components are spaces, can be summarised in the following theorem:

(vi) *The group* $\mathbf{G}^r(V)/\mathbf{G}^r_l(V)$ *is infinite cyclic, whenever* V *is a birational variety.*

If in (iv) we denote by $\mathbf{P}^N$ the ambient space of V and consider each divisor X_i as a partial intersection of a divisor on $\mathbf{P}^N \times \mathbf{P}^m$ with V, in such a way that the residue intersection has not all V as its projection, then one easily obtains:

(vii) *Any rational system belonging to a projective variety* V *can be cut on* V *by a suitable intersection-system of the ambient space, outside some semifixed components.*

It seems probable that, if the given rational system on V is effective, then the cutting system can also be selected so as to be effective: this has been proved, together with the preceding theorems, by SEVERI for a surface over the complex field.

We give in this note some complementary notices concerning SEVERI's notion of rational equivalence.

Two r-cycles X and $\overline{X}$ on a non-singular variety V are called *rationally equivalent* by SEVERI if they are expressible in the forms $X = \sum_{i=1}^{s} m_i X_i, \overline{X} = \sum_{i=1}^{s} m_i \overline{X}_i$, where the m_i are integers and the $X_i, \overline{X}_i$ are cycles belonging to one and the same "*complete intersection system*" on V: i. e. to the set of the intersection-cycles of the divisors belonging to $n - r$ given linear systems on V, as we have already stated in (VI, 10).

Hence it follows that the rationally vanishing r-cycles are those which admit an expression of the type $\sum_{i=1}^{s} m_i(X_i - \overline{X}_i)$: this is the zero-group of our equivalence, and in this form the definition is due to TODD [2]. Then SEVERI has proved {see [a], p. 83} that any two rationally equivalent cycles in this sense are totally contained in a (possibly virtual) complete intersection system, and, therefore, in a rational system.

Conversely, by th. (v) in (VI, 10) (which is also due to SEVERI), two cycles belonging to the same rational system are rationally equivalent in SEVERI's sense. It follows that *our notion of rational equivalence and that of* SEVERI *concide.* We add that SEVERI calls (and in this we shall follow him) an *equivalence system* any algebraic subsystem of r-cycles which is totally contained in a rational system: this system is then obviously compounded of cycles which are mutually equivalent in the rational equivalence. This gives together with th. (vi) in (VI, 10) the immediate consequence that the set of all effective r-cycles of a given degree in $\mathbf{P}^n$, or any algebraic subsystem of the set, is a system of equivalence.

We pass on to report the more significant results of topological and transcendental nature obtained by SEVERI in his researches on equivalence systems {for the general notions which we shall use see Ch. IX}.

Let Σ be an algebraic system, irreducible and d-dimensional, of r-cycles on a non-singular variety V defined over the complex field, and let $\mathbf{C}^d$ be the CHOW set of Σ. Then as a variable point of $\mathbf{C}^d$ describes a (topological) algebraic q-cycle Γ^q on the RIEMANN manifold of $\mathbf{C}^d (0 \leqq q \leqq 2\,d)$, the corresponding element of Σ describes a (topological) cycle $T(\Gamma^q)$ on the RIEMANN manifold of V. Now we give the definitions:

a) The system Σ is of *q-dimensional circulation zero* if $T(\Gamma^q) \sim 0$, for any q-cycle Γ^q.

b) The system Σ is of *q-dimensional algebraic circulation* if $T(\Gamma^q)$ is an algebraic cycle for any q-cycle Γ^q.

c) The system Σ is *without torsion* if from $\lambda\ T(\Gamma^q) \sim 0$, where λ is an integer, it follows that $T(\Gamma^q) \sim 0$ on V for any q-cycle Γ^q

Then SEVERI has proved {[20, 23]; see also [d]} that: any equivalence system satisfies (c) and, moreover, it satisfies (a) if q is odd and (b) if q is even. We do not know if these conditions are also sufficient for an algebraic system to be an equivalence system: nevertheless, this has been conjectured by SEVERI.

Moreover, if we call *system of pseudoequivalence* any algebraic system having some integral multiple which is an equivalence system, then the preceding conditions (a) and (b) are still satisfied and, in this case, they are also sufficient for the characterisation of a pseudoequivalence system. The existence of such systems depends on the existence of some topological torsion in V and in $\mathbf{C}^d$: SEVERI has proved that, if V is free from torsion, then any pseudoequivalence system is also an equivalence system.

Moreover if V is a surface, an algebraic series Σ is a pseudoequivalence series if and only if V is regular and with geometrical genus zero: this series, in its turn, is an equivalence series if and only if V is without torsion {see SEVERI [20]}. The interesting question of deciding the rationality or otherwise of such a surface is still unsolved.

Now let Σ be an algebraic series of order l, on a variety V, satisfying (a) and (b): it follows immediately that if ω^p is a p-form of the first species on V, then the form $\sum_{j=1}^{l} \omega^p_{(j)}$, where $\omega^p_{(j)}$ denotes the form ω^p calculated at the point P_j of a generic group of Σ, and which is also of the first species, has all its periods zero (since the p-cycles on V are or homologous to zero by (a) or are algebraic by (b) {see e. g. HODGE [a], p. 213}.

Consequently, by the fundamental theorem of the uniqueness of a harmonic form with given periods {see (IX, 4)}, we have necessarily $\sum_{j=1}^{l} \omega^p_{(j)} = 0$. By using the fact that, conversely, if $r = \dim V = 2$, any cycle giving zero-periods for any double integral of the first species is algebraic (criterion of LEFSCHETZ {see (X, 6)}), SEVERI has proved in [28], as it is quite easy to do, that *"the n. a. s. c. in oder that an algebraic series on a non-singular surface V should be a pseudoequivalence series, is that* $\sum_{j=1}^{l} \omega^p_{(j)} = 0, p = 1, 2,$ *for all the p-ple forms of the first species on V".*

This theorem gives an extension of the classical theorem of ABEL on the linear series on a curve. Evidently the attempt to extend such a result to a variety V, with $\dim V > 2$, depends on the possibility of obtaining a criterion for a p-cycle on V to be algebraic, which is, at present, an unsolved question {see some suggestions of HODGE in [4, 5, 8] and also (X, 6)}.

Finally we remark that, though the notion of a complete equivalence system of effective r-cycles raises very serious difficulties in the present state of the theory, SEVERI has been able to prove that, if a positive group of l-points on a surface is not contained in some group of irregularity {see (IX, 9)} then it determines uniquely a complete series with 1-dimensional circulation zero and of dimension $2l-q$, where q is the irregularity of the surface. If $2l \leqq q$ the series reduces to the given group and if this belongs to ∞^{i-1} groups of irregularity, then it belongs to some series of order l with 1-dimensional circulation zero and of dimension $\leqq 2l-q+i$ {see SEVERI [28]}: *this gives a theorem of* RIEMANN-ROCH *for such series* {see (X, 9)}.

VII. The Abelian Varieties from the Algebraic Viewpoint; and Related Questions

1. Jacobi Variety

The theory of the Abelian varieties from the transcendental standpoint is classical {this field has been recently reviewed by IGUSA [3], A. WEIL [4], CHOW [6]: see (XI, 2); the outstanding classical contributions are due to PICARD, POINCARÉ, ENRIQUES, CASTELNUOVO, SEVERI and LEFSCHETZ: see e. g. the report [b] of LEFSCHETZ}: the Abelian varieties considered in an abstract, purely algebraic manner, have been investigated only recently, giving rise to a beautiful chapter in algebraic geometry, whose fundamental ideas and general methods are chiefly due to A. WEIL {see the treatises [b, c] of A. WEIL} and whose applications are chiefly linked with the names of CHOW {see CHOW [6, 7, 8, 9]}, MATSUSAKA {see MATSUSAKA [3, 4, 5, 6, 7, 8]}, NÉRON {see NÉRON [1, 2]} and SAMUEL {see NÉRON and SAMUEL [1]}.

Among these the most important are the construction of the Jacobi variety of a curve or, more generally, that of the PICARD manifold of a variety, and the proof of the existence of a finite base for the group of rational equivalence on a variety over any field of definition, which reduces to a theorem of SEVERI in the classical case {see SEVERI [12, 14, 22, 34, 36]}.

In spite of its recent birth this theory covers a wide territory and we must in consequence restrict ourselves to a short description of the outstanding applications, omitting the general theory of the Abelian varieties, for which we refer the reader to WEIL's treatises.

Before proceding to the construction of the Jacobi variety of a curve, we recall a few definitions. A variety V is said to have *a composition law* when there exists a function f defined on $V \times V$ with values on V; and V is called *a group variety* if it is endowed with such a law, which must be everywhere defined on $V \times V$ and such as to induce a group structure on V. We now state, with A. WEIL {see A. WEIL [c], p. 28}, the

Def. *a. A projective algebraic variety V is called Abelian if it is a group variety.*

A theorem of CHEVALLEY {see A. WEIL [c], p. 25} affirms that

(i) *An Abelian variety is necessarily endowed with an Abelian group.*

Moreover we remark the less deep result {see A. WEIL [c], p. 18}:

(ii) *An Abelian variety, as indeed any group variety, is non-singular.*

We now begin with the following

Def. *b*. *Let C be a non-singular curve defined over a field k. A variety J(C) will be called a Jacobi variety of C if and only if:* (1) *J(C) is an Abelian variety defined over a field* $k' \supset k$; (2) *There is a homomorphism f, defined and rational over k', of* $\mathbf{G}_a(C)$ *over J(C) having as kernel* $\mathbf{G}_l(C)$: *this is called the canonical homomorphism*; (3) *The homomorphism f has the universal mapping property.*

Here (2) means that f preserves the rationality and the specialisation notion of cycles on C over any extension of k', contained in K, while (3) means that if g is any rational homomorphism of $\mathbf{G}_a(C)$ into any Abelian variety $\mathfrak{A}$, then g is the product of f and of a rational homomorphism of $J(C)$ into $\mathfrak{A}$. Hence follows easily the uniqueness of $J(C)$ to within a birational isomorphism, with respect to $J(C)$ as a group, if $J(C)$ exists.

A. Weil was the first to establish the existence of a Jacobi model {see A. Weil [c], Ch. V}, but in the form of a so-called abstract variety, a concept which has been devised by A. Weil chiefly to solve this problem: an abstract variety, roughly speaking, is a finite set of varieties on any one of which is given a frontier, which is an algebraic proper subset of the ambient variety, to be excluded from the given variety and, moreover, the varieties are related to one another by a set of suitable coherent birational transformations operating on the frontiers {see A. Weil [a], Ch. VII}.

Obviously this set of pieces of varieties has the disadvantage of having to be considered *in abstracto* apart from any embedding in a projective space: nevertheless A. Weil has been able to extend the usual properties of the projective varieties to these abstract varieties, which suffices to legitimise his solution.

Subsequently Chow proved the theorem {see Chow [8]; here the reader will find all proofs of the facts mentioned in this section}:

(iii) *We can always construct a projective model J(C), and in such a way that J(C) and the homomorphism f are defined over the same definition field k of the curve C.*

We proced to sketch the proof of (iii).

Let us put $\mathbf{D} = \mathbf{G}(C)/\mathbf{G}_l(C)$ and $\mathbf{D}_0 = \mathbf{G}_a(C)/\mathbf{G}_l(C)$: $\mathbf{D}_0$ is then the subgroup of $\mathbf{D}$ of the classes of linear equivalence of degree zero, and the classes of linear equivalence, but of degree n, form a coset $\mathbf{D}_n$ of $\mathbf{D}/\mathbf{D}_0$. In particular, any class of $\mathbf{D}_g$, $g =$ genus of C, contains generally, by the Riemann-Roch theorem, one and only one positive divisor: so that there is a one-to-one correspondence between the generic point of the g-ple product of C for itself, which we denote by $C^{(g)}$, and the classes of $\mathbf{D}_g$, except for some proper subvariety of $C^{(g)}$.

This is usually {see e. g. van der Waerden [9]} the starting point for the construction of the Jacobi variety: but one has to overcome the initial difficulty that the special classes of $\mathbf{D}_g$ give rise to exceptions, since any one of them contains an infinity of positive divisors. Moreover, if one tries to remove the exceptions, one finds a model which may be singular and which, as we shall see, therefore, could not yield the required construction: hence WEIL's solution by means of abstract varieties.

More simply, CHOW has considered the classes of $\mathbf{D}_n$, rather than of $\mathbf{D}_g$, when $n > 2g - 2$: it is well known that then there are no special classes. With any class $\mathbf{B}$ of $\mathbf{D}_n$ is associated on $C^{(n)}$ a subvariety $F(\mathbf{B})$ of dimension exactly equal to $n - g$. If the class $\mathbf{B}$ is rational over $k' \supset k$, then $F(\mathbf{B})$ is defined over k'.

Now let P' be a generic point of $C^{(n)}$ over k and $\mathbf{B}(P')$ the class, rational over $k(P')$, associated with P' in $\mathbf{D}_n$: then $F(\mathbf{B})$ is rational over $k(P')$ and its CHOW coordinates determine a point P, rational over $k(P')$ of a suitable projective space $\mathbf{P}^N$. Let X be the g-dimensional variety locus of P over k and let us consider the rational transformation $P = f(P')$ of $C^{(n)}$ onto X, defined over k.

CHOW proves that this is the CHOW variety of the involutory system whose generic member is $F(P)$, and that f is the associated incidence correspondence: this is sufficient, by CHOW's criterion in (VI, 6), to assure that X is non-singular, provided C be such, which in effect CHOW verifies.

Here it is implicit that X represents biunivocally the classes of $\mathbf{D}$, with $f(P') = f(Q')$, where P' and Q' are points of $C^{(n)}$, if and only if the C-divisors associated with P' and Q' are linearly equivalent on C.

Let us denote by $\mathbf{B}^*$ a class of $\mathbf{D}_n$, rational over k, certainly existent for suitable values of n. It will be associated with a point P^0 of X rational over k: then, if P^1 and P^2 are two points of X and $\mathbf{B}^1$ and $\mathbf{B}^2$ the associated classes of $\mathbf{D}_n$, we shall define the symbol $\mathbf{B}^1 + \mathbf{B}^2$ as the point P^3 of X associated with the class $\mathbf{B}^3 = \mathbf{B}^1 + \mathbf{B}^2 - \mathbf{B}^0$.

So we arrive at a composition law on X, and we can then demonstrate that this is a rational function on $X \times X$, defined over k, and uniquely determined at each point of X, and then, as X is non-singular, also defined at any point of X: moreover it is seen at once that this establishes an Abelian structure on X, having, as unit element P^0 and, as inverse $-P$ of each point P, the transform of the class $2\,\mathbf{B}^* - \mathbf{B}$, where $\mathbf{B}$ is the transform of P. Moreover there is a group of birational transformations of X onto itself, called "*translations*", given by the expression $P' = A + P$, where A is a fixed point anywhere on X.

Considering two systems of s divisors of degree n on C, $p_1, p_2, \ldots, p_s$ and $q_1, q_2, \ldots, q_s$, we have $\sum_1^s f(p_i) = \sum_1^s f(q_i)$, if and only if $\sum_1^s p_i \equiv \sum_1^s q_i$, where f is supposed to be defined directly on the divisors of degree n of C: this comes immediately from observing that there is a divisor p of B^* and two positive divisors p_0 and q_0, such that $\sum_1^s p_i \equiv (s-1)\,p + p_0$,

$\sum_1^s q_i \equiv (s-1)\,p + q_0$ and from using the above observation that if $f(p_0) = f(q_0)$, then $p_0 \equiv q_0$.

Therefore, taking $(n-1)$ generic points $x^{(1)}, x^{(2)}, \ldots, x^{(n-1)}$ independent over k of C, let x be a C-point generic over $k(x^{(1)}, x^{(2)}, \ldots, x^{(n-1)})$ and put, for any 0-cycle $p = \sum_1^s n_i(x^{(i)})$ in $\mathbf{G}_a(C)$, $f(p) = \sum_1^s n_i\, f(x^{(i)})$: then the above relation gives the conclusion that f is a rational homomorphism of $\mathbf{G}_a(C)$ into X, whose kernel is $\mathbf{G}_l(C)$, and, further, that it is defined over k and satisfies the universal property; hence we have $X = J(C)$.

When $n = 1$, f becomes the so-called canonical function of WEIL, defined over the field obtained from k by adjoining any point of C.

2. The Base for the Group of Algebraic Equivalence for Divisors

The subject with which we deal has been proposed and thoroughly investigated by SEVERI for the classical case, both by transcendental and algebro-topological methods; the abstract case, instead, has been treated by NÉRON and we shall here outline, as briefly as possible, the methods used by this author {see NÉRON [1, 2]}.

a) Let M be a generic point over k of a projective space $\mathbf{M}^n$ over k and $C(M)$ a curve defined over $K = k(M)$, supposed non-singular, embedded in a projective space $\mathbf{R}$ and of genus g. The curve $C(M) \times M$ will have a locus $\mathfrak{C}^{n+1} = \mathfrak{L}_k(C(M) \times M)$ in $\mathfrak{R} = \mathbf{R} \times \mathbf{M}$. We readily see that, by replacing C by a biregular birational model over K, C can be taken absolutely normal if one introduces a suitable extension of k.

We shall denote by $\mathbf{G}$, $\mathbf{G}_a$, $\mathbf{G}_l$ the usual groups related to C and by $\mathfrak{G}$, $\mathfrak{G}_a$, $\mathfrak{G}_l$ those referred to $\mathfrak{C}$; moreover we denote by $\mathfrak{H}_0$ the group of the $\mathfrak{C}$-divisors $\mathfrak{A}$ such that $\mathrm{pr}_M \mathfrak{A} = 0$, by $\mathfrak{H}$ the group of the $\mathfrak{C}$-divisors such that $\mathrm{pr}_M \|\mathfrak{A}\|$ is a set of dimension less than $n-1$ and we put $\mathfrak{H}_a = \mathfrak{H} + \mathfrak{G}_a$, $\mathfrak{H}_l = \mathfrak{H} + \mathfrak{G}_l$.

Setting $A' \times M' = \mathfrak{A} \cdot (C(M') \times M')$, where $\mathfrak{A}$ is a $\mathfrak{C}$-divisor and M' is a generic point of $\mathbf{M}$ over a definition field for $\mathfrak{A}$, and observing that $A' \times M' = \mathfrak{A} \cdot (\mathbf{R} \times M')$, we at once deduce that $\mathfrak{H}_0$ is also the group of the $\mathfrak{C}$-divisors which cut on $C(M')$ a divisor A' of degree zero and that, similarly, $\mathfrak{H}$ is the group of those which cut the zero-divisor on $C(M')$.

From this and the definitions, we deduce the inclusions

$$\mathfrak{G} \supset \mathfrak{H}_0 \supset \mathfrak{H}_a \supset \mathfrak{H}_l \supset \mathfrak{H}\,,$$
$$\mathfrak{H}_a \supset \mathfrak{G}_a \supset \mathfrak{G}_l\,,$$
$$\mathfrak{H}_a \supset \mathfrak{H}_l \supset \mathfrak{G}_l\,.$$

It is clear that $\mathfrak{G}/\mathfrak{H}_0$ is isomorphic to the group of integers. Putting $f(P \times M) = M$, for any $P \in C(M)$, we prove the theorem:

(i) *If* $\mathfrak{H}'$ *is the group of the* $\mathfrak{C}$*-divisors having the form* $\mathfrak{A} = f^{-1}(a)$, *where a is an* **M**-*divisor, then* $\mathfrak{H}' \subset \mathfrak{H}$, *and* $\mathfrak{H}/\mathfrak{H}'$ *is of finite type.*

Moreover, with the definitions

$$\mathfrak{H}' \cap \mathfrak{G}_a = \mathfrak{H}'_a, \quad \mathfrak{H}' \cap \mathfrak{G}_l = \mathfrak{H}'_l,$$

$\mathfrak{H}'/\mathfrak{H}'_a$ and $\mathfrak{H}'/\mathfrak{H}'_l$ are isomorphic to some quotient groups of the group of the integers.

The inclusion $\mathfrak{H}' \subset \mathfrak{H}$ follows from $||a|| = \mathrm{pr}_M||\mathfrak{A}||$, if $\mathfrak{A} = f^{-1}(a)$. For the rest, if $\mathfrak{A} = \Sigma \lambda_i \mathfrak{A}_i \in \mathfrak{H}$, let $M_i \times P_i$ be a generic point of $\mathfrak{A}_i$ over the field $k_i = \mathrm{def} A_i$, and let $a_i = \mathfrak{L}_{ki}(M_i)$ and $a = \Sigma \lambda_i a_i$, where $\dim a_i = n - 1$. The mapping $\mathfrak{A} \to a$ satisfies $f^{-1}(a) \to a$, and so is a homomorphism χ of $\mathfrak{H}$ onto **G(M)**. If $\mathfrak{A} \to 0$ and $a_i = \mathrm{pr}_M \mathfrak{A}_i$, it is either $\dim a_i < n - 1$ or $\dim a_i = n - 1$: the former case comes out of elements of the type $C \cap (\mathbf{R} \times M_i)$ of irregular dimension and also the latter but, instead, reducible.

Thus this occurs, in both ways, if and only if $\mathfrak{A}$ belongs to an algebraic proper subset of **M**, independent of $\mathfrak{A}$, and therefore the kernel $\mathfrak{K}$ of χ is a finite subgroup of $\mathfrak{H}$. Then, since $\mathfrak{H} = \mathfrak{H}' + \mathfrak{K}$ and also $\mathfrak{H}/\mathfrak{H}' \cong \mathfrak{K}/\mathfrak{K} \cap \mathfrak{H}'$, the group $\mathfrak{H}/\mathfrak{H}'$ is finite.

Finally, if $a \in \mathbf{G}_a(\mathbf{M})$, $f^{-1}(a) \in \mathfrak{H}'_a$ {see (VI, 7)} and if $a \in \mathbf{G}_l(\mathbf{M})$, $f^{-1}(a) \in \mathfrak{H}'_l$ {see (VI, 8)}: hence, if $\mathbf{G}'_a(\mathbf{M}) = f(\mathfrak{H}'_a)$ and $\mathbf{G}'_l(\mathbf{M}) = f(\mathfrak{H}'_l)$, we see that $\mathfrak{H}'/\mathfrak{H}_a$ and $\mathfrak{H}'/\mathfrak{H}_l$ are isomorphic to some quotient groups of $\mathbf{G}(\mathbf{M})/\mathbf{G}_a(\mathbf{M})$ and $\mathbf{G}(\mathbf{M})/\mathbf{G}_l(\mathbf{M})$ respectively, which, in their turn, are isomorphic to the groups of the integers.

Moreover we have:

(ii) *If* $\mathfrak{A} \in \mathfrak{G}$ *and if* M' *is a generic point of* **M** *over a field* k' *of definition for* $\mathfrak{A}$, *putting* $A' \times M' = \mathfrak{A} \cdot (\mathbf{R} \times M')$, *then it is* $\mathfrak{A} \in \mathfrak{H}_l$ *if and only if* $A' \equiv 0$ *on* $C' = C(M')$.

The necessity of the condition is obvious by (VI, 8). Let then $A' \equiv 0$ on C'. There will exist a function ϑ on C' defined over $K' = k'(M')$ with $(\vartheta) = A'$ and, if P' is generic for C' over K', putting $\omega(P' \times M') = \vartheta(P')$, we derive $(\omega) \cdot (C' \times M') = A' \times M'$: therefore $[(\omega) - \mathfrak{A}] \cdot (C' \times M') = 0$ and $(\omega) - \mathfrak{A} \in \mathfrak{H}$. Hence, since $(\omega) \in \mathfrak{G}_l$, it follows that $\mathfrak{A} \in \mathfrak{H}_l$.

b) To study the group $h = \mathfrak{H}_0/\mathfrak{H}_l$, we introduce the Jacobi variety $J = J(M)$ of the curve $C = C(M)$, defined, like $C(M)$, over K and embedded in a projective space **P**. The canonical function φ related to this model is then defined over an extension K' of K and we shall suppose, at first, that $K' = K$ {see (VII, 1)}.

Let $J^{n+g} = \mathfrak{L}_k(J \times M) \subset \mathbf{P} \times \mathbf{M}$: then it is $J \times M = \mathfrak{J} \cdot (\mathbf{P} \times M)$. Putting $J' \times M' = \mathfrak{J} \cdot (\mathbf{P} \times M')$, we define $J' = J(M')$, when the intersection-product is defined: we can suppose, as for $\mathfrak{C}$, that J is non-singular. Moreover, by using a suitable modification of the characterisation of the Jacobi variety, NÉRON shows that $J' = J(M')$, where J' is a non-singular variety for almost any M' of M, is the Jacobi variety of $C' = C(M')$, which is likewise a curve of genus g, and that the specialisation φ' of φ over $M \to M'$ is still a canonical function for J'.

Now let k' be a definition field for a $\mathfrak{C}$-divisor $\mathfrak{A}$ and M' a generic point of **M** over k': then the cycle A' on C' defined by $A' \times M' = \mathfrak{A} \times \times (\mathbf{R} \times M')$ is rational over $K' = k'(M')$. If $A' = \Sigma \lambda_i A'_i$, we put

$S(\varphi'(A')) = \Sigma \lambda_i \varphi'(A'_i) = Z'$: then, Z' being rational over the regular extension K' of k', the locus $\mathfrak{Z} = \mathfrak{L}_{k'}(Z' \times M') \subset \mathfrak{J}$ is defined, and we have: $(\mathfrak{Z} : \mathbf{M}) = 1$.

Now (ii) shows that to all elements of a class of h corresponds the same $\mathfrak{Z}$: conversely, taking a subvariety $\mathfrak{Z}^n$ of $\mathfrak{J}$ defined over k' with $\mathfrak{Z} \cdot (J(M') \times M')$ reduced to a point Z', we can find a C'-divisor A' such that $Z' = S(\varphi'(A'))$, so that, putting $\mathfrak{A} = \mathfrak{L}_{k'}(A' \times M')$, we have $\mathfrak{A} \in \mathfrak{H}_0$. Thus the set of the $\mathfrak{Z}$ and h are in one-to-one correspondence and in what follows we shall identify them: observe that the $\mathfrak{Z}$ form the section group of the variety $\mathfrak{J}$ fibred by the J'.

We shall now prove, with NÉRON, the following lemma:

(iii) *Let U be a variety and let $C_i (i = 1, \ldots, m)$ be non-singular curves of U with each point simple on U, and let J_i be the Jacobi variety of C_i and φ_i the associated canonical function. If k is a definition field for all these, we associate to any U-divisor A the points $Z_i = S(\varphi_i(A \cdot C_i))$, whenever $A \cdot C_i$ is defined for any i. Let $Z = Z_1 \times Z_2 \times \cdots \times Z_m$ be a point on $J = J_1 \times J_2 \times \cdots \times J_m$.*

Then Z is independent of the choice of A $(\operatorname{mod} \mathbf{G}_l(U))$ and, if $A \in \mathbf{G}_a(U)$, the mapping $A \to Z$ defines a homomorphism λ of $g_a = \mathbf{G}_a(U)/\mathbf{G}_l(U)$ onto the group (Ω) of the points of an Abelian subvariety Ω of J. Moreover, there exists a subgroub of g_a parametrised by an Abelian variety π, whose λ-transform into J is (Ω).

First by (VI, 9), each class of $\mathbf{G}(U)/\mathbf{G}_l(U)$ contains a cycle A such that $A \cdot C_i$ is defined for any i: consequently any such class has a well determined transform onto J.

Let I be the transform of $\mathbf{G}_a(U)/\mathbf{G}_l(U)$: then I is a subgroup of J, corresponding to $\mathbf{G}_a(U)/\mathbf{G}_l(U)$ by a homomorphism λ. If A represents a class of g_a, let us denote Z its transform by λ. There exists a non-singular curve V and a $(U \times V)$-divisor such that $A = \operatorname{pr}_U X \cdot (U \times (M_1 - M_2))$, where $M_1, M_2 \in V$ {see (VI, 8)}.

Let J_V, g_V and φ_V be respectively the Jacobi variety, the genus and the canonical function related to the curve V, all supposed defined over a field k_V. Then if Z_V is generic on J_V over k_V, we can find a positive V-divisor a_V of degree g_V, satisfying the $Z_V = S(\varphi_V(a_V))$, and, putting $a_V \to a_V^{(0)}$ over k_V with $M_2 \in \|a_V^{(0)}\|$, we define $\bar{a}_V = a_V - a_V^{(0)}$, $\bar{A} = \operatorname{pr}_U(X \cdot (U \times \bar{a}_V))$.

Now the set of all specialisations of $\bar{A}$ over k_V, among which are A and the zero-divisor of U, gives rise to a sub-group g'_a of g_a parametrised by J_V, i. e. g'_a is such that there is a subset $\mathbf{G}'_a$ of $\mathbf{G}_a(U)$ parametrised by J_V, and such that g'_a contains all and only all the classes which admit a representative in $\mathbf{G}'_a$.

Therefore $\lambda(g'_a)$ is the variety $\zeta = \mathfrak{L}_{k_V}(\bar{Z}) \subset J$, where $\bar{Z} = \lambda(\bar{A})$, and further we have established the existence of a homomorphism of J_V into J, whose total transform is still ζ, which is an Abelian subvariety of J. Hence: a point Z of I and the point Z_0, unit of J, can be always connected by an Abelian subvariety of J, which is the transform of a subgroup g'_a of g_a parametrised by another Abelian variety.

If then Ω is an Abelian subvariety of J with the maximal dimension among those varieties whose group (Ω) belongs to I, from what we have proved above, it follows at once that $(\Omega) = I$. Finally the last assertion of the theorem comes from applying the above conclusion to a point Z of J generic over k.

c) We now proceed to apply (iii) to our case. The mapping $\mathfrak{A} \to Z = S(\varphi(A))$, with $A \times M = \mathfrak{A} \cdot (C \times M)$ defines a homomorphism of $\mathfrak{G}_a/\mathfrak{G}_l$ onto the group (Ω) of an Abelian subvariety Ω of $J(M)$. By (ii), Z depends only upon the class of $\mathfrak{A}(\text{mod}\, \mathfrak{H}_l)$, and since every class of $h_a = \mathfrak{H}_a/\mathfrak{H}_l$ has a representative in $\mathfrak{G}_a$, this mapping defines a homomorphism of h_a onto (Ω). We prove:

(iv) *If $\mathfrak{A} \in \mathfrak{H}_a$ and if $Z = S(\varphi(A))$ the mapping $\mathfrak{A} \to Z$ is an isomorphism of h_a onto (Ω), when $K = K^*$.*

For the proof it will be sufficient to show that there is one and only one section $\mathfrak{Z}$ of $\mathfrak{Z}$, belonging to h_a, which contains $Z \times M$ with $Z \in \Omega$. To this end let M' be another generic point, but independent of M over k, and let $J' = J(M')$.

The transform Λ of h_a into the product $J \times J'$ is an Abelian variety which can be obtained as the transform of an Abelian variety Π parametrising a subgroup h'_a of h_a by (iii). On the other hand Λ is the graph of a birational transformation between Ω and the variety Ω', defined for J' as Ω for J: in fact, otherwise, if Z_0 and Z'_0 are the units of J and J' respectively, there is a point $\overline{Z}'_0 \neq Z'_0$ of J and of finite orders, such that $Z_0 \times \overline{Z}'_0 \in \Lambda$.

This point comes from a point of order s of Π, and then from an element of order s of h_a: but the number of the elements of order s of h_a is finite, as one can prove using the fact that the number of such elements of J is finite, and, since M is generic, the only one among these elements passing through Z_0 is the unit section Z_0.

Since, further, Λ can be supposed rational over $k(M, M')$, Z' is defined over $k(M, M', Z) = K(M', Z)$. Therefore $K(Z' \times M') = K(Z, M')$, and hence the variety $\mathfrak{O}$ locus of $Z' \times M'$ over K, or of $\Omega(M) \times M$ over k, is birationally equivalent to the product $\Omega \times \mathbf{M}$, which comes to saying that $\mathfrak{O}$ is a trivial fibred subvariety of J.

Now replace the generic point M' by any point $M_1 \subset \mathbf{M}$: then, for almost any $M_1 \subset \mathbf{M}$, the pair (Ω', Λ') has a unique specialisation (Ω_1, Λ_1) over $M' \to M_1$ with respect to K {see (VI, 2)}. From here we readily deduce that if M_1 is defined over $\bar{k}$ Ω_1 is a model of Ω defined over an algebraic finite extension of k, precisely the one $k_1 = k(M_1)$.

Lastly, if $K^* \neq K$, we can replace k by $k^* = \bar{k} \cap K^*$. Proceeding then for k^* as for k, we have $K^* = k^*(M^*)$ when M^* is generic for M^* over k^* and we consider the variety $\mathfrak{C}^*$ locus of $C(M) \times M^*$ over k^*. The inverse of the mapping $M^* \to M$ of $\mathfrak{C}^*$ onto $\mathfrak{C}$ consequently defines an isomorphism of $\mathfrak{H}_a/\mathfrak{H}_l$ into $\mathfrak{H}^*_a/\mathfrak{H}^*_l$, which is analogously defined with respect to k^*, and one can easily prove the theorem:

(v) *The group $\mathfrak{H}_a/\mathfrak{H}_l$ is isomorphic to the group of the points of an Abelian subvariety Ω of J.*

We end this subsection by proving that:

(vi) *The groups of the elements of finite order s of $\mathbf{G}(U)/\mathbf{G}_l(U)$ and $\mathbf{G}(U)/\mathbf{G}_a(U)$ are finite for any variety U.*

Elementary group properties and the theory of the birational transformations easily show that the theorem has a birationally invariant character: of which we avail ourselves to reduce U birationally to a fibred model of the type of $\mathfrak{C}$.

Then, in such a case, we have $\mathfrak{G} \supset \mathfrak{H}_0 \supset \mathfrak{H}_l \supset \mathfrak{H}' + \mathfrak{G}_l \supset \mathfrak{G}_l$. By (iii) $\mathfrak{G}/\mathfrak{H}_0$ is of finite type: further $\mathfrak{H}_l/(\mathfrak{H}' + \mathfrak{G}_l) \cong \mathfrak{K}/\mathfrak{K}_l$, where $\mathfrak{K}_l = \mathfrak{K} \cap (\mathfrak{G}_l + \mathfrak{H}')$, by a well known theorem. Since $\mathfrak{R}$ is finite, such is then the group $\mathfrak{H}_l/(\mathfrak{H}' + \mathfrak{G}_l)$. Moreover, by the theorem quoted above, $(\mathfrak{H}' + \mathfrak{G}_l)/\mathfrak{G}_l \cong \mathfrak{H}'/\mathfrak{H}'_l$: therefore by (iii) also $(\mathfrak{H}' + \mathfrak{G}_l)/\mathfrak{G}_l$ is finite.

It follows that the following subgroups of elements of order s are all finite: $|\mathfrak{G}/\mathfrak{H}_0|_s$, $|\mathfrak{H}_l/(\mathfrak{H}' + \mathfrak{G}_l)|_s$, and $|(\mathfrak{H}' + \mathfrak{G}_l)/\mathfrak{G}_l|_s$: the group $|\mathfrak{H}_0/\mathfrak{H}_l|_s = |h|_s$ is also finite by (b). Consequently one readily derives the same result for $|\mathfrak{G}/\mathfrak{G}_l|_s$ and similarly for $|\mathfrak{G}/\mathfrak{G}_a|_s$.

d) We are now in a position to condense NÉRON's proof {here we follow NÉRON [2]} of the theorem (due to SEVERI):

(vii) *The group* $\mathbf{G}(U)/\mathbf{G}_a(U)$ *is of finite type for any variety* U.

NÉRON begins by proving that U can be replaced by any unirational model $f(U)$, $\{f(U) : U = q\}$.

For this purpose it is sufficient to apply (vi) to the elements of order q and to use the lemma that if, $\bar{A} \in G(f(U))$, with $f^{-1}(\bar{A}) = 0$, then $f(f^{-1}(\bar{A})) = q\,\bar{A} + \bar{B}$ with $f^{-1}(\bar{B}) = 0$.

This is used to show that: (1) We can take a model for U of the type $\mathfrak{C}$; (2) We can replace $\mathfrak{C}$ by the $\mathfrak{C}^*$ defined over K^* as in (e): in fact the function $\beta(P \times M^*) = P \times M$, is such that β^{-1} has only a finite number of determinations; (3) Using (2) one can suppose that φ is defined over K and that the solutions Z_i, $i = 0, 1, \ldots, s^{2g} - 1$, of the equation $s \cdot Z_i = 0$ are rational over K, when $s > 1$ and s is prime with p.

After which NÉRON proves SEVERI's theorem from the following statements which will be demonstrated by the classical arithmetic method of infinite descent {this method has been used by A. WEIL and other authors in similar arithmetical researches: see NÉRON [1], p. 132}:

(viii) *Let sh be the group of the elements* $\mathfrak{Z}$ *of* $h = \mathfrak{H}_0/\mathfrak{H}_l$ *of the form* $s \cdot \mathfrak{Z}'$, $\mathfrak{Z}' \in h$; *then* h/sh *is finite.*

(ix) *Let* $\mathfrak{Z}_l^{(0)}$ *be some elements of h in finite number and* $\mathfrak{Z}$ *an element of h, and let us suppose that there is a sequence* $\mathfrak{Z}, \mathfrak{Z}', \ldots, \mathfrak{Z}^{(\nu)}, \ldots$ *in h such that, for any* ν *it is* $s\,\mathfrak{Z}^{(\nu)} = \mathfrak{Z}^{(\nu-1)} - \mathfrak{Z}_{l_\nu}^{(0)}$, *for a suitable index* l_ν. *Then if* ν *is sufficiently large* $Z^{(\nu)}$ *cannot represent an infinite number of classes of* $h/h_a \cong \mathfrak{H}_0/\mathfrak{H}_a$, *and the classes represented are independent of the choice of the first element* $\mathfrak{Z}$.

We first prove that the SEVERI theorem follows from these theorems. In fact, since by (viii) h/sh is finite, we can take for $\mathfrak{Z}_{l_\nu}^{(0)}$ some representative of each class of h/sh. Then, since, by (ix), $\mathfrak{Z}^{(\nu)}$ is congruent $(\operatorname{mod} h_a)$ to an element of a finite subset h_f of h_a, $\mathfrak{Z}$ is congruent, with the same modulus, to a linear combination of the $\mathfrak{Z}_{l_\nu}^{(0)}$ and of the elements of h_f. The group $h/h_a \cong \mathfrak{H}_0/\mathfrak{H}_a$ is, therefore, of finite type. This is also true for $\mathfrak{G}/\mathfrak{G}_a$, which is obvious if $n = \dim \mathbf{M} = 0$.

We proceed now by induction with respect to n. We have $\mathfrak{G} \supset \mathfrak{H}_0 \supset \mathfrak{H}_a \supset \mathfrak{G}_a + \mathfrak{H}' \supset \mathfrak{G}_a$ {see (VII, 2, c)}, $\mathfrak{H}_a/(\mathfrak{G}_a + \mathfrak{H}') \cong \mathfrak{H}'/\mathfrak{H}'_a$ and $(\mathfrak{G}_a + \mathfrak{H}')/\mathfrak{G}_a \cong \mathfrak{R}/\mathfrak{R}_a$, where $\mathfrak{R}_a = \mathfrak{R} \cap (\mathfrak{G}_a + \mathfrak{H}')$. But the groups $\mathfrak{G}/\mathfrak{H}_0$ and $\mathfrak{R}/\mathfrak{R}_a$ are finite and, since $\dim \mathfrak{M} < \dim \mathfrak{C}$, the group $\mathfrak{H}'/\mathfrak{H}'_a$ is also finite; also the latter is isomorphic to a quotient subgroup of $\mathbf{G}(\mathbf{M})/\mathbf{G}_a(\mathbf{M})$:

this being true also for $\mathfrak{H}_0/\mathfrak{H}_a$, it is also true for $\mathfrak{G}/\mathfrak{G}_a$. This proves SEVERI's theorem.

The following is a brief account of the main ideas of the proof of (viii) and (ix).

Let $\mathfrak{X}$ be a $\mathfrak{J}$-divisor, rational over k and such that if $X \times M = \mathfrak{X} \cdot (J \times M)$, then $X \neq 0$. Let Z_i, $i = 0, \ldots, s^{2g} - 1$, be the solutions of $s \cdot Z_i = 0$, which are rational over k, and put $X_i = X + Z_i$. Denote by ξ_i a function defined over K, to within a constant of K, by the relation $(\xi_i) = s \cdot (X_i - X)$ and set $\eta_i(Q \times M) = \xi_i(Q)$, where Q is generic on J with respect to K.

If $\mathfrak{X}_i = \mathfrak{L}_k(X_i \times M)$ we have $(\eta_i) = s \cdot (\mathfrak{X}_i - \mathfrak{X}_0) + \mathfrak{Y}_i$, where $\mathfrak{Y}_i \cdot (J \times M) = 0$, i. e. $\dim(\mathrm{pr}_M||\mathfrak{Y}_i||) \leqq n - 1$. If $\mathfrak{Z} \in h$ and $||\mathfrak{Z}|| \not\subset ||\mathfrak{X}_i||$, which one can suppose by taking $\mathfrak{E} \supset ||\mathfrak{X}_i||$, where $\mathfrak{E}$ is an algebraic proper subset of $\mathfrak{H}$, and selecting $||\mathfrak{Z}|| \not\subset \mathfrak{E}$ in fact this situation can always be realised by a suitable translation on $\mathfrak{J}$. Then η_i induces on $\mathfrak{Z}$ a function ζ_i for any i: if $Z \times M = \mathfrak{Z} \cdot (J \times M)$ we put $\vartheta_i(M) = \xi_i(Z)$.

Hence it follows that $(\vartheta_i) = \mathrm{pr}_M((\eta_i) \cdot \mathfrak{Z}) + a_i'$, where $||a_i'||$ is contained in a fixed subset of $\mathbf{M}$ and then $(\vartheta_i) = s(\mathrm{pr}_M((\mathfrak{X}_i - \mathfrak{X}_0) \cdot \mathfrak{Z})) + a_i$ where the a_i have the same property as the a_i'. Hence $(\vartheta_i) = s \cdot c_i + c_i'$, c_i and c_i' being $\mathbf{M}$-divisors with the c_i' belonging to a fixed finite subset. Thus $s c_i$ can belong to only a finite number of classes, mod $\mathbf{G}_l(\mathbf{M})$ and the same is true for the c_i {see (VII, 2, iii)}. Hence it follows easily that there exist functions α_i and α_i' on $\mathbf{M}$ with $\vartheta_i = \alpha_i^s \alpha_i'$, where the α_i' belong to a finite fixed set of functions.

It is then proved that the mapping $\mathfrak{Z} \to \vartheta_i$ is such that the functions associated with $\mathfrak{Z}_1$, $\mathfrak{Z}_2$ and $\mathfrak{Z}_1 + \mathfrak{Z}_2$ (say ϑ_{1i}, ϑ_{2i} and ϑ_i) satisfy a relation of the form $\vartheta_i/\vartheta_{1i}\vartheta_{i2} = \delta^s$ where δ^s is a function on $\mathbf{M}$ defined over a field k' of definition for $\mathfrak{Z}_1$ and $\mathfrak{Z}_2$. Let h_s be the set of the $\mathfrak{Z}$ for which we can take $\alpha_i' = 1$ for any i in the relation $\vartheta_i = \alpha_i^s \alpha_i'$.

By the above formula it follows that h_s is a group and that h/h_s is finite since the number of determinations of the α_i' is finite. If now one proves that $sh \supset h_s$, and then that $sh = h_s$, i. e. that $\mathfrak{Z} \in h_s$ is a n. a. s. c. for the existence of a $\overline{\mathfrak{Z}}$ of h with $s\overline{\mathfrak{Z}} = \mathfrak{Z}$, then (viii) is true: the proof is achieved by GALOIS theory.

We pass to (ix). We begin by demonstrating that the degrees of the transforms of the varieties $\mathfrak{Z}_p^{(\nu)}$ onto a projective model $\mathfrak{J}_p$ of $\mathfrak{J}$ are, for any large ν, less than a suitable integer, independent of Z: then since the positive cycles of given degree and dimension on $\mathfrak{J}_p$ are distributed in a finite number of components of the CHOW model the theorem (ix) will follow. The above assertion is proved as follows.

Let $\mathfrak{Z}^0$ be an element of h and let β be the correspondence on J given by: $Z \to sZ + Z^0$ with $Z^0 = \mathfrak{Z}^0 \cdot J$, and let γ be its extension to $\mathfrak{J}$. Now introduce the linear system of all the positive divisors of $\mathfrak{J}$ whose images on the projective model $\mathfrak{J}_p$ form the multiple of the system of the hyperplane sections of a sufficiently large order. Denote by $\mathfrak{X}$ a generic element of the system, which is free from fixed point and put $\overline{\mathfrak{X}} = \gamma^{-1}(\mathfrak{X})$.

Then we can prove that for each $\mathfrak{Z}$ of the form $\gamma(\overline{\mathfrak{Z}})$ with $\overline{\mathfrak{Z}} \in h$, one has $\mathrm{pr}_M(\overline{\mathfrak{X}} \cdot \overline{\mathfrak{Z}} - \mathfrak{X} \cdot \mathfrak{Z}) = a$, where a denotes an element of a finite subgroup of $\mathbf{G}(\mathbf{M})$ independent of $\overline{\mathfrak{Z}}$. From this and the theory of linear equivalence on Abelian varieties we have: $\mathrm{pr}_M((\mathfrak{Z} - (s^2 - 1)\,\overline{\mathfrak{Z}}) \cdot \mathfrak{X}) = b$, where b belongs to a finite subgroup of $\mathbf{G}(\mathbf{M})$.

Further one can see that the coefficients of the components of b are all bounded: it is sufficient to prove this for the components of the cycle $\overline{\mathfrak{X}} \cdot \overline{\mathfrak{Z}}$, with respect to the intersection of $\overline{\mathfrak{Z}}$ with a finite but large number of independent generic specialisations of $\overline{\mathfrak{X}}$, since the fixed part of $\overline{\mathfrak{X}}$ certainly gives components of bounded multiplicity.

This is finally obtained by using the following lemma, whose proof is based on SAMUEL's theory of irregular intersections.

Let A_j be positive U-divisors, finite in number, passing through a simple subvariety W of U and let α be an integer. Consider on U all the subvarieties V passing simply through W and such that W is a proper component, in each intersection $V \cap A_j$, of higher multiplicity than α in the intersection of V with each component of $\bigcap_j A_j$. Then the multiplicity of W in each $V \cap A_j$ admits a superior limit independent of V.

3. The First Picard Variety

We shall give here a short account of the construction of *the (first)* Picard *variety* of a variety V following the method of Néron and Samuel {see Néron and Samuel [1]}; this is somewhat different from Matsusaka's, which generalises more directly the construction of the Jacobi variety of a curve, that is to say, the Picard variety of the curve. We first state some definitions.

Given a variety V and two subgroups $\mathbf{G}_1$ and $\mathbf{G}_2$ of $\mathbf{G}(V)$, with $\mathbf{G}_1 \supset \mathbf{G}_2$, and another variety V' together with two subgroups $\mathbf{G}'_1$ and $\mathbf{G}'_2$ of $\mathbf{G}(V')$ with $\mathbf{G}'_1 \subset \mathbf{G}'_2$, a homomorphism φ of $\mathbf{G}_1/\mathbf{G}_2$ into $\mathbf{G}'_1/\mathbf{G}'_2$ will be called *rational over a field k* if, for any field $K \supset k$ and for any divisor $A \in \mathbf{G}_1$ rational over K, the transform of the class of $A (\operatorname{mod} \mathbf{G}_2)$ by φ contains a divisor of $\mathbf{G}'_1$ rational over K: the same definition is similarly applied to the case of a homomorphism of a quotient group into the group of an Abelian variety. The isomorphism φ will be called birational over k if φ and φ^{-1} are rational over k.

The authors now begin with the theorem:

(i) *If T is a birational correspondence between two normal projective varieties V, V' defined over a field k_0, there exists a birational isomorphism over k' of $\mathbf{G}_a(V)/\mathbf{G}_l(V)$ onto $\mathbf{G}_a(V')/\mathbf{G}_l(V')$, where k' is a finite algebraic extension of k_0.*

The proof is simple: in fact let p be the projection function which represents T, after the graph of T has been normalised. We prove that: $D = p^{-1} p(D)$, where $D \in \mathbf{G}_a(V)$ and the symbol $p^{-1}(D')$, $D' \in V'$, is defined since V is normal. In the first place we have $\overline{D} = p^{-1} p(\overline{D}) + E$, where $\overline{D}$ is generic for an algebraic system containing D and the zero-divisor and where further $E \in D$ and $\dim \operatorname{pr}_{V'}\|E\| < n-1$, with $n = \dim V$. Specialising $\overline{D} \to D$, $\overline{D} \to 0$, we have $D = p^{-1} p(D)$. Moreover, by the normality of V and V' we obtain $p^{-1}(D) \in \mathbf{G}_a(V')$ when $D \in \mathbf{G}_a(V)$ and $p(D') \in \mathbf{G}_a(V)$ when $D' \in \mathbf{G}_a(V')$, and similarly for the linear equivalence. Whence the result.

We shall state without proof the following theorem, of a rather elementary nature, based on projective properties of the hyperplane sections of a variety:

(ii) *Let V^n be projective and k-normal. There is an extension k' of k and a one-to-one correspondence defined over k' between V and another variety V' such that:* (1) *Either this belongs to the product $\mathbf{P}^1 \times \mathbf{P}^N$ of two projective spaces and moreover $\operatorname{pr}_{\mathbf{P}^1} V' = \mathbf{P}^1$ and the cycle $(M \times \mathbf{P}^N) \cdot V'$, for any point $M \in \mathbf{P}^1$, is reduced to a variety $W(M)$, normal when M is generic over k'*; (2) *Or it belongs to a product of two projective spaces $\mathbf{P}^N \times \mathbf{P}^{n-1}$*

and $\mathrm{pr}_{\mathbf{P}^{n-1}} V' = \mathbf{P}^{n-1}$; *and the cycle* $(M \times \mathbf{P}^N) \cdot V'$ *is not prime if and only if the point* M *of* $\mathbf{P}^{n-1}$ *belongs to a subset of dimension less than* $n-2$; *moreover, for any point* M *of* $\mathbf{P}^{n-1}$, $\dim (M \times \mathbf{P}^N) \cdot V' = 1$.

The first part of (ii) gives a model of V fibred by divisors and the latter a model fibred by curves, like the one used in the preceding section: obviously (ii) would be trivial, if one did not consider the regularity conditions on the fibres and which are really used in the demonstration of the following theorem:

(iii) *If* V *is either model of* (ii) *defined over* k, *then* $\mathbf{G}_a(V)/\mathbf{G}_l(V)$ *is birationally isomorphic over* k *to* $\mathfrak{H}_a/\mathfrak{H}_l$.

Using the symbols introduced in (VII, 2) and recalling that V is normal, we see at once that, for V, the group $\mathfrak{H}'$ of the divisors $p^{-1}(d)$, where p is the projection of V on $\mathbf{P}^1$ or $\mathbf{P}^{n-1}$ and d is a divisor respectively of $\mathbf{P}^1$ or $\mathbf{P}^{n-1}$, coincides with the group $\mathfrak{H}$. Moreover let χ be the canonical homomorphism of $\mathbf{G}_a/\mathbf{G}_l$ onto $\mathfrak{H}_a/\mathfrak{H}_l$. Put $\overline{\mathfrak{H}}_a = p^{-1}(\mathbf{G}_a(\mathbf{B})) \cong \mathbf{G}_a(\mathbf{B})$, $\overline{\mathfrak{H}}_l = p^{-1}(\mathbf{G}_l(\mathbf{B})) \cong \mathbf{G}_l(\mathbf{B})$ where $\mathbf{B}$ denotes $\mathbf{P}^1$ or $\mathbf{P}^{n-1}$. We have: $\overline{\mathfrak{H}}_l \subset \subset \mathfrak{H} \cap \mathfrak{G}_l$, $\overline{\mathfrak{H}}_a \subset \mathfrak{H} \cap \mathfrak{G}_a$.

The kernel of χ is $((\mathfrak{H} + \mathfrak{G}_l) \cap \mathfrak{G}_a)/\mathfrak{G}_l = ((\mathfrak{H} \cap \mathfrak{G}_a) + \mathfrak{G}_l)/\mathfrak{G}_l \cong \cong (\mathfrak{H} \cap \mathfrak{G}_a)/(\mathfrak{H} \cap \mathfrak{G}_l)$. The last group is isomorphic to a subgroup of a quotient group of $\mathfrak{H}'/\mathfrak{H}'_l$, and, therefore, of $\mathbf{G}_a(\mathbf{B})/\mathbf{G}_l(\mathbf{B})$, where, we recall, $\mathbf{B}$ is a projective space: consequently that kernel is zero and then χ is an isomorphism, which is easily seen to be birational over k.

By (i), (ii), (iii) and recalling (VII, 2, ii) we can state the following proposition, by virtue of the observation that the isomorphism considered in the last theorem is rational over a finite algebraic extension of the definition field of the variety under examination:

(iv) *Let* V^n *be a projective normal variety defined over a field* k: *then there exists an algebraic finite extension* k' *of* k *and an Abelian projective variety* $\mathfrak{P}^*(V)$, *defined over* k', *such that there is an isomorphism of* $\mathbf{G}_a(V)/\mathbf{G}_l(V)$ *onto* $\mathfrak{P}^*(V)$ *which is rational over* k'.

Néron and Samuel give another proof of this theorem which, though resting on the same fundamental facts, is more purely algebraic. The most salient feature of this alternative proof is that, declining the help of (VII, 2, ii), which is an explicitly geometric statement, the authors construct a one-to-one parametrisation of maximal dimension of the subgroups of I {see (VII, 2)}, transferring the Abelian structure of these to the variety U of the parameters.

But then the composition law remains fixed only almost anywhere on U, and one must pass to a normal model to make it everywhere regular: this is an idea which Chow has previously used to construct the Jacobi variety of a curve before having been able to give the direct proof referred to in (VII, 1).

The theorem (iii) then assumes the following definitive form:

(v) *With the same hypotheses of* (iv) *there exists a finite algebraic extension k'' of k and an Abelian projective variety $\mathfrak{P}(V)$, defined over k'', such that there is a birational isomorphism over k'' of $\mathbf{G}_a(V)/\mathbf{G}_l(V)$ onto $\mathfrak{P}(V)$.*

The varieties $\mathfrak{P}^*(V)$ and $\mathfrak{P}(V)$ of theorems (iv) and (v) respectively, are both called the (first) PICARD variety of V.

The proof of (v) requires a great deal of algebra. We point out that $\mathfrak{P}(V) = \mathfrak{P}^*(V)$, when the canonical function associated with the Jacobian of the fibre is defined over $k'(M)$ with M generic in $\mathbf{P}^{n-1}$, and where we naturally refer to case (2) of (ii): again, supposing this hypothesis satisfied, and, further, supposing V embedded in a product variety of the type $C^{n-1} \times \mathbf{P}^N$, in such a manner that $\mathrm{pr}_C V = C$, where C is any variety, the authors prove that, under suitable hypotheses, $\mathbf{G}_a(V)/\mathbf{G}_l(V)$ is isomorphic to the product $\mathfrak{P}^* \times \mathfrak{P}(C)$ where $\mathfrak{P}(C)$ is the PICARD variety as in (v) and $\mathfrak{P}^*$ is the Abelian variety of (iv): the rest of the theorem is now proved by proceeding from this to the general case by many transformation of varieties and other devices. We remark the following consequences of the theory outlined above:

(vi) *If X is a divisor of $\mathbf{G}_a(V)$ and X' a specialisation of X over a field $K \supset k'$* {see (iv)}, *then the point x' of $\mathfrak{P}^*$ corresponding to X' is a specialisation over $X \to X'$ with respect to K of the point x associated with X.*

The importance of the above fact arises from the possibility of parametrising the algebraic V-systems by the subvarieties of the PICARD variety of V; from this and (v) at once follows *the existence of the so-called* POINCARÉ *families of divisors on a normal variety, i. e. of an algebraic system of divisors in $\mathbf{G}_a(V)$ birationally related over k'' with $\mathfrak{P}(V)$, and such that the point of $\mathfrak{P}(V)$ corresponding to a divisor X by the birational correspondence is also the λ-transform of the class of X* $\bmod \mathbf{G}_l(V)$.

In order to construct such a family it is sufficient to consider a generic point x of $\mathfrak{P}(V)$ over k'', a divisor X of V rational over $k''(x)$ representing x, and the algebraic system having X as a generic element over k''.

4. The Total Maximal Algebraic Families

We shall now describe MATSUSAKA's researches which generalise to PICARD varieties CHOW's work on the Jacobi variety of a curve. Many of the classical concepts and properties, chiefly due to the Italian school, extend to abstract reference fields: among these are the notions of total maximal algebraic family and superficial irregularity, already investigated by SEVERI {see SEVERI [27]} for the classical case by means of purely algebraic methods.

Let V be a projective normal variety defined over a field k, which we take as the basic field, i. e. all the fields which will be considered contain k. If C is a generic 1-section of V with a suitable subspace of the ambient space, we suppose that $K = \mathrm{def}(V, W, J, \varphi)$, where W is a component of the CHOW set $\mathbf{C}$ of all V-divisors, J is the Jacobi variety of C and φ the associated canonical function (VII, 1). If X is a V-divisor whose CHOW point (x) lies on W, X and then $X \cdot C$ are rational over $K(X)$. We put $h(x) = S(\varphi(C \cdot X))$ {see (VII, 1)}; then one can show {see MATSUSAKA [4, (I)]}, using ZARISKI's main theorem on birational correspondences {see (I, 6)}, that:

(i) *The function $h(x)$ on W with values in J is defined at every normal point of W.*

Further:

(ii) *Let $\xi = h(x) \in J$ with (x) generic on W. Then the points of every component of $h^{-1}(\xi)$ correspond to linearly equivalent V-divisors and moreover $h^{-1}(\xi)$ is the associated variety $T(X)$ of the complete linear system $|X|$.*

The proof of (ii) comes easily from the fact that W is a component of C and from the following lemma, which is the simplest of the criteria for linear equivalence: it is due to A. WEIL and extends well known results of SEVERI for surfaces over the complex field:

(iii) *There is a finite set of divisors D_i on V, all algebraic over k, and algebraically independent on V, such that, if $X \cdot C \equiv 0$, where C is a 1-section of V generic over k, then X is linearly equivalent to some linear combination of the D_i.*

For the criteria of linear equivalence of SEVERI (proved in [6, 11]) see SEVERI [a], p. 191: they have been extended to higher varieties over the complex field by SEVERI in [11] and to abstract fields of reference by A. WEIL in [5, 6], by replacing the use of transcendental methods by the theory of applications of V into the second PICARD variety related to V {see (VII, 8)}.

The algebraic system on V parametrised by the component W of C will be called *a maximal family* and we shall denote such a family by $\{X\}$, where X is a generic divisor of this system.

Now the point $\xi \in J$ in (ii) will have a locus $\Lambda(X)$ over K, which, as W varies among the set of all the components of C, which are infinite, describes a set $\mathbf{S}$ of subvarieties of J, which has, as one readily sees, the properties:

(iv) *There is a maximal family $\{X\}$ of positive V-divisors such that $\Lambda(X)$, if necessary after a translation on J, is an Abelian subvariety $\mathfrak{B}$ of J, where $\mathfrak{B}$ is such that any subvariety of J all lying in* $\mathbf{S}$, *is, if necessary after a translation on J, a subvariety of B. Moreover $\overline{K} = \mathrm{def}\,\mathfrak{B}$.*

The result is substantially the same as that of (VII, 2, iii) due to NÉRON.

Now, let $K = \mathrm{def}(V, C, J, \varphi, W, U)$ where W is the CHOW variety of a maximal family $\{X\}$ as the one described above in (iv), while U is the CHOW variety of any algebraic irreducible system of positive V-divisors $\{Y\}$. Let x, y be independent generic points respectively of W

and U over K associated to X and Y, and let $Y \cdot C$ be defined. Then we put $h(y \times x) = S(\varphi(C \cdot X + C \cdot Y))$, which is a function on $U \times W$ with values in J. Taking into account that, after a translation, we may assume $B = \mathrm{pr}_J Z$, where Z is the graph of h on $U \times W \times J$, we easily prove that:

(v) *Every component of* $h^{-1}(\zeta)$, *where* $\zeta = h(y \times x)$, *has projection* U *onto* U.

Finally MATSUSAKA proves that:

(vi) *The maximal family described in* (iv) *is such that if* Y *is any* V*-divisor with* $Y \equiv 0$, *then* $Y \equiv X - X'$, *where* X *and* X' *are generic member of* $\{X\}$ *over* $K = \mathrm{def}(V, W) \supset \bar{k}$.

The proof is as follows. Let C be a generic 1-section of V over K and $K' = \overline{K'} = \mathrm{def}(C, J, \varphi) \supset K$. Then $K' = \mathrm{def}\, B$. Writing $Y = Y_1 - Y_2$ with $Y_1, Y_2 > 0$, assume, as one always can by (VI, 7), that Y_1 and Y_2 belong to one and the same maximal family of positive V-divisors $\{Y\}$, having U associate variety containing the CHOW points y_1 and y_2 of Y_1 and Y_2. Obviously by (VI, 1) $\bar{k} = \mathrm{def}\, U \subset K$.

Applying (v), if T is a component of $h^{-1}(\zeta)$, there will be in T points whose projections on U are respectively y_1 and y_2: such points may be written in the form $y_1 \times x_1$ and $y_2 \times x_2$.

Then, assuming U and W to be normalised and using (i), we have $\zeta = h(x_i \times y_i)$, $i = 1, 2$. Hence it follows easily that $h(y_i \times x)$ is a generic point of B over K', to within a translation on J, and so we can choose x_i as a generic point of U over K'. Then (ii) gives at once the required result.

Def. *a. A family on* V *such as the one considered in* (vi) *will be called a total or regular family.*

5. A Property of the Arithmetic Genus

We could now proceed to construct the first PICARD variety of V starting from the locus of the CHOW-point of $T(X)$ where $\{X\}$ is a total family on V, precisely as CHOW has done in the case where V is a curve. Unfortunately we are faced here with the same difficulty, namely that $\dim |X|$ is not generally constant over the specialisations of X: hence the representation may acquire exceptional features and, clearing these away, we may be left with singular models. Hence the necessity of proceeding with the theory of WEIL's abstract varieties or, better still, of looking for a total family on V having the desired character.

In the case where V was a curve it sufficed to apply the RIEMANN-ROCH theorem in order to obtain a non-special linear series, which fulfils our condition of regularity.

Before demonstrating the existence of such families for a variety V of any dimension, we point out that the PICARD variety when already constructed as an abstract variety gives rise also to a projective model in virtue of the following important statement of MATSUSAKA {see MATSUSAKA [6]}:

(i) *Let* U *be an abstract variety having a normal composition law, both defined over a field* k. *Then, if* U *is birationally equivalent over* k *to* *an Abelian variety, it is also birationally equivalent over k to an Abelian projective variety defined over k.*

In spite of this, the direct geometrical construction of a projective model of the PICARD variety maintains its importance since it allows a deeper insight into the structure of algebraic and linear equivalence groups on V. We begin with the following noteworthy fact, proved by KODAIRA on the complex field {see (XI, 3)} and by MATSUSAKA on any field {see MATSUSAKA [5]}:

(ii) *If X and X' are divisors on a non-singular variety V satisfying $X \equiv X'$, then $p_a(X) = p_a(X')$.*

The proof is by induction with respect to $r = \dim V$. In fact (ii) is true if $r = 1$. Supposing it to be true if $\dim V < r$, by (VI, 7, vii) we can assume X, X' to be positive V-divisors of one and the same maximal family $\{X\}$. If $\{A\}$ is any total V-family, let X and X' be rational over $K = \operatorname{def}(V, \{A\}, \{X\})$ and $\bar{X}, A$ denote independent generic members over K of the respective families.

By the results in (VII, 4) it is easily seen that $C_s + X + A \equiv Z$ and $C_s + X' + A \equiv Z'$, where C_s is a member of the s-ple of the system of hyperplane sections and Z, Z' are generic members over $\bar{K}$ of a maximal V-family containing the divisor $C_s + \bar{X} + A$, which is also total as $\{A\}$. This shows that $p_a(Z) = p_a(Z')$, since Z and Z' are generic specialisations of each other over $\bar{K}$.

But, for large s, there are varieties Y and Y' such that $Y \equiv C_s + X$ and $Y' \equiv C_s + X'$: therefore, by the invariance of p_a with respect to linear equivalence on V {see (V, 2, i)}, we have $p_a(Y + A) = p_a(Y' + A)$. Using (III, 7) and the fact that $\{A\}$ is total and thereby also $\{A + C_n\}$ for any n, we may assume A to be non-singular and $Y + A$, $Y' + A$ to be linearly equivalent to some non-singular varieties: let they be W and W' respectively. Then $p_a(W) = p_a(W')$.

Further we may assume that $|A|$ is an ample system {see (III, 6)} and then, interposing a birational biregular correspondence, we can treat $|A|$ as if it were a system of hyperplane sections, to which we apply the modular property (V, 2, ii): hence, over a suitable definition field of the varieties over which A is supposed generic, we have

$$p_a(W) = p_a(Y + A) = p_a(Y) + p_a(A) + p_a(Y \cdot A)\,,$$
$$p_a(W') = p_a(Y' + A) = p_a(Y') + p_a(A) + p_a(Y' \cdot A)\,.$$

Since $Y \equiv Y'$ on V, it is by (VI, 7, iii) $Y \cdot A \equiv Y' \cdot A$ on A and therefore $p_a(Y \cdot A) = p_a(Y' \cdot A)$. Consequently $p_a(Y) = p_a(Y')$, or $p_a(C_s + X) = p_a(C_s + X')$: whence the result by (V, 2, i) and the induction hypothesis.

6. Non-special Total Families

We begin with the following lemma of CASTELNUOVO {see CASTELNUOVO [a], p. 396}, which has been extended to the present case by MATSUSAKA {see MATSUSAKA [5]}.

(i) *Let X be a positive V-divisor rational over a field k and C be a generic hyperplane section of V over k. Then $|X + C_h| \cdot C$ is a complete system on C for all sufficiently large h.*

The proof is immediately achieved by calculating {use (28) of Ch. V and (V, 2, ii)} the dimensions of both the systems $|X + C_h| \cdot C$ on C and of $|X \cdot C + C_h \cdot C|$ on C. After this we prove:

(ii) *Let $\{X\}$ be a total family on V. Then* $\dim |X + C_m| = \dim |X' + C_m|$ *for large m and any two elements X, X' of $\{X\}$.*

The theorem is true if $\dim V = 1$, by the Riemann-Roch theorem. Suppose it to be true on any variety of dimension less than r.

Let X be a generic element of $\{X\}$ over $K = \mathrm{def}(\{X\})$. Then obviously if $x' \in W$, we have $\dim |X| \leqq \dim |X'|$, where X' is associated with x', and there is a proper subset S of W algebraic over K such that $\dim |X| = \dim |X'|$ if $x' \in W - S$ and otherwise $\dim |X| < \dim |X'|$. Applying this fact to the total family $\{X + C_m\}$ we find a sequence of exceptional subsets S_i, $i = 0, 1, \ldots, h, \ldots$

Suppose now h_0 to be such an integer that $|X + C_h| \cdot C$ is complete on C and that every divisor of a total family of C-divisors containing $C \cdot X$ has already the property stated in our theorem, both statements holding for $h \geqq h_0$, as may be supposed by our induction hypothesis. Then, if $x' \in W - S_h$ and the divisor X' corresponding to x' is such that the intersection-product is defined, we have:

$$\dim |X + C_{h+1}| \cdot C = \dim |X + C_{h+1}| - \dim |X + C_h| ,$$
$$\dim |X' + C_{h+1}| \cdot C = \dim |X' + C_{h+1}| - \dim |X' + C_h| .$$

By our choice of h, $\dim |X + C_{h+1}| \cdot C \geqq \dim |X' + C_{h+1}| \cdot C$ and, as $x' \in W - S_h$ we have: $\dim |X + C_h| = \dim |X' + C_h|$. Therefore $\dim |X + C_{h+1}| \geqq \dim |X' + C_{h+1}|$, and then necessarily $\dim |X + C_{h+1}| = \dim |X' + C_{h+1}|$. Repeating this process we have:

$$\dim |X + C_h| = \dim |X' + C_h|$$

for any $h \geqq h_0$ and $x' \in W - S_h$.

Now let $S_h = \sum_i U_i^{(h)}$ be the decomposition of the set S_h in its components over $\bar{K}$ and let $x_i^{(h)}$ be a generic point of $U_i^{(h)}$ over K, associated with the divisor $X_i^{(h)}$. We select an integer $h_1 > h_0$ such that, for any $h \geqq h_1$, we have applying (28) of Ch. V:

$$\dim |X + C_h| = (-1)^r \{p_a(V) + p_a(-X - C_h)\} - 1 ,$$
$$\dim |X_i^{(h_1)} + C_h| = (-1)^r \{p_a(V) + p_a(-X_i^{(h_1)} - C_h)\} - 1 .$$

Hence, by (i), we derive: $\dim |X + C_h| = \dim |X_i^{(h_1)} + C_h|$, so that, when the Chow point of X' is on $W - S_{h_1}$, we get: $\dim |X + C_h| = \dim |X' + C_h| = \dim |X_i^{(h_1)} + C_h|$. Hence by what has been proved above, S_{h_1} is a proper subset, with respect to each component, of S_{h_0}.

Consequently, for large h, S_h will be empty and our statement is demonstrated.

c) The starting point of the direct construction of the PICARD variety, together with some developments of the notion of total family is contained in the following theorem of MATSUSAKA:

(iii) *Two total V-families $\{X\}$ and $\{Y\}$ necessarily coincide when X and Y are algebraically equivalent.*

(iv) *If $\{X\}$ is any maximal family on V, there is a maximal family $\{X + C_m\}$ containing $X + C_m$ which is total whenever m is greater than some integer m_0 independent of the choice of $\{X\}$.*

(v) *There exists an integer m'_0 such that the total family of the preceding theorem is also non-special if $m > m'_0$, in the sense that every member of the family must belong to a complete linear system of constant dimension.*

The first assertion is obvious, for if X is a generic divisor of the maximal family $\{X\}$, then, supposing $\{Y\}$ also to be total, there exists a V-divisor $Y \in \{Y\}$ such that $Y \equiv X$ and thereby $|Y| = |X|$, so that the two families coincide.

The second follows immediately by supposing m'_0 to be selected in such a way that L_m is, for $m > m'_0$, complete and partially contains a generic divisor A of a total, arbitrarily chosen, family $\{A\}$: then the family $\{X_0 + C_m + A - A_0\}$ where $A_0 \in \{A\}$ and $X_0 \in \{X\}$ is also total and obviously contains $X_0 + C_m$.

Finally the third theorem comes from (ii).

Moreover, observing that any V-divisor algebraically equivalent to some member of a non-special family is also linearly equivalent to a suitable member of the family, since this is total, and that $\dim|X|$ is constant for any specialisation of a suitable generic member, we have:

(vi) *A non-special total family on V is complete with respect to algebraic equivalence, i. e. it contains all the positive V-divisors algebraically equivalent to any one element of the family.*

This is the reason why such families have also been called *complete*.

If X is a positive V-divisor rational over a field $K = \overline{K}$ and $\{A\}$ is a total family having A as generic member over $K = \text{def}(V, \{A\})$, denoting by A_0 a rational member of $\{A\}$ over K and proceeding as for (iv), we can prove that:

(vii) *N. a. s. c. in order that the V-divisor X should belong to a total family on V is that the system $|X + A' - A_0|$ should effectively exist for any specialisation A' of A over K.*

Hence defining, with SEVERI {see SEVERI [31]}, an arbitrary V-divisor as *linearly* effective if and only if its linear equivalence class contains some positive divisor, we at once obtain the condition {see for all that precedes in this section the criticism of SEVERI in [31, 41]}:

(viii) *Each divisor of a class of $\mathbf{G}(V)/\mathbf{G}_a(V)$ is linearly effective if and only if this class contains a total family.*

7. The First Picard Variety According to Matsusaka

We are now able to describe Matsusaka's construction of a projective model for the (first) Picard variety of a non-singular variety V, omitting the proofs which are rather long {see Matsusaka [4, II^0]}.

Let $\{X\}$ be a total family on V and let U be the associated variety of $\{X\}$. Then if k is the basic field, we have $\bar{k} = \operatorname{def} U$ and the locus W of the Chow point of the associated variety $T(X)$ of $|X|$ over k, is defined over k, provided $\{X\}$ contains a rational divisor over k. In this case $k = \operatorname{def} U$. Further one can always find a non-special total family containing such a divisor X_0: it is sufficient for this to consider a family containing the sum of a complete system of conjugates of a member of $\{X\}$. This also is a total family; and adding a suitable multiple to it, it gives rise by (v) to a non-special family evidently containing a divisor of the kind required.

Now if X_1 and X_2 are generic independent divisors over k of the family $\{X\}$ so obtained, since $\{X\}$ is total, there is a divisor X_3 in $\{X\}$ such that $X_3 \equiv X_1 + X_2 - X_0$, where X_0 is a rational divisor over k of the family. If M_i is the Chow point of $T(X_i)$, $i = 1, 2, 3$, then one can prove that the locus Z of $M_1 \times M_2 \times M_3$ over k is a subvariety of $W \times W \times W$ and that the projection of Z on the products of any two factors is regular: therefore Z defines on W a normal composition law which is moreover commutative and defined, as is Z, over k.

It seems most likely that W itself is non-singular, when one takes into account its homogeneous generation, but, in defect of a formal proof, Matsusaka proceeds by k-normalising W into a model W^*, proving then that W^* is absolutely normal and non-singular. This is sufficient in order that the above composition law, when transferred to W^*, should be everywhere defined on W^* itself. Clearly $W^* = \mathfrak{P}$ is isomorphic to $\mathbf{G}_a(V)/\mathbf{G}_l(V)$ as a group; further $\mathfrak{P}$ and its composition law are both defined over $k = \operatorname{def} V$. This projective variety is called a projective model of the Picard variety of V.

Moreover we have the following characterisation theorem for such a variety:

(i) *Defining the (first)* Picard *variety* $\mathfrak{P}$ *of the non-singular projective model* V *as follows:*

(1) $\mathfrak{P}$ *is isomorphic to* $\mathbf{G}_a(V)/\mathbf{G}_l(V)$.

(2) *There is a field* $K = \operatorname{def}(V, \mathfrak{P})$ *such that if* $Y \in \mathbf{G}_a(V)$ *is rational over a field* $K' \supset K$, *then its class on* $\mathfrak{P}$ mod $\mathbf{G}_l(V)$ *is rational over* K' *itself and such that if* y *is a point on* $\mathfrak{P}$ *rational over a field* $K'' \supset K$, *then there is a* V*-divisor rational over* K'' *belonging to* $\mathbf{G}_a(V)$ *and whose class on* $\mathfrak{P}$ *is represented by* y.

Then the variety $\mathfrak{P}$ *is determined uniquely to within an isomorphism.* The reader will notice the substantial identity of these results with those of Néron and Samuel stated in (VII, 3), save for the fact that these latter refer to the more general hypothesis that the variety V be only normal.

8. The Second PICARD Variety and the Superficial Irregularity

An Abelian variety $\mathfrak{A}$ is said to be *generated by a variety V* when there is a function f defined on V with values in $\mathfrak{A}$ and a finite number of simple points $x_1, x_2, \ldots, x_s$ such that $f(x_1) + f(x_2) + \cdots + f(x_s)$ is a generic point of $\mathfrak{A}$ over $K = \operatorname{def}(V, \mathfrak{A}, f)$.

Then MATSUSAKA has proved {see MATSUSAKA [3]} that *$\mathfrak{A}$ is generated by a generic 1-section C of V over K and by the function f_C induced on C by f.* Consequently $\dim \mathfrak{A} \leqq$ genus C, since the genus of C is the same for any such C: hence *there is an Abelian variety generated by V and having the maximal dimension q.*

Suppose K to be algebraically closed and consider q independent generic points $x_1, x_2, \ldots, x_q$ of C over K, putting $\xi = \sum_{i=1}^{q} f_C(x_i)$. There is a minimal field $K(z)$ of K, over which $\sum_{i=1}^{q} x_i$ is rational. Hence: $K(\xi) \subset K(z)$, and if $\dim \mathfrak{A} = q$, $K(z)$ is algebraic over $K(\xi)$. Putting $(K(z):K(\xi)) = d$, we have $\operatorname{pr}_{\mathfrak{A}} Z = d\,\mathfrak{A}$, where Z is the locus of $z \times \xi$ over K. This integer d *depends only on f*, and not on C: it is denoted by $d(\mathfrak{A}, f)$.

We state the following

Def. *a. The second* PICARD *variety $\mathfrak{A}$ of V is an Abelian variety satisfying the condition: there is a function φ defined on V with values in $\mathfrak{A}$, such that when f is a function defined on V with values in an Abelian variety $\mathfrak{B}$, then there is a homomorphism λ from $\mathfrak{A}$ to $\mathfrak{B}$ giving $f = \lambda\,\varphi + c$, where c is a constant.*

It is then possible to prove that, by referring to the previous notions, there is an Abelian variety $\mathfrak{A}$ of the maximal dimension q generated by V and by a suitable function g, such that, for any other similar pair $\mathfrak{A}^*$, g^*, we have: $d(\mathfrak{A}, g) \leqq d(\mathfrak{A}^*, g^*)$ and moreover:

(i) *This variety $\mathfrak{A}$ is, to within an isomorphism, the second* PICARD *variety of V.*

(ii) *The first and the second* PICARD *varieties of V are isogeneous, i. e. they have the same dimension q and there is a homomorphism of $\mathfrak{A}$ into $\mathfrak{P}$ and conversely.*

The absolute invariant of the birational class of V, given by the dimension q of both these varieties, if V is defined on the complex field coincides with the so-called *superficial irregularity* of V, a character known from transcendental and topological considerations, as we shall see later in (XI, 3): consequently the same name is used in the abstract case, where only the strictly algebraic characterisation of our number is significant.

The attribute "*superficial*" is justified by the following theorem {see MATSUSAKA [7]}, which translates into a suitable abstract language a famous theorem of CASTELNUOVO-ENRIQUES which was proved by transcendental methods in the complex case:

(iii) *The non-singular variety V,* $\dim V \geqq 3$, *and its sufficiently general divisor have the same* PICARD *variety to within an isomorphism.*

The algebraic proof of this theorem, which is not difficult, has been given by MATSUSAKA: it rests on the preceding abstract construction of the PICARD variety and on WEIL's extension to abstract fields of the criteria for linear equivalence. We point out that, as regards the purely dimensional aspect, (iii) had already been proved by SEVERI {see SEVERI [27]} over the complex field, with algebraic tools similar to those later used by MATSUSAKA.

The distinction between the first and the second variety of PICARD (which appears in earlier works of SEVERI) is fully realised over the complex field in IGUSA [3] and ANDREOTTI [1]: the universal map of a variety V into its second PICARD variety has been largely used by the last author as a fundamental tool of investigating and classifying irregular surfaces: see ANDREOTTI [1, 2].

Further developments of the results expounded in this chapter may be found in the important researches [4, 5] of BARSOTTI and [1, 2] of ROSENLICHT.

VIII. Theory and Applications of the Canonical Systems

1. Introduction

If the theory of rational equivalence resembled more closely that of linear equivalence, then the algebraic geometry of the higher dimensional varieties would probably be as simple as the classical geometry of curves. But such a simplification, if indeed attainable, is still a distant goal.

At present, however, there exists a considerable body of methods and facts, though not amounting to a general theory, which will be expounded in this chapter.

The earliest examples of invariant systems of equivalence, which, in what follows, are always to be understood in SEVERI's sense, i. e. as irreducible algebraic subsystems, proper or otherwise, of rational systems {see (VI, 10)}, are due to SEVERI himself; these are series of point-sets on a surface, e. g. the so-called SEVERI series {see (VIII, 3e); moreover see SEVERI [a], p. 236, and the report [a] of CONFORTO, where the reader may find much historical information}.

Following some extensive work on threefolds by B. SEGRE {see B. SEGRE [2, 5]} and some important suggestions of EGER {see EGER [1, 2, 3, 4]} (the latter inspired by transcendental methods), TODD {see TODD [5, 11, 14]} saw how to define, on any non-singular variety over the complex field, relatively invariant systems of equivalence of every dimension: these are known as the TODD canonical systems. Finally B. SEGRE {see B. SEGRE [12, 14]} has recently reviewed such a theory from a new standpoint, obtaining many important algebraic results and throwing light on far-reaching relations between this subject and recent topological developments.

After giving an alternative derivation of the canonical divisor-systems, due to SEVERI and TODD, we describe how TODD establishes the general notion of canonical systems; we then outline SEGRE's methods and, finally, sketch the chief applications of these theories. In this chapter we shall be concerned only with non-singular varieties defined over the complex field, so that this assumption can always be understood in the sequel.

2. A new Definition of the Canonical Divisors

{This new derivation was given by SEGRE for threefolds in [2] and by TODD for varieties of any dimension in [11], supposing the canonical system already defined; instead, as an independent definition, by SEVERI in [35], in the manner here described.}

Let L_i, $i = 0, 1, \ldots, h$, be $h + 1$ ample linear systems, distinct or not, on the non-singular v-dimensional variety V.

Consider in each system L_i a generic pencil F_i and denote by $J_h(F_0, F_1, \ldots, F_h)$, or more briefly by $J_h(F)$, the algebraic set on V consisting of the points of V which are singular for some member of F_0, if $h = 0$, or otherwise are such that the tangent spaces to the members of the pencils F_i through them are dependent. It is easily shown by analytic methods, and also by a geometrical inductive proof with respect to the values of h and v {see e. g. SEVERI [a], p. 200; this derives in effect also from the following proof}, that $J_h(F)$ is a non-singular subvariety of V, which, considered as a prime cycle, is called the Jacobian cycle of the $h + 1$ pencils F_i.

Putting $D_j = A_0 \cdot F_j$ and $E_j = B_0 \cdot F_j$, $j = 1, 2, \ldots, h$, where A_0 and B_0 are respectively a member and the base-variety of F_0, and denoting by $J_{h-1}(D)$ and $J_{h-1}(E)$ the Jacobian cycles respectively of the h pencils D_j or E_j on A_0 or B_0, we immediately get the relations:

$$A_0 \cdot J_h(F) = B_0 + J_{h-1}(D) \,, \qquad (h + 1 = v) \,, \tag{1}$$

$$A_0 \cdot J_h(F) = J_{h-1}(D) + J_{h-1}(E) \,, \quad (h + 1 < v) \,. \tag{2}$$

Let us first consider the case $h + 1 = v$ and then prove the theorem:

(i) *The cycle $J_h(F) - 2\,(A_0 + A_1 + \cdots + A_h)$, where $h + 1 = v$ and A_i is a member of F_i, is independent of the choice of the pencils F_i, $i = 0, 1, \ldots, h$, and is a canonical impure divisor on V.*

The theorem is elementarily known if $v = 1$: therefore we proceed by induction with respect to v, assuming the theorem for any non-singular variety of dimension less than v and proving the statement for V. From (1) and our inductive hypothesis we derive:

$$A_0 \cdot J_h(F) \equiv B + 2 \sum_{i=1}^{h} A_0 \cdot A_i + K' \,, \tag{3}$$

where K' is a canonical impure divisor on A_0.

From here it follows that the V-divisor $J_h(F) - 2\sum_{i=1}^{h} A_i$ cuts on A_0 a divisor linearly equivalent to $B + K'$ which is obviously independent of the choice of the pencils F_i, $i = 1, 2, \ldots, v$. Therefore, if we replace the pencils F_i by other pencils $\overline{F}_i$ satisfying the same hypotheses, we obtain with an obvious symbolism:

$$\left(\overline{J}_h(\overline{F}) - 2\sum_{i=1}^{h} \overline{A}_i\right) \cdot A_0 \equiv \left(J_h(F) - 2\sum_{i=1}^{h} A_i\right) \cdot A_0 .$$

Then, by applying a criterion of linear equivalence {see e. g. SEVERI [a], p. 190}, we have on V:

$$\overline{J}_h(\overline{F}) - 2\sum_{i=1}^{h} \overline{A}_i \equiv J_h(F) - 2\sum_{i=1}^{h} A_i ,$$

which proves the independence of the divisor $J_h(F) - 2\sum_{i=1}^{h} A_i$ of the pencils F_i, $i = 0, 1, \ldots, h$. Moreover (3) yields the formula on A_0:

$$K_1 \cdot A_0 \equiv K' - A_0 \cdot A_0 , \tag{4}$$

where we have put $K_1 = J_h(F) - 2\sum_{i=0}^{h} A_i$ and $A_0 \cdot A_0$ denotes a member of the characteristic system of L_0 on A_0.

We now go on to prove that $K_1 = K$, denoting by K an impure canonical cycle on V. For this, we project V onto a primal V', with only ordinary singularities, of a $(v + 1)$-dimensional projective space and assume, by using induction with respect to v, that the assertion is true if $\dim V < v$ since it is known to be true if $v = 1$: then we shall prove it if $\dim V = v$.

To this end we take as system L_0 the system of hyperplane section of V, which, as we already know does not alter K. Then, if n is the order of V', by (IV, 2e) and our inductive hypothesis, the adjoint forms of order $n - v - 1$ to V' cut on a generic member A_0 of L_0 canonical cycles K' residually to the singular double locus S of V'.

Hence, by (4), we derive the linear equivalence on $A : A_0 \cdot C \equiv \equiv K_1 \cdot A_0 + A_0 \cdot A_0$, and then, as before, the other equivalence on $V' : C \equiv \equiv A_0 + K_1$, where C denotes the intersection of V' with an adjoint form to V' outside S. Our statement now follows immediately by again using (IV, 2e).

3. Todd's Canonical Systems

{All this section follows TODD [11].}

With the same situation as in (VIII, 2), but with $h + 1 < v$.

a) We begin by defining *two operators:* the first, denoted by $^*X_h(M)$, will be defined on the set of all *prime non-singular* $(h + i)$-dimensional cycles, $i = 1, 2, \ldots, v - h$; the second will be a linear operator denoted by φ_h and defined, instead, on the group of all h-cycles on V, which are free from singular components, $h = 0, 1, \ldots, v$.

The definitions are as follows

$$*X_h(M^{h+i}) = J_h^{(i)}(F^{(i)}) - \varphi_h \left\{ \prod_{j=0}^{h} (1 + A_j^{(i)})^2 - 1 \right\}, \qquad (5_i)$$

$$\varphi_h(X^k) = X^k, \qquad \text{if } k = h, \qquad (6)$$

$$\varphi_h(X^k) = 0, \qquad \text{if } k < h, \qquad (7)$$

$$\varphi_h(M^{h+i}) = *X_h(M^{h+i}), \quad i = 1, 2, \ldots, v - h. \qquad (8_i)$$

In (5_i), $J_h^{(i)}$ and $F_j^{(i)}$ have, on the prime non-singular cycle M^{h+i}, the same definition as the analogous symbols defined on V in (VIII, 2) while the $A_j^{(i)}$ denote, instead, almost any member of the ample system $L_j^{(i)}$, $i = 1, 2, \ldots, v - h$: this is necessary in order to be able to suppose that the argument of φ_h in (5_i) is a cycle free from singular components; it is also sufficient, as we may prove by a straightforward application of Bertini's theorems.

We explain that, in the symbolic product of (5_i), all meaningless terms are to be omitted.

We now observe that (5_1) defines directly $*X_h(M^{h+1})$, since, in this case, $\varphi_h\left\{ \prod_{j=0}^{h} (1 + A_j^{(1)})^2 - 1\right\}$ reduces, by (6) and (7), to $2(A_0^{(1)} + A_1^{(1)} + \cdots + A_h^{(1)})$, and, thereby, we again find the canonical system on M^{h+1} as defined in (VIII, 2). After this (8_1) acquires significance, and then (5_2) and so on. Nevertheless it is clear that (5_i) does not define uniquely the cycle $*X_h(M^{h+i})$: if $i = 1$ we already know that, by allowing the $A_j^{(1)}$ and the pencils $F_j^{(1)}$ to vary in their respective linear systems $L_j^{(1)}$, the cycle $*X_h(M^{h+1})$ describes on M^{h+1} a linear system; precisely the impure canonical system on M^{h+1} {see (VIII, 2)}.

Generally there are two reasons for this indeterminateness: (1) *The choice of the members $A_j^{(i)}$ of the systems $L_j^{(i)}$*; (2) *The choice of these systems themselves, and of the respective pencils $F_j^{(i)}$.*

b) We now prove a theorem which deals with (1):

(i) *If the pencils $F_j^{(i)}$ are fixed, then the cycle $*X_h(M^{h+i})$ is determined on M^{h+i} to within a rational equivalence.*

We sketch Todd's proof. Let $X_j^{(i)}$, $j = 1, 2, \ldots, r$, be a prime non-singular $(h + i - 1)$-cycle on M^{h+i} such that the intersection-product $X_1^{(i)} \cdot X_2^{(i)} \cdot \cdots \cdot X_r^{(i)}$ is defined on M^{h+i} and is, moreover, a prime non-singular cycle on M^{h+i}. Then $*X_h(X_1^{(i)} \cdot X_2^{(i)} \cdot \cdots \cdot X_r^{(i)})$ is also defined: since the operator φ_h is linear, our statement will obviously be true if we prove that, for any integer r, this h-cycle on M^{h+i} is determined to within a rational equivalence, when the $X_j^{(i)}$ are fixed to within a linear equivalence on M^{h+i}.

Now we can readily reduce the problem to the case $r = 1$, which we proceed to consider. We suppose, accordingly, that $X^{(k)} \equiv \overline{X}^{(k)}$ and we prove that $*X_h(X^{(k)}) \equiv *X_h(\overline{X}^{(k)})$, where $X^{(k)}$ and $\overline{X}^{(k)}$ are prime and non-singular and where the equivalences are on M^{h+k}. The assertion is certainly true if $k=1$, since then $\dim X^{(1)} = h$ and so ${}^*X_h(X^{(1)}) = X^{(1)}$. We assume the theorem for any $i < k$, and prove it when $i = k$. Put $S_j^{(k-1)} = X^{(k)} \cdot F_j^{(k)}$ and $\bar{S}_j^{(k-1)} = \bar{X}^{(k)} \cdot F_j^{(k)}$.

Then from (5_{k-1}) follow both the equivalences, on $X^{(k)}$ and on $X^{(k)}$ respectively:

$$ {}^*X_h(X^{(k)}) \equiv J_h^{(k-1)}(S^{(k-1)}) - \varphi_h\left\{\prod_{j=0}^{h}(1 + B_j^{(k-1)})^2 - 1\right\}, $$

$$ {}^*X_h(\bar{X}^{(k)}) \equiv J_h^{(k-1)}(\bar{S}^{(k-1)}) - \varphi_h\left\{\prod_{j=0}^{h}(1 + \bar{B}_j^{(k-1)})^2 - 1\right\}, $$

where $B_j^{(k-1)}$ and $\bar{B}_j^{(k-1)}$ are suitable members of ample linear systems, on $X^{(k)}$ and on $\bar{X}^{(k)}$ respectively, containing the pencils $S_j^{(k-1)}$ or $\bar{S}_j^{(k-1)}$. By our inductive hypothesis we can suppose that the $B_j^{(k-1)}$ and the $\bar{B}_j^{(k-1)}$ are cut on $X^{(k)}$ and $\bar{X}^{(k)}$ respectively by the same $A_j^{(k)}$ of an ample linear system $L_j^{(k)}$ on M^{h+k}.

Therefore the cycles furnished by the operator φ_h in the above equivalences are rationally equivalent on a certain variety arising as intersection of the $A_j^{(k)}$ and hence also on M^{h+k} {see (VI, 8)}. Finally, since $X^{(k)} \equiv \bar{X}^{(k)}$ there exists a pencil $\mathfrak{T}$ which contains both these cycles and, if D_j is the pencil cut out by $F_j^{(k)}$ on the base-variety of T, we derive from (VIII, 2) the relation:

$$ J_h^{(k-1)}(S^{(k-1)}) \equiv X^{(k)} \cdot J_{h+1}(\mathfrak{T}; F^{(k)}) - J_h(D), $$

which yields immediately the conclusion that, when $X^{(k)}$ varies in T then also $J_h^{(k-1)}(S^{(k-1)})$ varies in a system of equivalence {see (VI, 8)}. With this the proof is evidently complete.

c) The following theorem of Todd deals with (2) in (a).

(ii) *The h-cycle* ${}^*X_h(M^{h+i})$ *on* M^{h+i} *is defined to within a rational equivalence by the cycle* M^{h+i} *only.*

Here we outline Todd's proof, which is by induction on i: the statement is true when $i=1$ by (VIII, 2). We shall suppose it true when $i < k$, deducing that it holds when $i = k$, where, we recall, $i = 1, 2, \ldots, v - h$.

Let $F_{h+1}^{(k)}$ be another pencil of divisors on M^{h+k}, selected under the same conditions as the $F_j^{(k)}$, $j = 0, 1, \ldots, h$; and let $J_{h+1}^{(k)}(F^{(k)})$ be the Jacobian cycle relative to the $h+2$ pencils $F_j^{(k)}$ on M^{h+k}, which may be supposed prime and non-singular.

Denoting by $D_{jl}^{(k-1)}$ and $E_{jl}^{(k-2)}$ the pencils cut on $A_l^{(k)}$ and $B_l^{(k)}$ respectively, where the former is a non-singular member of $F_l^{(k)}$, $l \neq j$, and the latter is the base-variety of $F_l^{(k)}$, also non-singular, let us, suppose, as we may, that the respective Jacobian cycles on $A_l^{(k)}$ or on $B_l^{(k)}$ are also h-cycles which are prime and non-singular.

A simple calculation resting on our inductive hypothesis, on (5_{k-1}) and on (5_{k-2}), gives the following equivalences on $A_l^{(k)}$ and on $B_l^{(k)}$ respectively, and therefore on M^{h+k} also:

$$J_{h,l}^{(k-1)}(D^{(k-1)}) \equiv \varphi_h\left[(A_l^{(k)} \cdot \overset{h}{\underset{j=0}{\Pi}}{}' (1 + A_j^{(k)})^2\right],$$

$$J_{h,l}^{(k-2)}(E^{(k-2)}) \equiv \varphi_h\left[(B_l^{(k)} \cdot \overset{h}{\underset{j=0}{\Pi}}{}' (1 + A_j^{(k)})^2\right],$$

where the apex over the symbolic products denotes that the value $j = l$ has to be excluded.

Moreover, from (VIII, 2) we have

$$J_{h+1}^{(k)}(F^{(k)}) \cdot A_l^{(k)} \equiv J_{h,l}^{(k-1)}(D^{(k-1)}) + J_{h,l}^{(k-2)}(E^{(k-2)}) .$$

Now let J be the Jacobian cycle of the pencils cut on $J_{h+1}^{(k)}(F^{(k)})$ by the $h+1$ pencils $F_j^{(k)}$: this is an h-cycle which by (VIII, 2) may be written in the following form:

$$J \equiv {}^*X_h\left(J_{h+1}^{(k)}(F^{(k)}) + 2\sum_{j=0}^{h} J_{h+1}^{(k)}(F^{(k)})\right) \cdot A_j^{(k)} ,$$

and which may be also alternatively expressed, as TODD shows (with a rather delicate analysis which we shall omit), in the form:

$$J \equiv Z + J_h^{(k)}(F^{(k)}) + 2\sum_{l=0}^{h} J_{h,l}^{(k-2)}(E^{(k-2)}) + J_{h,h+1}^{(k-2)}(E^{(k-2)}) ,$$

where Z is an h-cycle on $J_{h+1}^{(k)}(F^{(k)})$ symmetrically related to the pencils $F_j^{(k)}$, $j = 0, 1, \ldots, h+1$.

From the above three relations we obtain:

$$\begin{aligned} J_h^{(k)}(F^{(k)}) &\equiv J - 2\sum_{l=0}^{h} J_{h,l}^{(k-2)}(E^{(k-2)}) - J_{h,h+1}^{(k-2)}(E^{(k-2)}) - Z \\ &\equiv {}^*X_h(J_{h+1}^{(k)}(F^{(k)}) + 2\sum_{l=0}^{h} J_{h,l}^{(k-1)}(D^{(k-1)}) - J_{h,h+1}^{(k-2)}(E^{(k-2)}) - Z. \end{aligned}$$

Using now the expressions for $J_{h,l}^{(k-1)}(D^{(k-1)})$ and $J_{h,l}^{(k-2)}(E^{(k-2)})$ we find after some calculations:

$$\begin{aligned} J_h^{(k)}(F^{(k)}) - \varphi_h\left\{\prod_{j=0}^{h}(1 + A_j^{(k)})^2 - 1\right\} &\equiv {}^*X_h(J_{h+1}^{(k)}(F^{(k)})) - Z + \\ &+ \varphi_h\left\{2\sum_{j=0}^{h+1} A_j^{(k)} \cdot \overset{h+1}{\underset{l=0}{\Pi}}{}' (1 + A_l^{(k)})^2 - \prod_{j=0}^{h+1}(1 + A_j^{(k)})^2 + 1\right\}, \end{aligned}$$

where evidently the cycle on the right hand side is symmetrically related to the $h+2$ pencils $F_j^{(k)}$ and the cycle on the left hand side does not depend on $F_{h+1}^{(k)}$ and therefore on none of the $F_j^{(k)}$.

d) After what we have proved, we may conclude that *there are on V v well determined classes of rational equivalence, having as representatives the cycles* $*X_h(V)$, $h = 0, 1, \ldots, v-1$: these classes are *the* TODD *canonical classes on* V.

It is moreover obvious by the definition that these classes are *relative invariants* of the model V among all the *non-singular* models of the rational function field on V. The behaviour of canonical classes under irregular birational transformations is, instead, little known; as we shall see later in (VIII, 9).

It is now possible to extend, with TODD, the adjunction property, which is classical for divisors on V {see (IV, 5)}, to the case of cycles of any dimension. To this end we first establish the following formula:

$$*X_{h-1}(B) \equiv B \cdot *X_h(B) + *X_h(V) , \tag{9}$$

where B is a prime non-singular divisor on V isolated or not with respect to the linear equivalence {see MACPHERSON [1]}; the proof is by a straightforward induction resting on (5_{v-h}) and on (VIII, 2).

The cycle:

$$A_h(B) \equiv *X_h(B) + *X_h(V) , \tag{10}$$

is called an adjoint h-cycle to B and its equivalence class is called *the adjoint class of B on V.*

Obviously we have

$$*X_{h-1}(B) \equiv B \cdot A_h(B) . \tag{11}$$

We remark that, if L is almost any $(h+1)$-dimensional linear subsystem of an ample linear system on V, and if the pencils F_j, $j = 0, 1, \ldots, h$, before considered, are all distinct and belong to L, then the Jacobian cycle $J_h(F)$ coincides with the so-called *Jacobian cycle* $J_h(L)$ of L: this is associated, by definition, with the algebraic subset of V, at whose points the tangent spaces to the members of L through one of them have a common $(v-h)$-dimensional space.

From (5_{v-h}), setting $\varphi_h(1) = *X_h(V)$, comes easily the equivalence

$$J_h(L) \equiv \varphi_h((1+A)^{h+2}) \equiv \sum_{i=0}^{h+2} \binom{h+2}{i} *X_h(A^{[i]}) , \quad A^{[0]} = V , \tag{12}$$

where A is almost any member of L.

This proves incidentally that $J_h(L)$ can be denoted also by $J_h(A)$, since it depends only on the linear class of A. In a first attempt to define canonical systems, this relation was taken by TODD as a starting point; it still maintains a considerable importance in many questions.

e) We give here a few applications of (12) {see B. SEGRE [12], p. 104}. To begin with, if $h = 0$, (12) becomes the following:

$$J_0^{(0)}(A) \equiv *X_0^0(V) + 2\, *X_0^0(A) + *X_0^0(A^{[2]}) ,$$

which, if $\delta, c_v, c_{v-1}, c_{v-2}$ denote respectively the numbers of points in the point-groups appearing in its expression, gives the enumerative relation

$$\delta = c_v + c_{v-1} + c_{v-2}\,. \tag{13}$$

If V is a curve of genus p and A a V-divisor of degree a, we have $c_{v-1} = a$ and $c_{v-2} = 0$, so that (13) becomes

$$c_1 = \delta - 2a = 2p - 2\,.$$

If, instead, V is a surface and A is one of its divisors of degree a and genus p, we have $c_1 = 2p - 2$, $c_0 = a$. Then (13) furnishes

$$c_2 = \delta - 4p - a + 4\,,$$

where the character $\delta - 4p - a$ is the well known ZEUTHEN-SEGRE *invariant* I_2 {see C. SEGRE [1]}; thereby, we get:

$$c_2 = I_2 + 4\,. \tag{14}$$

$*X_0^0(V)$ is, in this case, a group of the series of equivalence first discovered by SEVERI and now called *the* SEVERI *series* {see SEVERI [18, 19] and SEVERI [a], p. 236}; the relative invariant character of this series and its order follows immediately.

More generally, (13) may be written in the form:

$$c_v = I_v + (-1)^v \cdot 2v\,, \tag{15}$$

where $I_v = \delta - 2c_{v-1} - c_{v-2} - (-1)^v \cdot 2v$ is, by definition, the ZEUTHEN-SEGRE *invariant of* V.

This proves that TODD's work generalises, at least in the case of the canonical zero-dimensional class or SEVERI's class, the classical proof {see C. SEGRE [1]; ENRIQUES [a], p. 167} of the relative invariance of the ZEUTHEN-SEGRE invariant, which was previously restricted to the enumerative aspect alone.

f) We state here the following theorem due to B. SEGRE {see B. SEGRE [12], p. 105}:

(iii) *If h and s are any integers such that* $0 \leqq h \leqq v - 1$ *and* $s > v - h$, *we have*

$$\begin{aligned}*X_h(V) \equiv \sum J_h(A_i) - \sum J_h(A_i + A_j) + \sum J_h(A_i + A_j + A_l) - \\ - \cdots + (-1)^{s-1} J_h(A_1 + A_2 + \cdots + A_s)\,,\end{aligned}$$

where $A_1, A_2, \ldots, A_s$ are prime non-singular cycles such that the above Jacobian cycles are defined and where the sums are extended over all the simple combinations of class 1, 2, ..., s.

The proof of the theorem is a straightforward formal calculation deriving from (12).

From it we may deduce *a formal definition of a Jacobian h-cycle $J_h(B)$ for any divisor B on V.*

To this end, we consider an integer $t \geqq v - h$ and select a divisor A on V such that the linear systems $|i\,A|$, $|i\,A + B|$, $i = 1, 2, \ldots, t$, admit a Jacobian h-cycle; then, in particular, each such system will be effective and at least $(h + 1)$-dimensional. Substituting in (iii) $s = t + 1$, $A_i = A$, $i = 1, 2, \ldots, s - 1$, $A_s = B$, we easily derive the following expression for $J_h(B)$:

$$J_h(B) \equiv {}^*X_h(V) + \sum_{i=1}^{t} (-1)^i \binom{t}{i} (J_h(iA) - J_h(iA + B)) . \tag{16}$$

This defines $J_h(B)$ to within a rational equivalence, since, as it is not difficult to show, the cycle on the right hand side does not depend upon t or A. In particular, for $B \equiv 0$, we find that

$$J_h(0) \equiv X_h(V) . \tag{17}$$

We point out that (iii) could also be taken as the starting point for the definition of the canonical systems: as we have already mentioned in (d).

4. Introduction to Segre's Theory

We shall now review the recent state of the theory of canonical equivalence classes, which is due to B. Segre. We begin with some generalities concerning equivalence groups.

a) We put

$$\mathbf{A}(V) = \Gamma(V)/\mathbf{G}_l(V) = \sum_{m=0}^{v} \mathbf{A}^m(V), \tag{18}$$

where $v = \dim V$ and

$$\mathbf{A}^m(V) = \mathbf{G}^m(V)/\mathbf{G}_l^m(V) . \tag{19}$$

By (I, 9) and (VI, 9) $\mathbf{A}(V)$ is *a commutative ring with respect to the intersection-product:* its unit element is the transform y of V by the group-homomorphism

$$f : \Gamma(V) \to \mathbf{A}(V) . \tag{20}$$

We shall denote by a *Greek letter*, e. g. μ, *any homogeneous element of* $\mathbf{A}(V)$ *and by the corresponding* Roman *letter, m, its dimension:* the notion of a homogeneous class is obviously justified by observing that if a class of $\mathbf{A}(V)$ has a homogeneous representative cycle, then so also has any other cycle of the same class. The ring $\mathbf{A}(V)$ is graduated with respect to the dimension by (18), and we have the following homogeneous group-homomorphisms

$$f_m : \mathbf{G}^m(V) \to \mathbf{A}^m(V) . \tag{21}$$

Denoting by ω^m the zero of $\mathbf{A}^m(V)$ and by O the zero of $\mathbf{A}(V)$, we find from (18) the following graduation of the zero: $O = \sum_{m=0}^{v} \omega^m$.

Now let V' be a non-singular subvariety of V: we can define for V' all groups and rings or other characters already defined for V, which we shall denote by the same symbols as before, but endowed with apices. We consider *the inclusion maps*

$$i: \mathbf{A}(V') \to \mathbf{A}(V)\,, \tag{22}$$

$$i_m: \mathbf{A}^m(V') \to \mathbf{A}^m(V)\,, \quad (m \leqq v' = \dim V')\,. \tag{23}$$

If $\mu = i_m(\mu')$, $\mu' \in \mathbf{A}^m(V')$ and $\mu \in \mathbf{A}^m(V)$, we shall write $\mu = \mu'_V$ and also $\mu' = \mu_{V'}$, instead of $\mu' = i_m^{-1}(\mu)$: observe that the latter operation is, when defined, generally many-valued {see (VI,8)}. Finally we denote by $\mu \cdot V'$, $\mu \in \mathbf{A}^m(V)$, the element $\mu' \in \mathbf{A}^{m'}(V')$, $m' = m + v' - v$, such that $\mu' = f'_m(M')$, being $M' = M \cdot V'$: M is any cycle representing μ and such that $M \cdot V'$ is defined on V. Obviously μ' does not depend on M and we have so defined, by the formal intersection-product properties, *a homomorphism*

$$\varphi_m: \mathbf{A}^m(V) \to \mathbf{A}^{m'}(V'), \quad m' = m + v' - v\,. \tag{24}$$

By repeated applications of intersection-products and of the preceding definitions one could now write down a great number of relations between the elements of $\mathbf{A}(V)$: since these follow most easily from the intersection theory transferred into $\mathbf{A}(V)$ by f, we shall omit them.

b) {See B. SEGRE [12], p. 13.} Now let P be a non-singular subvariety of V and consider the sequence $\{\pi\} = \pi_0, \pi_1, \ldots, \pi_i, \ldots$, where $\pi_i \in \mathbf{A}(P)$, $P \in \pi_0$, and, further, $\dim \pi_i = p_i = (\dim P) - i = p - i$: with the usual convention, we put $\pi_i = 0$, for $i > p$. Such a sequence, which is finite, will be called *a sequence of* $\mathbf{A}(P)$ *or of support* P.

We associate with $\{\pi\}$ the element $\sum_{i=0}^{p} \pi_i \in \mathbf{A}(P)$ or, better, *the formal power series* $(\pi, x) = \sum_{0}^{\infty} \pi_i \cdot x^i$, which is in fact a polynomial.

The sequence $\{\overline{\pi}\}$, still of support P, related to the series $(\pi, -x)$ will be called *the alternating sequence* of $\{\pi\}$, while the other $\{\widetilde{\pi}\}$ related to the series $(\pi, x)^{-1}$, which is defined since (π, x) has the first term equal to the unit of the ring $\mathbf{A}(P)$, will be called *the inverse sequence* of $\{\pi\}$.

We omit the obvious relations between all these sequences, merely remarking that we shall write formal series over the ring $\mathbf{A}(P)$ by replacing P by 1.

c) {See B. SEGRE [12], p. 15.} A very useful sequence is the sequence $\{y\}$ related to the series over $\mathbf{A}(V)$ given by $(y^{(s)}, x) = \prod_{h=1}^{s} (1 + \alpha_h x)$, where $\alpha_1, \alpha_2, \ldots, \alpha_s$ are classes of $\mathbf{A}^{v-1}(V)$: this will be called *the intersection sequence of the* α's.

In such a case we have $(y^{(s)}, x) = \sum_0^\infty y_i^{(s)}(\alpha) \cdot x^i$, where $y_0^{(s)}(\alpha) = V = 1$ and, if $i > 0$, $y_i^{(s)}(\alpha) = \sum \alpha_{j1} \cdot \alpha_{j2} \cdot \cdots \cdot \alpha_{ji}$, the sum being extended over all simple dispositions. It is clear that $y_i^{(s)}(\alpha) = 0$ when $i > v$ or $i > s$.

It is also immediately seen that, if $(y^{(s)}, x)^{-1} = \sum_0^\infty \tilde{y}_i^{(s)}(\alpha)\, x^i$, then $\tilde{y}_0^{(s)}(\alpha) = 1$ and, if $i > 0$, $\tilde{y}_i^{(s)}(\alpha) = \sum \alpha_{j1} \cdot \alpha_{j2} \cdot \cdots \cdot \alpha_{ji}$ where now the sum is extended over all dispositions: moreover, if $i > v$, we have $\tilde{y}_i^{(s)}(\alpha) = 0$.

d) {See B. SEGRE [12], p. 17.} Let us take two sequences $\{\pi\}$ and $\{\chi\}$ with the supports respectively at the *non-singular* subvarieties P and Q of V and with (π, x), (χ, x) as related series.

The series which is the formal product of (π, x) and (χ, x) does not determine a sequence, unless the intersection-product $P \cdot Q$ be defined on V. If, instead, we consider the series $(\pi, x)_V = (\pi_V, x)$ and $(\chi, x)_V = (\chi_V, x)$ related to the sequences $\{\pi\}_V = \{\pi_V\}$ and $\{\chi\}_V = \{\chi_V\}$, which are obtained from the given ones by the inclusion map i of (a), then we can always associate with their formal product, to be evaluated in $\mathbf{A}(V)$, a sequence of $\mathbf{A}(V)$: therefore *the set of all sequences having the support at any non-singular subvariety P of V, is, after inclusion in $\mathbf{A}(V)$, a multiplicative semi-group* whose unit element is the sequence $y, 0, 0, \ldots$

On the other hand it scarcely needs pointing out that, instead, the sequences of $\mathbf{A}(P)$, where P is a fixed non-singular subvariety of V, form a *multiplicative group*, whose unit element is the sequence $\pi, 0, 0, \ldots$

e) {See B. SEGRE [12], p. 19.} Let P be a non-singular subvariety of V and let α_i, $i = 1, 2, \ldots, s$, be an element of $\mathbf{A}(V)$ of dimension a_i, where $p = \dim P \geqq r = a_1 + a_2 + \cdots + a_s - (s-1)\, v$, such that there are representative cycles A_i of each α_i intersecting on V simply along P and regularly along any other component of their intersection-set.

As we see at once, the equivalence class χ in $\mathbf{A}(V)$ of this intersection cycle outside P, remains defined, and we shall write $\chi = (\alpha_1 \cdot \alpha_2 \cdots \alpha_s)^P$. It would be interesting to examine whether the class χ depends on the equivalence class π of P only, and how this operation could be extended to one without exceptions on $\mathbf{A}(V)$.

The element $\varepsilon(P; \alpha) \in \mathbf{A}^{v-s}(V)$ defined by

$$\varepsilon^{(s)}(P; \alpha) = \alpha_1 \cdot \alpha_2 \cdots \alpha_s - (\alpha_1 \cdot \alpha_2 \cdots \alpha_s)^P \tag{25}$$

is called *the rational equivalence of P with respect to the* α's, while the equivalence class χ is called, following B. SEGRE, *the symbolic product* of the α's. We remark the obvious properties:

(i) *If $\mu \in \mathbf{A}^m(V)$ and $\mu_i' = \mu \cdot \alpha_i$, $i = 1, 2, \ldots, s$, we have $(\mu \cdot \alpha_1 \cdot \alpha_2 \cdot \cdots \cdot \alpha_s)^P = (\alpha_1' \cdot \alpha_2' \cdots \alpha_s')$ whenever the symbolic product on the left is defined.*

(ii) *If* $p = v - 1$, *the class* χ *depends only upon the class* π *of* P *in* $\mathbf{A}(V)$ *and, if* $\alpha_i = \pi + \beta_i$, *then* $\chi = \beta_1 \cdot \beta_2 \cdots \beta_s$.

Applying to V a monoidal transformation with centre P, we may deduce from (ii) the theorem:

(iii) *If* $\chi = (\alpha_1 \cdot \alpha_2)^P$, *then* $\chi \cdot P \in \mathbf{A}^{q-1}(P)$.

5. The Covariant Sequence

We are now able to describe a fundamental concept due to B. SEGRE {see B. SEGRE [12], p. 25 and B. SEGRE [14], p. 139}, namely that of *covariant sequence of a non-singular subvariety P of V*.

a) Let $\alpha_1, \alpha_2, \ldots, \alpha_v$ be elements of $\mathbf{A}^{v-1}(V)$, such that the symbolic product $(\alpha_1 \cdot \alpha_2 \cdots \alpha_s)^P$ is defined for any $v - p \leqq s \leqq v$: it is clear that this supposition is legitimate and, further, that the α's may be selected in an infinity of ways.

The sequence $\{\pi(V)\}$ defined by writing

$$\varepsilon^{(s)}(P; \alpha) = \sum_{i=0}^{s+p-v} \pi_i(V) \cdot y^{(s)}_{s+p-v-i}(\alpha)\,, \quad s = v - p, \ldots, v\,, \tag{26}$$

has support P and will be called *the covariant sequence relative to the subvariety P of V*: this definition is justified by the following fundamental theorem {see B. SEGRE [14], p. 141}:

(i) *The sequence* $\{\pi(V)\}$ *defined by* (26) *does not depend on the* α*'s or on the non-singular representative* P *of the class* π.

We proceed to outline the proof. Assume a monoidal model $V' = T(V)$, such that T, which, following B. SEGRE {see B. SEGRE [11, 13]; see also (II, 2)}, will be here called a *dilatation*, has the centre at P on V and dilates each point of P into a linear space L'^{v-p-1} of V'. Then $P' = T(P)$ is a $(v-1)$-dimensional subvariety of V' locus of ∞^p spaces L'.

We recall that T is everywhere regular on V outside P and that T^{-1}, which contracts P' into P, has no fundamental points on V' and is regular outside P': moreover any subvariety $U \subset P$ is such that $U' = T(U)$ is an irreducible subvariety whether of V' or of P: We have $\dim U' = u' = v - 1 - p + u$, since U' is a locus of ∞^u spaces L'.

Our theorem may now be considered an obvious consequence of the following:

(ii) *If* $V' = T(V)$, *where* T *is as described above, we have the relation*

$$\pi_i(V) = (-1)^{v-p+i+1}\, T^{-1}(\pi'^{[v-p+i]})\,, \tag{27}$$

where $\pi' \in \mathbf{A}(V')$ *is the class represented by* P'.

The proof is given in the following subsections.

b) To begin with, if $\pi \in \mathbf{A}^{v-1}(V)$, from (26) it follows immediately that $\pi_i(V) = (-1)^i\, \pi^{[i+1]}$. This proves both (i) and (ii) in this case.

We find, incidentally, *that then* $\{\tilde{\pi}(V)\} = \pi, \pi^{[2]}, 0, 0, \ldots$, which is a consequence of the obvious relation $(1 + \pi x)^{-1} = \sum_0^\infty (-1)^i \pi^{[i]} x^i$.

c) We shall now suppose that $p < v - 1$ and therefore $v \geqq 2$ and $l = \dim L' = v - p - 1 > 0$. We begin by considering the first term $\pi_0(V)$, and proving (ii) for $i = 0$. Since two spaces L' of P' are *skew*, if L'^{l-i} is a subspace of L', we have $(\lambda'^{l-i} \cdot \lambda'^{l-j}) = 0$, $i, j = 0, 1, \ldots, l$, in $A(V')$, where λ'^{l-i} is the class of L'^{l-i}.

d) If A is a subvariety of V not containing P and if $A' = T(A)$, we have $T(P \cdot A^{[k]}) = P' \cdot A'^{[k]}$, where $P \cdot A^{[k]}$ is supposed to be defined, to within a rational equivalence, on V. Therefore the components of any cycle on V' of the type $P' \cdot A'^{[k]}$ are loci of spaces L': hence such a cycle is *null* if $k > p$, since, in such a case, so is $P \cdot A^{[k]}$.

e) If $A \equiv B$, and if B contains P simply, where A and B are subvarieties of V belonging to the same irreducible algebraic system having A as generic member, then the specialisation of $A' = T(A)$ over the $A \rightarrow B$, gives rise to the equivalence on V': $A' \equiv B' + P'$, where $B' = T(B)$. This yields the formulae:

$$(-1)^s \chi' \cdot \pi'^{[s]} = \chi' \cdot (\beta' - \alpha')^{[s]}, \tag{28}$$

$$(-1)^s \pi'^{[s+1]} = \pi' \cdot (\beta' - \alpha')^{[s]}. \tag{29}$$

f) Since B passes simply through a generic point of P, it is $\lambda'^l \cdot \beta' = \lambda'^{l-1}$ and more generally, with analogous hypotheses, $\lambda' \cdot \beta'_1 \cdots \beta'_s = \lambda'^{l-s}$: hence we get $\lambda' \cdot \beta'^{[l]} = \lambda'^0$, where λ'^0 is the class of $A(V)$ related to a point of L'.

From (28), using (c) and the above result, we derive, by formal development of the member on the right, $(-1)^s \chi' \cdot \pi'^{[s]} = \chi' \cdot \beta'^{[s]} = \lambda'^{l-s}$. Now if we take into account (d), we find that any class $\pi'^{[s+1]} \cdot \alpha'^{[k]}$ of $\mathbf{A}(V')$, $0 \leqq s < l$, $k > 0$, contracts, when we apply T^{-1} to some one of its representatives, i. e. it is transformed into a class of $\mathbf{A}(V)$ of lower dimension.

By applying this to the member on the right of (29) for $s = l$, we deduce that all the terms of the development are contracted except the first. Therefore $(-1)^l T^{-1}(\pi'^{[l+1]}) = T^{-1}(\pi' \cdot \beta'^{[l]})$; since from (f) it follows that $\pi' \cdot \beta'^{[l]}$ is associated with a single point of L', we have $(-1)^l T^{-1}(\pi'^{[l+1]}) = \pi = \pi_0(V)$, which proves (27) and then (ii) for $i = 0$.

g) The rest of the proof is *by induction with respect to the index* of the terms of the sequence $\{\pi(V)\}$. Supposing our statement holds for all $\pi_j(V)$, $j = 0, 1, \ldots, i - 1$, we prove it for $j = i \leqq p$.

Let $\alpha_1, \alpha_2, \ldots, \alpha_s$, $s = l + i$, $v - p \leqq s \leqq v - 1$, be elements of $\mathbf{A}^{v-1}(V)$, such that $\chi = (\alpha_1 \cdot \alpha_2 \cdots \alpha_s)^P$ is defined, and let $\alpha'_k \in A(V')$ be the class containing the transform of a variety A_k belonging to α_k and not containing P. Put $\beta'_k = -(\pi' - \alpha'_k)$, $k = 1, 2, \ldots, s$: then, if $\varrho = \chi \cdot P$ we have $\varrho = T^{-1}(\pi' \cdot \beta'_1 \cdots \beta'_s)$. Therefore we can write: $\varrho = (-1)^s T^{-1}(\pi' \cdot (\pi' - \alpha'_1) \cdots (\pi' - \alpha'_s))$.

Applying (d) we observe that in the development on the right of this expression, the terms containing $\pi'^{[h]}$, with $h < l - 1$ can be eliminated: hence, and from our inductive hypothesis, we find, after some calculations, that $\varrho = (-1)^{l+i} T^{-1}(\pi'^{[l+i+1]}) + \sum_{j=0}^{i-1} \pi_j(V) \cdot y^{(s)}_{i-j}(\alpha)$.

Finally the proof follows from this and the following fact, which is easily demonstrated, and which derives from the definition of $\{\pi(V)\}$:

(iii) *If* $\chi = (\alpha_1 \cdot \alpha_2 \cdots \alpha_s)^P$ *and* $\varrho = \chi \cdot P$, *then*

$$\varrho = \pi_i(V) + \sum_{j=0}^{i-1} \pi_j(V) \cdot y^{(s)}_{i-j}(\alpha),$$

where $i = s - v + p + 1$.

In fact it is sufficient to compare the value for ϱ given by (iii) with the value calculated above.

6. The Algebra of Covariant Sequences

Among covariant sequences there exist many relations {see B. SEGRE [12], p. 32}, whose ultimate *raison d'être* is in the algebra of the ring $\mathbf{A}(V)$. The following theorems indicate some of these: we shall suppose that the symbols used are well defined and we shall denote by $\{\pi_V(M)\}$, where M is any non-singular subvariety of V containing P, the sequence obtained from $\{\pi(M)\}$ by including its elements in $\mathbf{A}(V)$.

(i) *If* $\pi \in \mathbf{A}(M)$, *where* M *is a non-singular subvariety of* V *of class* μ, *we have* $\{\pi(V)\} = \{\pi_V(M)\} \cdot \{\mu(V)\}$.

(ii) *If* $\pi' = \pi \cdot y'$ *where* π, $y' \in \mathbf{A}(V)$, *we have* $\{\pi'(V')\} = \{\pi(V)\} \cdot V'$ *where* V' *is a non-singular subvariety of* V *representing* y'.

(iii) *If* $\pi = \mu \cdot \nu$, *with* μ, $\nu \in \mathbf{A}(V)$, *we have* $\{\pi(V)\} = \{\mu(V)\} \cdot \{\nu(V)\}$.

All these theorems have been stated and proved by B. SEGRE {see B. SEGRE [12], pp. 32, 34, 35}.

The proof of (i) has been obtained by B. SEGRE with the additional hypothesis that there exist non-singular hypersurfaces of V passing simply through the variety M representing the class μ. This hypothesis has been eliminated later by VESENTINI in [4], where the author gives also a new definition of SEGRE's covariant sequence independent of SEGRE's operation in (VIII, 4e), together with a new proof of th. (i) in (VIII, 5).

For the proof we merely remark that, for $m = \dim \mu = v - 1$, (i) arises immediately from (VII, 5b) and from the relations $(\pi_i(M))_V = \pi_i(V) + \pi_{i-1}(V) \cdot \mu$. These may be proved directly from the definition of $\{\pi(V)\}$ and, in this case, lead at once to our statement: the proof then proceeds by induction.

Theorem (ii) follows immediately by induction with respect to the index of the sequence, since it is certainly valid for the first term $\pi' = \pi \cdot y'$ and by then using the definition of $\{\pi(V)\}$. Finally (iii) derives at once from (i) and (ii) and is also valid for more than two factors.

We consider now some important consequences of these theorems.

a) {See B. SEGRE [12], p. 37.} If $\{\pi(V)\}$ is defined *we have:* $\tilde{\pi}_{v-p}(V) = \pi^{[2]}$, and if $p > v/2$, $v - p < i \leqq p$, $\tilde{\pi}_i(V) = 0$ also. This follows from the definitions and from (i).

b) {See B. SEGRE [12], p. 39.} Suppose that we have been able to express $\pi_0(V) = \pi$, $\pi_1(V)$, ..., $\pi_{i-1}(V)$ by operations only in $\mathbf{A}(V)$: observe that this is obvious for the first term, but not for the others, since the definition of $\{\pi(V)\}$ exceeds $\mathbf{A}(V)$.

Then: (1) Either $i = v - p$ and from (a) we derive at once $(\pi_{v-p})_V = -\pi^{[2]} - \sum_{j=1}^{v-p} (\pi_j(V) \cdot \tilde{\pi}_{v-p-j}(V))_V$; (2) Or $v - p < i \leqq p$ and from (a) it follows $\tilde{\pi}_i(V) = 0$, and therefore $(\pi_i(V))_V = -\sum_{j=1}^{v-p} (\pi_{i-j}(V) \cdot \tilde{\pi}_j(V))_V$; (3) Or else $i < v - p$, and then there exist $s = v - p - i > 0$ classes of $\mathbf{A}^{v-1}(V)$, say $\alpha_1, \alpha_2, \ldots, \alpha_s$, such that $\mu = \alpha_1 \cdot \alpha_2 \cdots \alpha_s$ contains a variety M passing simply through P. Hence P, embedded in M, is in the first case above and we can suppose we have determinded $\pi_i(M)$. By (i), written in the form $\{\pi_V(M)\} = (\{\pi_M(V)\} \cdot \{\tilde{\mu}_M(V)\})_V$ and by the formula $\tilde{\mu}_i(V) = \mu \cdot y_i^{(s)}(\alpha)$, which we can easily obtain from (ii) and the definition of $\tilde{\mu}_i(V)$, it follows that $(\pi_i(V))_V = (\pi_i(M))_V - \sum_{j=1}^{i} \pi_{i-j}(V) \cdot y_j^{(s)} \cdot (\alpha)$.

We conclude with the interesting statement that $\pi(V)$ *can always be constructed by operating only in* $\mathbf{A}(V)$: we could now extend the notion of covariant sequence to the case when P is an arbitrary cycle on V, and also obtain a new demonstration of the covariant character of $\{\pi(V)\}$ with respect to rational equivalence.

c) {See B. SEGRE [12], p. 40.} The question treated in (b) leads to the wider problem: assuming that N is a subvariety of V deduced from varieties $M_1, M_2, \ldots, M_s$ by operating on the rings $\mathbf{A}(M_i)$ and $\mathbf{A}(V)$, we require to express $\nu(V)$ as a function only of the sequences $\mu^{(i)}(V)$, where ν and $\mu^{(i)}$ are the equivalence classes of N and M_i respectively in $\mathbf{A}(V)$. B. SEGRE has solved this problem in the case where $\nu \in \mathbf{A}^{v-1}(V)$ is of the type $\nu = \mu^{(1)} \pm \mu^{(2)}$: we leave out these developments which are rather complicated.

7. The Canonical Sequence

We now outline the definition and properties of SEGRE's *canonical systems* {see B. SEGRE [12], p. 89}.

a) {See B. SEGRE [12], p. 84.} If V and V' are two non-singular varieties and $\{y\}$, $\{y'\}$ are two sequences respectively on V and V', we see at once that the classes of $\mathbf{A}(V \times V')$ of the type $\zeta_j = \sum y_i y'_l$ $i + l = j$, $j = 0, 1, \ldots$, give rise to a sequence of support $V \times V'$: this sequence will be called *the product of the given sequences*, and similarly for more than two factors.

It is then easy to extend the calculus of cycles on the product-varieties to a calculus of sequences on these same varieties: we merely remark that the operation of multiplication is *commutative with those of alternation and inversion, and that, if* $\pi \in \mathbf{A}(V)$, $\pi' \in \mathbf{A}(V')$ *and* $\eta = \pi \times \pi' \in \mathbf{A}(V \times V')$, *then* $\{\eta(V \times V')\} = \{\pi(V)\} \times \pi'(V')$.

b) {See B. SEGRE [12], p. 91.} We begin with the following definition, where "*projection*" is to be understood in a straightforward manner: *If V is non-singular and δ is the class of the diagonal Δ of $V \times V$ in* $\mathbf{A}(V \times V)$, *the projection on V of the sequence $\{\delta(V \times V)\}$ will be called the canonical sequence of V and denoted by $\{y^*\}$.*

The definition is justified by (VI, 8) and we may immediately prove *the relative invariance* of $\{y^*\}$ by transformations between non-singular models. We give now the following important properties {see SEGRE [12], pp. 92, 94, 95} whose proof is deducible from the preceding formal relations connecting covariant sequences:

(i) *If* $\pi \in \mathbf{A}(V)$, *then* $\{\pi^*\}_V = \{\overline{\pi}(V)\} \cdot \{y^*\}$.

(ii) *If* $V = V_1 \times V_2 \times \cdots \times V_s$, *then* $\{y^*\} = \{y_1^*\} \times \{y_2^*\} \times \cdots \times \{y_s^*\}$, *where y_i is the unit class of* $\mathbf{A}(V_i)$.

(iii) *If δ is the class of the diagonal of the s-ple product* $R = V \times V \times V \times \cdots \times V$ *in* $\mathbf{A}(R)$ *and if ϱ is the unit class of* $\mathbf{A}(R)$, *then* $\{\delta^*\}^{[s]} = \delta \cdot \{\varrho^*\}$, *which is a particular case of* (ii).

(iv) From (i) and from the formula $\widetilde{\pi}_j(V) = \pi \cdot \widetilde{\overline{y}}_j^{(v-p)}(\alpha)$, with $\pi = \alpha_1 \cdot \alpha_2 \cdots \alpha_{v-p}$, $\alpha_i \in \mathbf{A}^{v-1}(V)$, we deduce that $\pi_i^* = \pi \cdot \sum_{j=0}^{i} \widetilde{\overline{y}}_j^{(v-p)}(\alpha) \cdot y_{i-j}^*$. In particular, for $p = v - 1$, and then $\pi = \alpha \in \mathbf{A}^{v-1}(V)$, we obtain $\alpha_i^* = \alpha \cdot \sum_{j=0}^{i} \alpha^{[j]} \cdot y_{i-j}^* = \alpha \cdot (y_i^* + \alpha_{i-1}^*)$.

We give now another definition {see SEGRE [12], p. 96}:

The class $\alpha_{(i)} = y_i^* + \alpha_{i-1}^*$, belonging to $\mathbf{A}^{v-i}(V)$, will be called *the adjoint i-class* to the $(v-1)$-dimensional class α, which must contain at least a variety, and also to any its model A.

Using (iv) we derive at once the following *adjunction theorem:*

(v) *The adjoint i-class on V to a prime non-singular V-divisor A, cut on A canonical classes of index i*, i. e. $\alpha_i^* = \alpha \cdot \alpha_{(i)}$.

Finally we prove briefly the following theorem {see B. SEGRE [12], p. 113} which justifies the previous definitions:

(vi) *If $\{y^*\}$ is the canonical* SEGRE *sequence and if we denote by $\{{}^*y\}$ the sequence related to the* TODD *canonical classes, represented by the cycles* ${}^*X_i(V)$ *on V, then we have* $\{y^*\} = \{{}^*y\}$.

For the proof we begin by showing that the TODD canonical sequence also satisfies (i) and then, supposing V to be a linear space S containing the non-singular variety P, we have the two relations: $\{\pi^*\}_S = \{\overline{\pi}(s)\} \cdot \{\sigma^*\}$, $\{{}^*\pi\}_S = \{\overline{\pi}(S)\} \cdot \{{}^*\sigma\}$, where σ is the unit class of $\mathbf{A}(S)$.

Therefore the theorem need be proved only for a linear space S {see B. SEGRE [12], p. 109}. By using (16), we find that the degree of $^*\sigma_i$ is $(-1)^i \binom{s+1}{i}$ and then, since the group $\mathbf{G}^h(S)/\mathbf{G}^h_a(S)$ is infinite cyclic by (VI, 10, iv), we get the expression: $^*\sigma_i = (-1)^i \binom{s+1}{i} \sigma^{v-i}$, where σ^{v-i} is the class of a $(v-i)$-dimensional subspace of S.

It remains only to prove that $\sigma_i^* = (-1)^i \binom{s+1}{i} \sigma^{v-i}$. This is accomplished by formal calculations proceeding by induction with respect to the difference $s - i$.

The first step is for $s - i = 0$: one must then show that $(-1)^s \sigma_s^* = (s+1)\,\sigma^0$ which states that the member on the left is a group of $s+1$ points.

For this it is sufficient to observe that this group, by the definition of $\{\sigma^*\}$, has the same degree as the group Δ [2] in the product $S \times S$. Since the diagonal Δ belongs to the irreducible algebraic system of s-cycles on $S \times S$ which represent the homographies of the space S into itself and since, as is well known, a generic member of this system has just $s+1$ united points, simple and distinct, the above degree is precisely $s+1$.

We omit the rest of the inductive proof, which is merely formal.

8. Some Applications

We devote this section to describing very briefly some facts related to the theory of the canonical sequences.

a) As we have already stated in (VIII, 7b) the zero-dimensional term of $\{y^*\}$ is given {see B. SEGRE [12], p. 92}, with the symbols of the first definition of (VIII, 7, b), by $y_v^* = (-1)^v \cdot \mathrm{pr}_V \delta^{[2]}$: this follows at once from (VIII, 6a) and (VIII, 7b).

When $v = 1$ this result is the so-called principle of correspondence on a curve, or, rather, one of its forms for correspondences endowed with valence {see e. g. GODEAUX [a]}.

If, instead, $v = 2$, we have *a functional interpretation of the* SEVERI *series* {see (VIII, 2e)}, which is usually expressed by saying that *the* SEVERI *series is the characteristic series of the identity*, i. e. of Δ, *on the product of the surface with itself*, a result whose statement goes back to COMESSATTI {see SEVERI [a], pp. 304, 311; COMESSATTI [1]}.

b) {See B. SEGRE [12], p. 115.} By (15) and the fact that $\delta^{[2]}$ coincides with the characteristic of EULER-POINCARÉ of V, we have immediately the relation

$$b_v = I_v + 2(-1)^v (v-1) + 2 \sum_{i=1}^{v-1} (-1)^{v-i-1} b_i ,$$

where b_i is the i-th BETTI number of V.

This is the classical theorem of ALEXANDER which yields *a topological interpretation of the* ZEUTHEN-SEGRE *invariant* {see ALEXANDER [1]}.

It follows that, if v is odd, and hence b_v even by a well known result of LEFSCHETZ {see (IX, 6 a) and LEFSCHETZ [c], p. 272}, *I_v and the degree of the canonical series are both even* {see B. SEGRE [3]}.

c) The identification of the canonical classes in the sense of TODD and in the sense of SEGRE obviously implies *the coincidence of the respective adjoint classes whenever these are defined.*

We remark also that, by using (VIII, 7, iv), we can extend the notion of canonical classes to any class $\pi \in \mathbf{A}(V)$, even if it is free from prime and non-singular representatives: we shall not deal with this extension {see B. SEGRE [12], p. 101}, which, if discussed in details, would be rather delicate.

d) {See B. SEGRE [12], p. 102.} If π is a characteristic class of any index of the canonical system $\xi = v_1^*$ on V, the canonical cycles associated with a cycle P representing π, and having well defined canonical classes, all belong to the set $\mathbf{A}^*(V)$ obtained from $\mathbf{A}(V)$ by adjoining to it all the defined canonical classes of its elements, and which is called by SEGRE *the canonical extension* of $\mathbf{A}(V)$.

This follows from (VIII, 7, iv): in fact if $\pi = \xi^{[s]}$, $1 \leqq s \leqq v - 1$, we obtain $\pi_i^* = \pi \cdot \sum_{j=0}^{i} \binom{s+j-1}{j} \xi^{[j]} \cdot y_{i-j}^*$, $0 \leqq i \leqq v - s$.

In particular, when $i = 1$, this gives $\pi_1^* = (s+1)\,\xi^{[s+1]}$, which for $s = v - 1$, when P is the characteristic curve of ξ, furnishes $\pi_1^* = v \cdot \xi^{[v]}$. Hence follows the equality $2(\Omega_1 - 1) = v\,\Omega_0$ between the linear genus and the degree of ξ, having as particular cases classical formulae respectively of NOETHER and PANNELLI for $v = 2, 3$ {see B. SEGRE [3]; this relation is a particular case of the formulae of MAXWELL-TODD referred in (V, 9)}.

We remark that, if v is odd, then Ω_0 is *even*.

e) {See B. SEGRE [12], p. 115.} If S is a projective space, A a hyperplane of S and α its class in $\mathbf{A}(S)$, it follows from (VIII, 7) that $y_i^* = (-1)^i \binom{s+1}{i} \alpha^{[i]}$. If $\gamma \in \mathbf{A}^{s-1}(S)$, where $s = \dim S$, and if c is the degree of γ, we have $\gamma = c\,\alpha$.

From (VIII, 7, iv) we have $\gamma_i^* = \sum_{j=0}^{i} \gamma^{[i+1]} \cdot \sigma_{i-j}^*$ and hence, by using the preceding expression, we find: $\gamma_i^* = \gamma(i)\,\alpha^{[i+1]}$, where $\gamma(i)$ is the degree of the canonical cycles of index i of the class α.

As we easily find from TODD's definition, this is given by: $\gamma(i) = \sum_{j=0}^{i} (-1)^{i-j} \binom{s+1}{i-j} c^{j+1}$, $i = 0, 1, \ldots, s-1$.

If, more generally, $\pi = \alpha_1 \cdot \alpha_2 \cdots \alpha_s$, with $\alpha_i \in \mathbf{A}^{s-1}(S)$, we start from the formula $(\sigma^*, x) = (1 - \alpha\,x)^{s+1}$: this is true for any $\alpha \in \mathbf{A}^{s-1}(S)$ and follows from the above formulae for y_i^*. By taking into account that $\pi_i(S) = \pi \cdot \tilde{\sigma}_i^{(s)}(\alpha)$, we readily obtain the relation:

$$\begin{aligned}(\pi^*, x) &= \pi(1 - \alpha\,x)^{s+1} \prod_{h=1}^{s} (1 - \alpha_h\,x)^{-1} \\ &= \pi(1 - \alpha\,x)^{s+1}(1 - a_1\,\alpha\,x)\,(1 - a_2\,\alpha\,x), \ldots, (1 - a_s\,\alpha\,x)\,,\end{aligned}$$

where $\alpha_i = a_i\,\alpha$.

This important result furnishes *the canonical classes of any class* π *which is an intersection of several* $(s-1)$*-dimensional classes of* $\mathbf{A}(S)$ *and therefore of any variety of* S, *by the* SEVERI *theorem in* (VI, 10), provided the canonical sequence has been extended as we have described in (VIII, 4).

In particular there follows the classical result {see also GODEAUX [1], and EGER [4]}, which we have already used {see (V, 8)}, that *the canonical divisors of the variety* P *in* S *are cut on* P *by the forms of* S *of degree equal to* $N=\sum_{i=1}^{s} a_i - s - 1$, *where* $a_i =$ *degree* α_i.

9. The Behaviour of the Canonical Systems under Birational Transformations

The problem of determining the transformation law for canonical classes under arbitrary birational transformations is far from solved: at present we have only some partial results, due to B. SEGRE and TODD {see B. SEGRE [5, 14]; TODD [7, 10, 15]}, for the case where the transformation is a dilatation of V with centre at a subvariety of V.

We shall now describe the principal methods used in such researches.

a) {See B. SEGRE [14], p. 146.} In the first place, from (VIII, 4, ii) and from SEGRE's definition of canonical classes it follows easily that, if T is a dilatation of V with centre at the non-singular subvariety P of V, then we have $y_i'^{*} = T(y_i^{*}) + (\lambda_i')_{V'}$, where $V' = T(V)$, y and y' are the units of V and V', and where λ_i' is a suitable class of $\mathbf{A}^{v-i}(P')$, where $P' = T(P)$: this reduces the problem to that of finding the class λ_i'.

b) {See B. SEGRE [14], p. 147; see also TODD [7].} The simplest case is that in which $p = \dim P = 0$, and therefore P' is a space $S'^{v-1} \subset V'$.

In fact we have $\lambda_i' = a_i\, \sigma'^{v-i}$, where σ'^{v-i} is the class associated with a subspace S'^{v-i} and belonging to $\mathbf{A}(P')$. On the other hand, by (VIII, 7, vi) and by (VIII, 7, iv) we find $\pi_i'^{*} = (-1)^i \binom{v}{i} \sigma'^{v-i-1}$ and $\pi_i'^{*} = \pi' \cdot \sum_{j=0}^{i} \pi'^{[j]} \cdot y_{i-j}'^{*}$.

Therefore: $(-1)^i \binom{v}{i} \sigma'^{v-i-1} = \pi'^{[i+1]} + \sum_{j=0}^{i-1} \pi'^{[j+1]} \cdot y_{i-j}'^{*}$. But $\pi' \cdot T(y_i^{*}) = 0$: hence, by taking into account (a) and (VIII, 5), it follows easily, by intersecting with π', that: $\pi'^{[i+1]} = (-1)^i \sigma'^{v-i-1}$.

From this and the above result, we have the recursive formula: $a_i = (-1)^{i-1}\left\{\binom{v}{i} - \binom{v}{i-1}\right\} = (-1)^{i-1} \lambda_i^{v}$.

c) {See B. SEGRE [14], p. 148.} Another tractable case is $i = 1$. Then $\lambda_1' \in \mathbf{A}^{v-1}(P')$ and therefore $\lambda_1' = l\,\pi'$, where l is an integer. Hence $y_1'^{*} = T(y_1^{*}) + l\,\pi'$. We now prove that $l = v - p - 1$.

In effect this is true when $p = 0$ by (b) and is also obvious when $p = v - 1$, since T is then biregular. We proceed by induction with respect to v, supposing $p < v - 1$ and fixed.

Let $\beta \in \mathbf{A}^{v-1}(V)$ with a representative B containing P simply. If $\overline{\pi}' = \pi' \cdot \beta'$, the inductive hypothesis gives $\beta_1'^{*} = \overline{T}(\beta_1^{*}) + (v - p - 2)\,\overline{\pi}'$, where $\overline{T}$ is the dilatation of B into $T(B)$ with centre at P. From the above expression for $y_1'^{*}$ we have: $y_1'^{*} \cdot \beta' = \overline{T}(y_1^{*} \cdot \beta) + l\,\overline{\pi}'$.

Now let $\alpha = \beta$, with $\alpha' = \beta' + \pi'$ and apply the adjunction formula of (VIII, 7, v) giving $\beta_1'^{*} = y_1'^{*} \cdot \beta' + \beta'^{[2]}$, and $\beta_1^{*} = y_1^{*} \cdot \beta + \alpha \cdot \beta$. From this we derive $\overline{T}(\beta_1^{*}) = \overline{T}(y_1^{*} \cdot \beta) + \alpha' \cdot \beta' = y_1'^{*} \cdot \beta' - l\,\overline{\pi}' + {}$ $+ \beta'^{[2]} + \beta' \cdot \pi' = \beta_1'^{*} - (l - 1)\,\overline{\pi}'$, and hence the result.

d) For a dilatation with any p and i, TODD has *conjectured* {see TODD [15], p. 99} by extrapolation the following expression for the cycle λ_i':

$$\lambda_i' = -\sum_{r=t}^{p} \gamma_{r,i}' \cdot (\pi'^{[i-p+r]})_{P'}\,, \quad t = \max(0, p - i + 1)\,,$$

where

$$\gamma_{r,i}' = \sum_{s=0}^{p-r} \lambda_{v-i-r-s+1}^{v-p-s}\, \overline{\overline{\pi}}_s(V) \cdot \pi_{p-r-s}^{*}\,,$$

and

$$\lambda_n^m = \binom{m}{n} - \binom{m}{n-1}\,, \quad \text{if } n > 0\,; \quad \lambda_n^m = 0\,, \quad \text{if } n \leqq 0\,.$$

Observe that $\gamma_{r,i}'$ is a locus of ∞^r spaces L'^{v-p-1} of P', so that $\gamma_{r,i}' \in A(P')$.

These formulae have been proved by TODD for $i \leqq v - p$ and for any i, if $v - p = 2$, and by B. SEGRE if $i = 1, 2, 3$, for any v and p. The general proof depends, according to TODD, on the extension to rational equivalence of some criteria known to hold for linear equivalence.

Without insisting on these demonstrations, which depend on complicated inductions, we proceed instead to outline a different and interesting proof of TODD's formula, when $i = v$ and $v > 2p$, which is due to B. SEGRE {see B. SEGRE [14], p. 153} and which could probably be extended to more general cases.

e) In the first place if $i = v$ TODD's formula, as one easily shows, becomes

$$\lambda_v' = (-1)^{v-p}[(v - p - 1)\,\pi_p^{*}]\,,$$

where the symbol $[k\,G]$, k being an integer and G a group of g points on P, denotes a group of $k\,g$ points on P', formed by g groups of k points, each on the space L'^{v-p-1} arising from a point of G: by $[k\,\gamma]$ we denote, instead, the equivalence class of $k\,G$.

Now let $U = P \times P'$ and $U' = P' \times P'$, and let Δ and Δ' be the respective diagonal varieties of these products. From (VIII, 8 a) we have $\pi_{v-1}'^{*} = (-1)^{v-1}\,\mathrm{pr}_{P'}\,\delta'^{[2]}$ and $\pi_p^{*} = (-1)^p\,\mathrm{pr}_P\,\delta^{[2]}$.

Moreover T induces between U and U' a correspondence F, which associates with each point $Q \times Q$ of $U \times U$ a SEGRE variety $A = T(Q) \times \times T(Q) \subset U'$, where $T(Q)$ is a space L'^{v-p-1}.

Further $\delta^{[2]} = F^{-1}(\delta'^{[2]})$, while each point of $(\Delta'^{[2]})_U$ is the transform by F of all the points belonging to a characteristic group $(\Delta_A^{[2]})_A$, where Δ_A denotes the diagonal of A: on the other hand we know that, for $l = v - - p - 1$, such a group is equivalent to the group of the united points of an almost any homography in an l-space {see (VIII, 7 b)}. Hence we derive $\pi'^*_{v-1} = (-1)^l[(l+1)\,\pi^*_p]$.

Since $v > 2\,p$, we can find a prime non-singular divisor B of V passing simply through P, and hence, by using the same symbols as in (c), we have by this formula, $\overline{\pi}'^*_{v-2} = (-1)^{v-p}[(v-p-1)\,P^*_p]$.

It remains only to prove that $\overline{\pi}'^*_{v-2} = \lambda'_v$. For this consider V embedded in a pencil on a non-singular variety C^{v+1}, in such a way that the base-variety B of the pencil is non-singular and contain P simply. Extend T to a dilatation T^* of C into C', still with centre at P: then C' will contain a pencil of varieties V', which will have a certain base B'.

The varieties $W' = V' \times V'$ and $N' = \overline{P}' \times \overline{P}'$, where $\overline{P}'$ is a representative of $\overline{\pi}'$ will lie on the variety $R' = C' \times C'$.

Now let $\overline{V}'$ be a generic element of the pencil on C' and let $\overline{\overline{V}}'$ be the generic element of any equivalence system containing on $\overline{W}' = \overline{V}' \times \overline{V}'$ the variety $\overline{V}'$, considered as the diagonal variety. If $\overline{\overline{P}}' = \overline{\overline{V}}' \cdot N'$, we have $\overline{\overline{P}}' \equiv \overline{P}'$, and therefore we obtain for the respective equivalence classes: $\overline{\pi}'_{N'} \cdot \overline{\overline{\pi}}'_{N'} = \overline{\pi}'^{[2]}_{N'} = (-1)^v\,\overline{\pi}'^*_{v-2}$. Let $\overline{\overline{V}}' \to M'$ over $\overline{V}' \to V'$: then $M' \subset W'$ and $M' \equiv V'$. Thus $y'_{W'} \cdot \mu'_{W'} = y'^{[2]}_{W'} = (-1)^v\,y'^*_v$.

Our statement is now proved at once by taking into account the manner in which the correspondence induced by T^{-1} operates between W' and $W = V \times V$.

10. Irregular Intersection Problems

The preceding theory has been applied by TODD and SEGRE {see B. SEGRE [2, 12]; TODD [8, 11]} to the so-called problems of irregular intersection, which, first considered by SEVERI {see SEVERI [1]}, may be essentially typified by the following two.

Let $M_1 \perp (M_2 \perp (\ldots \perp M_s)\ldots) = P + Q$, where M_i, $i = 1, 2, \ldots, s$, are subvarieties of V, and where the cycle Q is of regular dimension $q = \sum_{i=1}^{s} m_i - v - 1$, $m_i = \dim M_i$; while the cycle P may be irregular, with $p \geqq q$ {see (I, 9)}.

Then we require to express, in terms of the covariant sequences of the M_i and P on V, supposing that they exist and that P is a non-singulare subvariety of V, the following equivalence classes of $\mathbf{A}(V)$: (1) *The* *so-called rational equivalence of P with respect to the μ_i, where μ_i is the class of M_i, defined by the formula* {see (VIII, 4e)}

$$\varepsilon^{(s)}(P;\mu) = \mu_1 \cdot \mu_2 \cdots \mu_s - (\mu_1 \cdot \mu_2 \cdots \mu_s)^P .$$

(2) *The class $\varrho = \chi \cdot P$, where χ is the symbolic product of the μ_i with respect to P.*

These problems have at present been solved only when the varieties M_i are *non-singular* and contain *simply* the non-singular variety P, always supposing the ambient V to be non-singular also: extensions have been obtained only for very particular cases. With the above hypotheses the following theorems of B. SEGRE are valid {see B. SEGRE [12], p. 51; preceding particular cases are in TODD [11]}:

(i) $\varepsilon(P;\mu) = \pi_t^0$, *where π_t^0 is the term of index $t = p - q$ of the following sequence supported by P:* $\{\pi^0\} = \pi_P(M_1) \cdot \pi_P(M_2) \cdots \pi_P(M_s) \cdot \tilde{\pi}_P(V)^{[s-1]}$.

We shall merely sketch the proof for the case $s = 2$, taking $M_1 = M$, $M_2 = N$. First one easily shows that $\{\pi^0\}$ may be also written in the form: $\{\pi^0\} = (\pi_M(V) \cdot \tilde{\mu}(V))_N \cdot \tilde{\nu}(V)$. Therefore the formula to be proved, which is obvious for $t = 0$, becomes one of the following:

$$(\varepsilon(P;M,N))_N = \sum_{i=0}^{t} \sum_{j=0}^{i} \big((\pi_i(V))_M \cdot (\tilde{\mu}_{i-j}(V))_M\big)_N \cdot (\tilde{\nu}_{t-i}(V))_N,$$

$$(\varepsilon(P;M,N))_N = \sum_{i=0}^{t} (\pi_i(M))_N \cdot (\tilde{\nu}_{t-i}(V))_N ,$$

$$(\varepsilon(P;M,N))_P = \sum_{i=0}^{t} \sum_{j=0}^{t-i} (\pi_i(M))_P \cdot (\pi_j(N))_P ,$$

which are all of interest, and where the second is obtained by using (VIII, 6, i).

The proof begins by distinguishing between the four cases: (1) $p = m$, (2) $n = v - 1, p < m$, (3) $n < v - 1, p = m - 1$, (4) $n < v - 1, p < m - 1$. In the first case $\pi = \mu$ and hence $\chi = (\mu \cdot \nu)^P = 0$. Moreover $\{\pi(M)\} = \{\mu(N)\} = \mu, 0, 0, \ldots$, so that the proof of our statement is reduced to showing that $(\mu \cdot \nu)_V = \mu \cdot (\tilde{\nu}_{v-n}(V))_V$ or $(\mu \cdot \nu)_N = \mu_N \cdot (\nu_V^{[2]})_N$, which is certainly true because $\mu = \pi \in \mathbf{A}(N)$. In the second case $\dim \pi = \dim \chi = m - 1$ and $t = 0$, and the theorem is obvious. The third case may be treated by induction with respect to $v - n$. It is in effect true when $v - n = 1$. The induction process is effected by using a prime divisor on V which passes simply through P, and by the theorems of (VIII, 6). The last case is treated, instead, by induction with respect to $m - p$, the theorem still being true if $m - p = 1$, and by using the same methods as in the third case.

The following theorem answers question (2).

(ii) *If $\chi = (\mu_1 \cdot \mu_2 \cdots \mu_s)^P$ and $\varrho = \chi \cdot P$, we have $\varrho = \pi_{t+1}^0$ with respect to the same sequence of* (i).

We mention that, if in $\{\pi^0\}$, the covariant sequences are expressed in terms of the canonical sequences by (VIII, 7 b), we find at once an invariant form for (i) and (ii).

The problems (1) and (2) may be generalised by requiring the determination of the covariant sequences of some representative Q of χ and of $R = P \cdot Q$ in terms of the M_i, P and their canonical cycles: this may be useful for the calculation of certain characters of Q or R. The solution has been obtained by SEGRE {see B. SEGRE [12], p. 62} *for* $\chi(V)$, *when* $p = q$, *or when* $p > q$ *for some special cases, and in any case for* $\varrho(Q)$, *with the elegant formula:* $\varrho_i(Q) = \pi^0_{i+i+1}$.

The problems of irregular intersection are quite characteristic of algebraic geometry, since they have at present no analogue in topology. We add the conjecture of B. SEGRE, according to which any such problem is totally solvable only by the covariant sequences of the varieties under examination: a fact which is true for any case considered up to now. Moreover such problems have never been investigated from the abstract algebraic standpoint, apart from some general researches on foundations {see BARSOTTI [3]}.

11. Miscellaneous Results

We shall describe here some applications of the above theory.

a) {See B. SEGRE [12], p. 67.} Take $\pi_i(V) = [\pi(V)]^j, j = v - p + i$, where $[\pi(V)]^j$ is regarded as a symbolic power, null if $j < v - p$ or $j > v$, and let $f(\pi) = \sum_{j=0}^{n} \mu_j \, \pi^{[j]}$ be a symbolic polynomial in π, with $\mu_j \in \mathbf{A}(V)$: we put $f(\pi(V)) = \sum_{j=0}^{n} \mu_j [\pi(V)]^j$, so that, if $\dim \mu_j = r + j$, with r independent of j, we have $\dim f(\pi(V)) = r$. Thus the definition of $\{\pi(V)\}$ becomes $\varepsilon(P; \alpha) = (\alpha_1 + \pi) \cdot \cdots \cdot (\alpha_s + \pi)$.

Now if the α_i admit some representative A_i containing P with the multiplicity $k_i \geqq 1$, and such that $A_1 \cap A_2 \cap \cdots \cap A_s$ is *regular* outside the component P, one may then show that among the representatives of $\varepsilon(P; \alpha)$ there is also the cycle $E = (A_1 + k_1 P) \ldots (A_s + k_s P)$, which solves the problem *of finding the rational equivalence of* P *with respect to the* α_i *and with respect to the given multiplicities* k_i {see also the particular cases in ARCHBOLD [1]; BARKER [1]}.

b) If $2p \geqq v$, one may show, as SEVERI has done {see SEVERI [1]}, that *almost any member* P *of an algebraic maximal system on* V *has necessarily a locus* $\mathfrak{D}$ *of double points*, which is pure and of dimension $d = 2p - v$. If δ is the class in $\mathbf{A}(V)$ of $\mathfrak{D}$, we have in $\mathbf{A}(V)$ the relation: $\delta = \pi^{[2]} - \tilde{\pi}_{v-p}(V)$, as TODD {see TODD [11]} and SEGRE {see B. SEGRE [12], p. 72; special cases of this formula are in THULLEN [1]; B. SEGRE [4]; YOAXALL [1]} have proved.

Observe that if, in particular, $\mathfrak{D}$ is rationally null, then we find again the result of (VIII, 6 a). By using (VIII, 7 b, i), we can also write $\delta = \pi^{[2]} - (-1)^{v-p} \sum_{i=0}^{v-p} \pi^*_i \cdot y^*_{v-p-i}$.

c) If $p < m \leqq 2p - 1$, $m < v$, any subvariety of V passing through P generally has *a singular locus* $\mathfrak{S}$, which, besides the locus $\mathfrak{D}$ double for P (generally empty if $2p < v$), contains *a locus* $\mathfrak{R}$ *simple for* P *with* $\dim \mathfrak{R} = 2p - m - 1$, as has been proved by SEVERI and SEGRE {see SEVERI [1], p. 74 or [e], p. 54; B. SEGRE [12], p. 72; SEVERI considers the case in which V is a linear space and P, M are complete intersections of forms}:

SEGRE shows that, if ϱ is the class of $\mathfrak{R}$ and if $m < v - 1$, then $\varrho = (\tilde{\pi}_{m-p+1}(M))_V$, and, instead, if $2p > m = v - 1$, $\varrho = (\tilde{\pi}_{v-p}(M))_V - \delta$ where δ is as above in (b). Also these relations may be translated by using the canonical sequences (VIII, 7 b, i).

d) {See B. SEGRE [12], p. 118.} If $p = \sum_{j=1}^{h} i_j$, it is clear that $(\pi^*_{i_1} \cdot \pi^*_{i_2} \cdot \cdots \cdot \pi^*_{i_h})$ is a zero-dimensional class $\xi_{i_1 i_2 \cdots i_h}$ where the classes π^*_j are, as usual, the canonical classes of the subvariety P of V. Put $x_{i_1 i_2 \cdots i_h} = \deg \xi_{i_1 i_2 \cdots i_h}$. One may prove without difficulty that if $\sum c_{i_1 i_2 \cdots i_h} \xi_{i_1 i_2 \cdots i_h} = 0$, and then $\sum c_{i_1 i_2 \cdots i_h} x_{i_1 i_2 \cdots i_h} = 0$ where the $c_{i_1 i_2 \cdots i_h}$ are integers independent of P and the sums are extended to all partitions of p, then $c_{i_1 i_2 \cdots i_h} = 0$, i. e. the ξ's and the x's are *linearly independent.*

In particular, by intersecting suitably the elements of the sequences $\{\pi^*\}$ and $\{\pi(V)\}$, one can find various independent equivalence series. SEGRE has shown {see B. SEGRE [12], p. 119} that on a surface there are exactly five independent equivalence series.

IX. The Algebraic Varieties as Complex-analytic Manifolds

1. The Complex Manifolds and CHOW's Theorem

We begin with this chapter the transcendental and topological treatment of algebraic varieties defined over the complex field, omitting the more familiar part of this subject which one may find in well-known treatises {see HODGE [a]; DE RHAM-KODAIRA [a]; DE RHAM [a]; LICHNEROWICZ [a]; SEGRE [c]; moreover see the works: GARABEDIAN-SPENCER [1, 2]; KODAIRA [1]; ECKMANN [1, 2]; ECKMANN-GUGGENHEIMER [1]; GUGGENHEIMER [1]; HODGE [10]}.

The concept of a complex-analytic manifold (complex manifold) generalises the well known definition of a RIEMANN surface in the abstract sense.

A *complex manifold* $\mathfrak{M}^n$ of complex dimension n, is a HAUSDORFF space with each point P of which is associated a neighbourhood $U(P)$ which is mapped topologically onto a subdomain of the Euclidean space of the complex variables $z^1, z^2, \ldots, z^n$, the local coordinates in $U(P)$, in such a way that the relation between two such systems in the intersection of their neighbourhoods is given by complex-analytic functions (pseudo-conformal mapping). The manifold $\mathfrak{M}$ is called *compact* if it is a compact topological space; and we shall consider only such manifolds.

It is obvious that the projective space $\mathbf{P}^n$ has a natural complex structure and it is easily proved that a non-singular algebraic variety in $\mathbf{P}^n$ is a compact complex manifold.

Chow has proved {see Chow [2]}, conversely, the following theorem:

(i) *Any compact complex manifold embedded in* $\mathbf{P}^n$ *is an algebraic variety.*

Chow's theorem is actually stronger, proving that every compact analytic set E embedded in $\mathbf{P}^n$, i. e. a set which is given locally by the solutions of a set of holomorphic equations, is algebraic.

The proof may be reduced {these simplifications are due to Samuel [2]} to the case $n = 2$, by using inductively the fact that if the $\mathbf{P}^{n-1}$ of $\mathbf{P}^n$ containing the origin O, intersect an algebroid manifold V of $\mathbf{P}^n$ along algebraic varieties, then V is algebraic. But for an algebroid curve C of P, through O, which is not algebraic, it is easily proved that there exists, for any integer N, an algebraic cycle Γ of order n such that $i(O; C \cap \Gamma) \geqq n\,N$. The proof is then completed by showing that if, locally at O, C is contained in a compact analytic set of P, then $i(O; C \cap \Gamma) \geqq n\,N$ for some N.

More elegant proofs have been given by Stein {see Cartan [c]: exposé XIV par K. Stein} and Serre {see Cartan [c]: exposé XIX par J. P. Serre; see also Cartan [1], where previous works of P. Thullen are quoted}. Stein's proof starts from the theorem:

Every cone in $\mathbf{A}^n$ *is analytic at its vertex, if it is so everywhere outside the vertex itself.*

It proceeds by embedding the $\mathbf{P}^n$ in $\mathbf{A}^{n+1}$ and projecting the given set E from a point O of $\mathbf{A}^{n+1}$. The cone $O(E)$ so obtained is then analytic at O, and, therefore, a convenient neighbourhood of the vertex O may be represented by a set $f_j(x) = 0$, $j = 1, 2, \ldots, s$, where the $f_j(x)$ are holomorphic with respect to the homogeneous coordinates x in $\mathbf{P}^n$. Developing $f_j(x)$ into a homogeneous polynomial series, we show that all polynomials appearing are null, by taking into account the structure of the cone at the vertex O: thus E appears as an intersection of algebraic divisors of $\mathbf{P}^n$.

Serre's demonstration is based on the theory of *coherent ideal stacks* {see (X)}, and, precisely, on the theorem that such a stack in $\mathbf{P}^n$ is always generated by the homogeneous polynomials of the stack.

Chow's theorem may be readily extended to *any analytic compact set on the products of more projective spaces*. Since a meromorphic transformation of an algebraic projective variety into a projective space $\mathbf{P}^n$ has a graph which can be thought of as an analytic compact subset of the projective space, namely the product of $\mathbf{P}^n$ by the ambient space of V, we have the statement {see Chow [2]}:

(i) *An everywhere meromorphic function on an algebraic projective variety is necessarily a rational function.*

2. HERMITE's and KÄHLER's Metrics

Let $\mathfrak{M}^n$ be a compact complex manifold and let $z^1, z^2, \ldots, z^n$ be local coordinates on it. We shall here briefly recall some general facts, omitting proofs which may be found in the standard treatises.

a) Putting

$$z^\alpha = x^{2\alpha-1} + \sqrt{-1}\, x^{2\alpha}, \quad \alpha = 1, 2, \ldots, n, \tag{1}$$

we introduce *real* local coordinates $x^1, x^2, \ldots, x^{2n}$, which clearly determine a *real analytic structure on* $\mathfrak{M}$.

Since the Jacobian of a pseudoconformal transformation between local coordinates is always positive, $\mathfrak{M}$ is an *orientable* manifold, and we shall select as natural orientation the one related to the ordering $x^1, x^2, \ldots, x^{2n}$.

b) A positive definite *Hermitian metric* on $\mathfrak{M}^n$ is given by the expression $ds^2 = 2 \sum_{\alpha,\beta=1}^{n} g_{\alpha\beta^*}\, dz^\alpha\, d\bar{z}^\beta$ with $g_{\alpha\beta^*} = \bar{g}_{\beta\alpha^*}$: such a metric can be always introduced on $\mathfrak{M}$.

The Hermitian metric is called *a* KÄHLER *metric* if $g_{\alpha\beta^*}$ satisfy the relations:

$$\frac{\partial g_{\gamma\beta^*}}{\partial z^\alpha} - \frac{\partial g_{\alpha\beta^*}}{\partial z^\gamma} = 0, \qquad \frac{\partial g_{\alpha\gamma^*}}{\partial \bar{z}^\beta} - \frac{\partial g_{\alpha\beta^*}}{\partial \bar{z}_\gamma} = 0. \tag{2}$$

Then, as KÄHLER has shown {see KÄHLER [3]}, there is a local function $N(z, \bar{z})$ such that

$$g_{\alpha\beta^*} = \frac{\partial^2 N}{\partial z^\alpha\, \partial \bar{z}^\beta}. \tag{3}$$

Passing to real coordinates x^i by (1), ds^2 can be written in the form

$$ds^2 = \sum_{j,k=1}^{2n} g_{jk}\, dx^j\, dx^k, \quad g_{jk} = g_{kj}, \tag{4}$$

so that $\mathfrak{M}^n$ may be considered as a real $2n$-dimensional orientable RIEMANN manifold with a positive definite metric.

c) Let $\zeta_0, \zeta_1, \ldots, \zeta_n$ be homogeneous coordinates in $\mathbf{P}^n$, normalised by $\sum_{i=0}^{n} \zeta_i \bar{\zeta}_i = 1$. Then define $X^h = \sqrt{2}\, \zeta_h \bar{\zeta}_h$, $X^{hk} = \zeta_h \bar{\zeta}_k + \bar{\zeta}_h \zeta_k$, $Y^{hk} = i(\zeta_h \bar{\zeta}_k - \bar{\zeta}_h \zeta_k)$ and assume X^h, X^{hk}, Y^{hk} as orthogonal Cartesian coordinates in a Euclidean space $\mathbf{E}^N$, $N = (n+1)^2$. Thus we have defined a real locus $\mathbf{P}^{*2n}$ in $\mathbf{E}^N$ which is a biregular real model of the complex space $\mathbf{P}^n$: then any algebraic variety V^r of $\mathbf{P}^n$ admits a real biregular model V^{*2r} in $\mathbf{P}^{*2n}$.

The manifold V^{*2r} is called *the* RIEMANN *manifold* of V^r and, it is more precisely known as the MANNOURY model {see HODGE [a], p. 150}.

If we take as metric on V^{*2r} the one induced by the Euclidean metric of the embedding space $\mathbf{E}$, then we find, by a simple calculation, that ds^2 has precisely the KÄHLER form as in (b), with

$$N = \frac{1}{2\pi} \log \sum_{k=0}^{n} \left| \frac{\zeta_k}{\zeta} \right|^2, \tag{5}$$

where $\zeta = \sum_{i=0}^{n} \lambda_i \zeta_i$ is an arbitrary linear form in the ζ_i, associated with the hyperplane at infinity of $\mathbf{P}^n$.

Therefore *the algebraic non-singular varieties of* $\mathbf{P}^n$ *are* KÄHLER *manifolds.*

d) {See e. g. HODGE [a], Ch. II.} We shall associate with the complex manifold $\mathfrak{M}^n$ *exterior differential forms* of the type $\varphi^p = \sum_{i_1 < i_2 < \cdots < i_p} a_{i_1 i_2 \cdots i_p}$, $dx^{i_1} \wedge dx^{i_2} \wedge \cdots \wedge dx^{i_p}$, where $x^1, x^2, \ldots, x^{2n}$ are real local coordinates on $\mathfrak{M}$ and the coefficients $a_{i_1 i_2 \cdots i_p}$ are *complex-valued functions* of the x^i of class C^∞. Applying usual summation conventions, the p-form φ^p may be also written as $\varphi^p = 1/p!\, a_{i_1 i_2 \ldots i_p}\, dx^{i_1} \wedge dx^{i_2} \wedge \cdots \wedge dx^{i_p}$.

Then referring to the real RIEMANN structure of $\mathfrak{M}$, we can define the *derived* $(p+1)$-form $d\varphi$ and the *adjoint* $(n-p)$-form $*\varphi$.

These linear operators satisfy the relations: (1) $d(\varphi \wedge \psi) = d\varphi \wedge \psi + (-1)^p \varphi \wedge d\psi$; (2) $d^2\varphi = 0$.

Selecting a sufficiently fine triangulation of $\mathfrak{M}$, such that each p-simplex is contained in the existence domain of a local system (z), we can define integration of a p-form φ on a p-simplex and, after, by linearity, on a p-chain C^p, both of class one at least, with complex or real coefficients.

There is the well-known *formula of* STOKES which states that

$$\int_{b\,C^{p+1}} \varphi = \int_{C^{p+1}} d\varphi, \tag{6}$$

where φ is a p-form and C^{p+1} is a differentiable $(p+1)$-chain having $b\,C^{p+1}$ as boundary.

Recalling that a p-cochain γ^p is a *linear function* $\gamma^p(C^p)$ on the group of p-chains and that the coboundary operator $\beta\gamma^p$ is defined by $\beta\gamma^p(C^{p+1}) = \gamma^p(b\,C^{p+1})$, whence $\beta^2\gamma^p = 0$, we can infer from this that the linear functional $\int_{C^p} \varphi$ is a p-cochain and, by using STOKES's theorem (6), that this cochain has for coboundary $\int_{C^p} d\varphi$.

Hence we derive a certain analogy between the theory of exterior differential p-forms and that of the p-cohomology on $\mathfrak{M}$. This is the starting point of DE RHAM's concept of a current on $\mathfrak{M}$, which we shall now describe {see DE RHAM-KODAIRA [a], Ch. II or DE RHAM [a], Ch. III}.

3. The Currents

{See the account in KODAIRA [1], p. 816.}

Let $\mathfrak{D}$ be a fixed but arbitrary subdomain of $\mathfrak{M}^n$ and let $\{\varphi\}$ be the linear space of the $(n-p)$-forms of class C^∞, null on $\mathfrak{M} - \mathfrak{D}$.

Then a *p-current of dimension* $n - p$ on $\mathfrak{D}$ is a linear functional $T[\varphi]$ defined on $\{\varphi\}$ which converges to zero when φ varies in a sequence (φ_h) for $h \to \infty$, where every φ_h is null outside the same compact subset of $\mathfrak{D}$ covered by a local system (x), and where, for $h \to \infty$, every $\partial^s \varphi_h / \partial x^p\, \partial x^q \ldots \partial x^r$, $s = 0, 1, \ldots$, converges uniformly to zero.

If ψ is a locally integrable p-form, then

$$\psi[\varphi] = \int_{\mathfrak{M}} \psi \wedge \overline{\varphi} \tag{7}$$

defines a current in $\mathfrak{M}$, which will be identified *with the form* ψ.

Similarly a $(2n-p)$-*chain* C may be thought of as a current by taking

$$C[\varphi] = \int_C \overline{\varphi}\,. \tag{8}$$

A p-current is *zero* at the point $p \in \mathfrak{D}$ if there is a neighbourhood U_p of p such that $T[\varphi] = 0$ for all φ vanishing outside U_p. The *carrier* $|T|$ of a current is the complement of the largest open set in $\mathfrak{D}$ on which $T = 0$.

By (7) and (IX, d) we get at once $d\psi[\varphi] = (-1)^{p+1}\, \psi[d\varphi]$, $*\psi[\varphi] = (-1^p)\,\psi[*\varphi]$; generally we shall define $dT[\varphi] = (-1)^{p+1}\, T[d\varphi]$, $*T[\varphi] = (-1)^p\, T[*\varphi]$ respectively as the *exterior derivative* dT and the *adjoint current* $*T$ of a p-current.

Then we have $ddT = 0$, $**T = (-1)^p\, T$.

If T is a $(2n-p)$-chain C, we find: $dC = (-1)^{p+1}\, bC$, so that the derived chain is, apart from sign, the boundary of C.

We now introduce *the operators* δ and Δ defined respectively as $\delta = -*d*$ and $\Delta = d\delta + \delta d$, the former being the so-called *codifferential* satisfying $\delta^2 T = 0$ and the latter *the extended* LAPLACE-BELTRAMI *operator*.

Next we put $(T \wedge \psi)\,[\varphi] = T[\psi \wedge \varphi]$ where ψ is a q-form of class C^∞ on $\mathfrak{M}$, which is called *the exterior product*, and moreover $(T, \varphi) = \overline{(\varphi, T)} = T[*\varphi]$, which is, instead, *the scalar product* of T and a p-form φ of class C^∞ on $\mathfrak{M}$. If T is a p-form ψ of class C^∞, we have $(\psi, \varphi) = \int_M \psi \wedge *\overline{\varphi}$ and then $(\varphi, \varphi) > 0$.

A p-current is said to be *regular* at a point $p \in \mathfrak{D}$, if T coincides with a p-form of class C^∞ in some neighbourhood of p. The set of all singular, i. e. non-regular, points of T is called *the singular set* $|T|_s$ of T.

By using this notion the scalar product of two arbitrary currents S, T, with $|S|_s \cap |T|_s$, empty, may be defined by writing: $(S, T) = (\varphi, \psi) + (\varphi, T - \psi) + (S - \varphi, \psi)$ where φ and ψ are p-forms of class C^∞

such that $|T - \psi| \cap |S - \varphi|$ is empty. One shows that (S, T) is independent of the choice of φ and ψ, and that $(dS, T) = (S, \delta T)$, $(\Delta S, T) = (S, \Delta T)$.

4. The Fundamental Existence Theorems

{See HODGE [a], Ch. II.}

We shall recall here some results of analysis, chiefly due to DE RHAM and HODGE, which are of importance for algebraic geometry.

a) A p-current is called *closed* or *coclosed* respectively if $dT = 0$ or $\delta T = 0$. It is said to be *homologous to zero*, $T \sim 0$, if there is a $(p-1)$-current S such that $T = dS$: then T is called *the boundary* of S. It is clear that, if $T \sim 0$, then T is closed. If $T = dS$, T is said *to cobound* S and then T is consequently coclosed.

We can now construct in an obvious manner *homology and cohomology groups* for the set of p-currents.

A current T is called *harmonic* in $\mathfrak{D}$ if $\Delta T = 0$ in $\mathfrak{D}$: a harmonic p-form h in all the compact variety $\mathfrak{M}$ satisfies, everywhere in $\mathfrak{M}$, $dh = \delta h = 0$, i. e. *h is closed and coclosed.* Such a form is frequently called of *the first species.*

These forms generate a vector space $\mathbf{H}^p$ of *a finite dimension* on the complex field, which follows, for instance, as an application of FREDHOLM's theory of integral equations.

If $h_1, h_2, \ldots, h_s$ is an orthonormal basis of $\mathbf{H}^p$, i. e. $(h_i, h_j) = \delta_{ij}$, the p-form $HT = \sum_i (h_i, T)\, h_i$ is harmonic and is called *the harmonic component* of the current T. We have $dH = Hd = 0$; $\delta H = H\delta = 0$, $*H = H*$, and $(HS, T) = (S, HT)$.

The following results are due to DE RHAM:

(i) *There is one and only one linear operator G which transforms every p-current T defined in $\mathfrak{M}$ into a $(p-1)$-current GT such that* (1) $\Delta GT = G\Delta T = T - HT$; (2) $HGT = GHT = 0$.

(ii) *Let T be a p-current in $\mathfrak{D} \subseteq \mathfrak{M}$. If ΔT or dT and δT are regular in $p \in \mathfrak{D}$, so is also T in p.*

The operator G is *real, self-adjoint* and *commutable with d and δ.*

b) From (i) we derive: $T = \Delta GT + HT = d\,\delta\,GT + \delta\,d\,GT + HT$; this decomposition of T in the sum of three currents, one homologous to zero, an other cohomologous to zero and the last equal to a harmonic form, is unique.

It follows that *a closed (coclosed) current T is homologous (cohomologous) to one and only one harmonic form HT:* hence we have the far-reaching result that *the vector space of the $(2n-p)$-homology group* $\mathbf{H}_{2n-p}(\mathfrak{M}, \mathbf{C})$ *and of the p-cohomology group* $\mathbf{H}^p(\mathfrak{M}, \mathbf{C})$ on $\mathfrak{M}$ ($\mathbf{C}$ = complex field) *are both isomorphic to the vector space* $\mathbf{H}^p$ *of the harmonic p-forms of the first species on* $\mathfrak{M}$.

Especially cohomology and homology groups of $\mathfrak{M}$ have *finite dimension:* moreover they are *dually related* by the adjoint operator * on the group $\mathbf{H}^p$, giving $*\mathbf{H}^p = \mathbf{H}^{2n-p}$, so that if b_p, the so-called *p-number of* BETTI, is the dimension of $\mathbf{H}^p$, we have $b_p = b_{2n-p}$, as is well known also from the POINCARÉ duality.

c) From the decomposition formula in (b) it follows that, if T is closed, then $T \sim HT$ and, if $T = dS$, then $S = \delta GT$: moreover S is unique, since a current cohomologous to zero and closed is null. Further a closed current T is homologous to zero if and only if $HT = 0$.

d) Calling *"period"* of a closed p-form φ on a p-cycle Γ^p the value of the integral $\int_\Gamma \varphi = \Gamma[\varphi]$, we have the following theorem of HODGE:

(iii) *There is one and only one harmonic p-form on $\mathfrak{M}$ having arbitrarily assigned periods on p homologically independent p-cycles.*

The uniqueness follows from (b) and the existence by taking a harmonic form h such that $(H\,\Gamma_i, h) = p_i$ where Γ_i, $i = 1, 2, \ldots, p$, are the p-cycles and p_i the periods. Then $*h$ is harmonic and we have $\int_{\Gamma_i} *h = (\Gamma_i, h) = (H\,\Gamma_i, h) = p_i$.

e) DE RHAM has proved that, if C^{2n-p}, C^p are two arbitrary chains such that the boundary of either does not intersect the other, then the topological intersection number $I(C^{2n-p}, C^p)$, or KRONECKER *index*, is given by: $I(C^{2n-p}, C^p) = (C^{2n-p}, *C^p)$.

If, in particular, the chains are cycles, we have: $I(\Gamma^{2n-p}, \Gamma^p) = (\Gamma^{2n-p}, *\Gamma^q) = \int_{\Gamma^p} \varphi^p = \Gamma^p[\varphi]$, where $\varphi^p = H\,\Gamma^{2n-p}$, and then $\varphi \sim \sim \Gamma^{2n-p}$.

5. The Complex Operators

{See HODGE [a], Ch. IV; A. WEIL [1]; ECKMANN-GUGGENHEIMER [1]; GUGGENHEIMER [1]; GARABEDIAN-SPENCER [2]; HODGE [10].}

a) Let φ be a p-form. Putting $2\,x^{2\alpha-1} = z^\alpha + \bar{z}^\alpha$, $2\,x^{2\alpha} = i(\bar{z}^\alpha - z^\alpha)$, φ assumes the following structure

$$\varphi = \sum_{r+s=p} \varphi^{r,s} \tag{9}$$

with

$$\varphi^{r,s} = \sum_{r+s=p} \frac{1}{r!}\frac{1}{s!} a_{\alpha_1\alpha_2\cdots\alpha_r\beta_1^*\beta_2^*\cdots\beta_s^*}\, dz^{\alpha_1} \wedge \cdots \wedge dz^{\alpha_r} \wedge d\bar{z}^{\beta_1} \wedge \cdots \wedge d\bar{z}^{\beta_s}. \tag{10}$$

We can now define the complex operator C by the relation

$$C\varphi = \sum_{r+s=p} (-1)^{r-s} \varphi^{r,s}. \tag{11}$$

Then, if

$$\omega = i \sum_{\alpha,\beta=1}^{n} g_{\alpha\beta^*}\, dz^\alpha \wedge d\bar{z}^\beta, \tag{12}$$

which is a real two-form associated with the Hermitian metric on M we define Λ *by*

$$\Lambda\,\varphi = (-1)^p\, {}^*\omega \wedge {}^*\varphi\,. \tag{13}$$

C, Λ are *the* HODGE-WEIL *operators*; they possess the following properties:

$$C\,C\,\varphi = (-1)^p\,\varphi\,,\quad \Lambda\,d - d\,\Lambda = C^{-1}\,\delta\,C\,. \tag{14}$$

Further, for a p-current T, we put $C\,T[\varphi] = (-1)^p\,T[C\,\varphi]$. A p-form $\varphi^{r,s}$ is called pure of type (r, s), and this definition may be extended to a current T by writing T as a symbolic differential form, whose coefficients are well determined distributions in the sense of L. SCHWARTZ. Therefore we shall denote by $P_t T$ *the pure current of type* $(p-t, t)$ additively contained by T. The operator $P_t T$, which is a so-called *projector,* is commutable with Δ, H and G.

b) A p-form of type $(p, 0)$ is called *holomorphic,* or *meromorphic,* at a point $p \in \mathfrak{M}$, if their coefficients at p are holomorphic or meromorphic, and similarly for a domain $\mathfrak{D}$.

A meromorphic p-form is said to be *regular* at p, if it is holomorphic at p: otherwise it is considered as singular at p.

A *p-differential* in a domain $\mathfrak{D}$ is a p-form meromorphic in $\mathfrak{D}$ and closed whenever it is regular: the p-differential is called *holomorphic,* if it is so as a p-form.

Let φ be such a form defined on all $\mathfrak{M}$ and observe that $C\varphi = i^p\varphi$, $\Lambda\varphi = 0$, and also that $\delta\varphi = i^{-p}\,\delta C\varphi = i^{-p}\cdot C(\Lambda d - d\Lambda)\,\varphi$. Hence an arbitrary holomorphic p-differential on $\mathfrak{M}$, is closed and coclosed in all $\mathfrak{M}$. Consequently it is a harmonic form of the first species and it is therefore called a *p-differential of the first kind.*

c) Let T be a *pure closed p-current of type* $(p, 0)$ in $\mathfrak{M}$. As in (b) we prove that T is coclosed in $\mathfrak{M}$. Then by (IX, 4 a, ii) $T\cdot$is a p-form on $\mathfrak{M}$, which will be of type $(p, 0)$ and endowed with holomorphic coefficients $a_{i_1 i_2 \dots i_p}$ being $d\,T = 0$, and therefore $\partial\, a_{i_1 i_2 \dots i_p}/\partial \bar{z}_\beta = 0$. Hence *$T$ is a p-differential of the first kind.*

6. The HODGE-ECKMANN Theory

{See HODGE [a], Ch. IV; HODGE [8, 9]; ECKMANN [1, 2].}

We shall proceed to explain some important results deriving from the sole supposition that the manifold $\mathfrak{M}$ is endowed with a KÄHLER metric. These are due to A. WEIL and, chiefly, to HODGE and ECKMANN: when in particular $\mathfrak{M}$ is algebraic and non-singular, these properties had been, in part, discovered by LEFSCHETZ.

a) By (IX, 5 a) we find that $\mathbf{H}^p = \sum_{r+s=p} \mathbf{H}^{r,s}$ where $\mathbf{H}^{r,s}$ denotes the group of pure harmonic p-forms of type (r, s). If $b^{r,s}$ denotes the rank of $\mathbf{H}^{r,s}$ the BETTI number b^p is given by $b^p = \sum_{r+s=p} b^{r,s}$. Taking into account that, if φ is harmonic of type (r, s), then the conjugate form $\overline{\varphi}$ is still harmonic and of type (s, r), *we infer that* $b^{r,s} = b^{s,r}$.

Therefore, *if p is odd, b^p is even:* for algebraic varieties this is a well-known result of LEFSCHETZ.

Moreover, *the two-form ω defined in* (12) *is, together with any power ω^k, $k = 1, 2, \ldots, n$, closed and coclosed,* as one readily proves, and so is *harmonic.* Since $\omega^k \neq 0$, $k \leq n$, $\omega^k \not\sim 0$ and so $b^{2k} \geq 1$.

b) We take here into consideration the isomorphism between $\mathbf{H}^p(\mathfrak{M},\mathbf{C})$ and $\mathbf{H}^p$, which are considered as groups on the complex field, and denote by the same symbol φ a closed form and the corresponding cohomology class.

The dual homology class $D\,\varphi$, where D is the duality operator on $\mathfrak{M}$, is given by $\varphi \cap \mathfrak{M}$, where $\mathfrak{M}$ is the fundamental $2\,n$-cycle of the oriented manifold $\mathfrak{M}$ and $\cap$ denotes the CECH-WHITNEY cup-product.

Let us call the classes $D\,\omega^k = \omega^k \cap M = Z_{2(n-k)}$, of topological dimension $2(n-k)$, *the principal homology classes of* $\mathfrak{M}$. Clearly, writing Z for $Z_{2(n-1)} = D\,\omega$, $Z_{2(m-2)}$ is the intersection $Z \otimes Z$, and so on.

If $\mathfrak{M}$ is the complex projective space $\mathbf{P}^n$, the principal classes Z_{2q} are numerical multiples of the projective subspaces $\mathbf{P}^q$ of $\mathbf{P}^n$, as one easily derives from (IX, 2 c).

Now let $\mathfrak{M}'$ be a complex manifold of complex dimension $n' < n$, analytically and regularly embedded into $\mathfrak{M}$: then the KÄHLER metric of $\mathfrak{M}$ induces a KÄHLER metric on $\mathfrak{M}'$ and the form ω' associated with the latter metric is the one induced onto $\mathfrak{M}'$ by ω. We have $D\,\omega' = D(\omega' \cap \mathfrak{M}') = Z(\mathfrak{M}')$ and, on $\mathfrak{M}$, $\omega \cap \mathfrak{M} = D\,\omega \otimes \mathfrak{M} = Z(\mathfrak{M}') \otimes \mathfrak{M}$. Hence $Z(\mathfrak{M}') = Z(\mathfrak{M}) \otimes \mathfrak{M}'$.

Thus *the principal classes of $\mathfrak{M}'$ are the intersections of $\mathfrak{M}'$ with the ones of $\mathfrak{M}$.*

For us is particularly important the case of a complex manifold $\mathfrak{L}$ regularly embedded in $\mathbf{P}^n$, which is certainly algebraic, by CHOW's theorem (IX, 1): the principal classes of $\mathfrak{L}$ are now the plane sections of the manifolds and hence they are *algebraic.*

c) Let us call *effective* a p-form φ, $p \leq n$, if $\omega^* \varphi = 0$ or, which is the same if $\Lambda\,\varphi = 0$, and *simple of class k* if $\varphi = \omega^k \wedge \psi$ where ψ is effective.

Then if φ is of class k we have the HODGE formula:

$$^*\varphi = u\,\omega^{n-r} \wedge C\,\varphi\,, \tag{15}$$

where u is *a numerical non-zero factor depending on n, p and k.*

If $p \leq n-2$, it follows that $\omega^*(\omega \wedge \varphi) = u'\,\omega \wedge \omega^{n-p-2} \wedge C\,\omega \wedge \wedge\,\varphi = u''\,{}^*\varphi$ with u' and u'' numerical non-zero factors. Therefore:

(i) $\omega \wedge \varphi = 0$, *if and only if* $\varphi = 0$,

and, by a simple inductive argument,

(ii) *Every p-form has a unique decomposition into a sum of single forms of class $k = 0, 1, \ldots, q = [p/2]$, as the*

$$\varphi = \sum_0^q {}_i\, \omega^i \wedge \psi_i\,. \tag{16}$$

(iii) *The operator $\omega \wedge$ is an isomorphism of the group of all p-forms into the group of all $(p+2)$-forms.*

d) The operators Δ and $\omega^k \wedge$ are *permutable:* hence, taking into account the commutativity of Δ with $*$, C, Λ, it follows that all results in (c) are still true when they are referred to the group $\mathbf{H}^p$, *which for $p \leqq n$, admits,* therefore, *the following structure:*

$$\mathbf{H}^p = \sum_p^{p-2q} {}_i\, \omega^i \wedge \mathbf{H}_0^{p-2i}\,, \tag{17}$$

where $\mathbf{H}_0^{p-2i}$ is *the group of effective harmonic $(p - 2i)$-forms.*

Moreover $\omega \wedge$ is an *isomorphism* of $\mathbf{H}^p$ into $\mathbf{H}^{p+2}$: it follows that $b^p - b^{p-2}$ is the rank of $\mathbf{H}_0^p$ or, more generally, $b^{p-2k} - b^{p-2k-2}$ the one of the group $\mathbf{H}_k^p$ of the harmonic p-forms of class k.

e) Since $*\omega^k = u \cdot \omega^{n-k}$, $D^*\,\omega^k$ is, except for a constant non-zero factor, the principal homology class Z_{2k} of $\mathfrak{M}$. Each homology class z_p of $\mathfrak{M}$, $p \leqq n$, may be written as {see (IX, 4 b)}

$$Z_p = D^*\,\varphi = D^*\left(\sum_0^q {}_k\, \omega^k \wedge \psi_k\right).$$

Then, apart from a constant non-zero factor,

$$\begin{aligned} D^*\,\omega^k \wedge \psi_k &= D(C\,\omega^k \wedge \psi_k \wedge \omega^{n-p}) = C\,\omega^k \wedge \psi_k \wedge \omega^{n-p} \cap \mathfrak{M} \\ &= C\,\omega^k \wedge \psi_k \cap (\omega^{n-p} \cap \mathfrak{M}) \\ &= C\,\omega^k \wedge \psi_k \cap D\,\omega^{n-p} = C\,\omega^k \wedge \psi_k \cap Z_{2p}\,. \end{aligned}$$

Therefore we obtain the theorems:

(iv) *All homology classes of dimension $p \leqq n$ lie in the principal class Z_{2p}, in the sense that they are cut on Z_{2p} by some cycle of $\mathfrak{M}$.*

(v) *Each homology class z_p, $p \leqq m$, has a unique decomposition into a sum of the type*

$$Z_p = \sum_0^q {}_k\, Z_p^{[k]}\,, \quad q = [p/2]\,, \tag{18}$$

where

$$Z_p^{[k]} = D^*\,\omega^k \wedge \psi_k \tag{19}$$

lies on $Z_{2(p-k)}$, but not, generally, on Z_{2l}, $l < p - k$.

The cycles in $z_p^{[0]}$ are called *effective p-cycles* and those in $z_p^{[k]}$ are called *simple p-cycles of class k.*

This is LEFSCHETZ's classification of p-cycles on $\mathfrak{M}$, which has been obtained by this author, by using direct topological tools, for algebraic varieties only: the above proof, valid for any KÄHLER variety, is due to ECKMANN.

The LEFSCHETZ results, however, are stronger since they use integral homology and not, as here, complex homology, though this refinement may be attained within the frame work of harmonic integrals, by means of the fact that *the principal homology class Z is, for algebraic varieties, an integral class, as follows immediately from* (IX, 6b): moreover, as we shall see in (X, 10, i), the last property is characteristic of such KÄHLER varieties {see HODGE [9]; KODAIRA [6]}.

7. HODGE's Birational Invariants

{See HODGE [8, 9]; HODGE [a], Ch. IV.}

We shall now consider how the above results depend *on the* KÄHLER *metric* with which our complex manifold is endowed.

a) From (IX, 4 e) and the definition of the operator D^* follows the useful formula:

$$D^* \psi^p [\varphi^p] = (\varphi^p, \psi^p) , \tag{20}$$

giving the period of the harmonic form φ^p on the p-cycle $D^* \psi^p$.

It is not difficult to prove that *harmonic forms of different type or of different class have zero scalar-product,* i. e. they are *orthogonal.* The former fact follows immediately by counting the number of the differentials dz^α and $d\bar{z}^\alpha$ appearing in the scalar product $\varphi^p \wedge *\psi^p$ and the latter using the reduction formula {see GUGGENHEIMER [1]}

$$(\varphi^{p-2} \wedge \omega, \psi^p) = (\varphi^{p-2}, \Lambda \psi^p) , \tag{21}$$

besides the definitions.

Hence and from (20) we derive that the period of the harmonic p-form φ on the cycle Γ^p vanishes, *unless the cycle and the form have the same type and class,* where the type and the class of the cycle Γ^p are defined as equal to those of the form ψ^p such that: $\Gamma^p = D^* \psi^p$.

b) Harmonic forms, and hence the corresponding classification of the cycles on the manifold $\mathfrak{M}$, depend on the choice of the KÄHLER metric, which, on a complex manifold admitting such a metric, is obviously not unique. However, as HODGE has proved, there are some characters of the manifold which are invariant under this choice.

Let us call $\varrho^{r,s}$ the rank of $\mathbf{H}_0^{r,s}$, i. e. of the group of harmonic effective p-forms of types (r, s). Then, by (IX, 6 d)

$$\sum_{r+s=p} \varrho^{r,s} = b^p - b^{p-2} , \tag{22}$$

and, as we find by taking conjugate forms,

$$\varrho^{r,s} = \varrho^{s,r} . \tag{23}$$

We shall now prove that the numbers $\varrho^{r,s}$ are invariant with respect to the KÄHLER metric.

For this, let $\mathfrak{C}^p$ be an independent basis for the p-cycles in the given KÄHLER ω-metric and let $\mathfrak{F}'^p$ be an independent basis for harmonic p-forms in a different KÄHLER ω'-metric both on $\mathfrak{M}$: clearly $\mathfrak{C}^p$ and $\mathfrak{F}'^p$ consist of b^p elements.

Let $\Gamma_i^p \in \mathfrak{C}^p$ and $\varphi_j'^p \in \mathfrak{F}'^p$, $i, j = 0, 1, \ldots, b^p$. Then, as in (a) we may prove that $\Gamma_i^p[\varphi_j'^p] = 0$, if the cycle and the form are of different type. But the matrix $\|\Gamma_i^p[\varphi_j'^p]\|$, $i, j = 0, 1, \ldots, b^p$, is not singular, since it is the same as the matrix $\|\Gamma_j'^{2n-p} \cdot \Gamma_i^p\|$, where $\Gamma_j'^{2n-p} = D\,\varphi_j'^p$, of two dual independent basis for cycles on $\mathfrak{M}$: on the other hand this is possible if and only if the number of forms in $\mathfrak{F}'^p$ of type (r, s) is the same as the number of cycles in $\mathfrak{C}^p$ of the same type.

Consequently:

$$\sum_{h=0}^{\min(r,s)} \varrho'^{r-h,\,s-h} = \sum_{h=0}^{\min(r,s)} \varrho^{r-h,\,s-h},$$

where $\varrho'^{r,s}$ is the analogous character of $\varrho^{r,s}$ for the ω'-metric. By applying this formula to the dimension $p-2$, we obtain

$$\sum_{h=1}^{\min(r,s)} \varrho'^{r-h,\,s-h} = \sum_{h=1}^{\min(r,s)} \varrho^{r-h,\,s-h},$$

and, then, by subtracting, we find $\varrho'^{r,s} = \varrho^{r,s}$.

Hence it follows easily that the numbers $\varrho^{r,s}$ are *birational invariants of an algebraic non-singular variety under biregular transformations:* in particular, by taking harmonic p-forms of type $(p, 0)$ i. e. the p-differentials of the first kind {see (IX, 5 b)}, which are obviously all effective, it follows that the number $\varrho^{p,0}$ *is simply the geometric genus* $p_g(V)$ of the algebraic variety V as it has been defined in (IV, 4): thus its *relative invariance is again established.*

8. Miscellaneous Results

a) Let φ be an arbitrary simple differential of the first kind on $\mathfrak{M}$. Then the integral $F(z) = \int^z \varphi$ is a *many-valued holomorphic function* on $\mathfrak{M}$ with $\varphi = dF$. Such a function is known as a *simple* PICARD *integral of the first kind.* If φ is simple and holomorphic but not of the first kind, the same integral, and the differential form itself, is called *of the second or of the third kind* according as it is *locally single-valued or not.*

For $n = 1$, i. e. on a complex KÄHLER curve, PICARD integrals are usually called Abelian integrals.

There is an important classical relation which, for algebraic varieties, goes back to CASTELNUOVO and SEVERI, between the first BETTI number b^1 and the number q of linearly independent differentials of the first kind {see CASTELNUOVO-ENRIQUES [2]; the proof given here is due to HODGE [a], p. 201; see also WEIL [1]}: it is found at once by taking $p = 1$ in the formula (22) of (7 b). Thus $b^1 = 2q$, where $q = \varrho^{1,0} = \varrho^{0,1}$.

We shall prove elsewhere {see (XI, 3)} that q coincides with *the first irregularity,* or superficial irregularity, of $\mathfrak{M}$, which has been defined in (VII, 8), when $\mathfrak{M}$ is algebraic.

b) Let $\{\varphi_{h,j}\}$ denote an independent basis for harmonic p-forms of a fixed class h and type $(p-j, j)$, $j = 0, 1, \ldots, p$, on an algebraic variety V, and let ω^j be the period matrix of the $\varphi_{h,j}$ and α the intersection matrix relative to the cycles $D^* \varphi_{h,j}$, $j = 0, 1, \ldots, p$.

Then a direct calculation establishes *the matrix relation*

$$\omega^j \alpha_{-1}^{-1} \overline{\omega}_{-1}^j = i^p (-1)^{p+h+j} \mu^j \tag{24}$$

where μ^j is *a positive definite Hermitian matrix* {see HODGE [a], p. 195; HODGE [9]}.

Especially important is the case $p = 1$, $h = 0$ and $j = 0$; here we deduce that the matrix $i\, \omega\, \alpha_{-1}^{-1} \overline{\omega}_{-1}$ is positive definite Hermitian and, since α_{-1}^{-1} is a skew-symmetric integral matrix, ω *is a* RIEMANN *matrix*, where now ω is *the period matrix of* PICARD *integrals of the first kind:* this is a well-known classical result related to the transcendental construction of the PICARD variety of $\mathfrak{M}$ {see (XI, 2)}.

c) If $h = 0$, the matrix α considered in (b) is the intersection matrix of the effective p-cycles; it is a square matrix of order $b^p - b^{p-2}$, symmetric when p is even.

The signature of α, which is obviously *a topological invariant of* $\mathfrak{M}$, is given, HODGE has proved {see HODGE [a], p. 201; HODGE [8, 9]}, by

$$(\varrho^{p,0} + \varrho^{p-2,2} + \cdots + \varrho^{0,p}, \quad \varrho^{p-1,1} + \varrho^{p-3,3} + \cdots + \varrho^{1,p-1}), \tag{25}$$

if $\mathfrak{M}$ is algebraic.

The proof follows readily from the matrix relation in (b) with $j = 0$. If (in particular) $\mathfrak{M}$ is a surface, the first signature character is $2\, p_g(V) + 1$, where $p_g(V) = \varrho^{2,0}$.

Thus the geometric genus of a surface is *a topological character* and, by (a), so also is $q - p_g(V)$, which, as we shall later see in (XI, 3), is equal to the arithmetic genus of the surface.

9. CHERN's Classes as Canonical Classes

We shall here establish an important relation between the concept of *covariant sequence of* B. SEGRE *and that of characteristic sequence in the topological sense of* CHERN.

For the definition of CHERN's classes see e. g. CHERN [1, 3]; STEENROD [a], p. 210; WU WEN-TSUN [a]. The proofs of this section are in KODAIRA [2], p. 319 and in VESENTINI [1, 2]. See also HODGE [11], and CHERN [3]. We take occasion to remark that EGER proved in [2] that a p-fold integral which is the direct product of p PICARD integrals of the first species is generally stationary along an algebraic set of p — 1-dimensions, belonging to a $(p-1)$-dimensional cycle of the canonical system on V. In particular, for $p = 1$, we obtain the SEVERI series, which was first defined in such a manner by SEVERI in [18] and then called the irregularity series of the variety. It is probable that the p-fold integrals of the first kind attached to V, even if of general type, are related in some manner to such a canonical system. Little is known on this subject for the case $p \neq 1$, $\dim V - 1$. See HODGE [2, 5]. Related works, which may be perhaps useful for this purpose, are: DE FRANCHIS [2]; KUNDERT [1, 2]; VESENTINI [3, 4, 5].

a) We begin by recalling that, if $\mathfrak{B}$ denotes a *unitary* $(2n-1)$-*sphere bundle* over a compact complex manifold $\mathfrak{M}$ and if $Y'_h = W_{n,n-h}$, $h = 0, 1, \ldots, n-1$, denotes *the complex* STIEFEL *manifold*, whose elements are the orthogonal h-frames in a complex n-space, then *the* $2q$*-th characteristic class of* $\mathfrak{B}$, $q = 1, 2, \ldots, n$, *in the sense of* CHERN, is the characteristic class, in the sense of HINDERNIS, of the bundle $\mathfrak{B}_{q-1}$ associated with $\mathfrak{B}$ whose fibre is Y'_{q-1}.

We denote it by $c_{2q}(\mathfrak{B})$: this is a $2q$*-cocycle* belonging to the $2q$-th cohomology group of $\mathfrak{M}$ with coefficients in *the first non-zero homotopy group of* Y'_{q-1}, which is the infinite cyclic group π_{2q-1}. It appears as *the obstruction cocycle* when we proceed to extend a continuous section of $\mathfrak{B}_{q-1}$ defined over the $(2q-1)$-dimensional skeleton of a complex $\mathfrak{R}$ covering $\mathfrak{M}$ to one over the $2q$-dimensional skeleton of the same complex. We shall moreover put $c_0(\mathfrak{M}) = 1$, where 1 denotes the unit of the group $\mathbf{H}^0(\mathfrak{M}, \mathbf{Z})$ over the integrals $\mathbf{Z}$.

The polynomial

$$c(x) = \sum_0^n {}_q\, c_{2q}\, x^q = \{c(\mathfrak{B}), x\}\,, \tag{26}$$

is called *the characteristic* CHERN *polynomial* associated with $\mathfrak{B}$ and sometimes it is more convenient to consider it as an infinite power series whose coefficients after the $(n+1)$-th are all zero.

If, in particular, $\mathfrak{B}$ is *the tangent sphere bundle to* $\mathfrak{M}$, then the characteristic classes of $\mathfrak{B}$ coincide with the so-called *characteristic classes or* CHERN *classes of* $\mathfrak{M}$.

We recall that if we write

$$C_{2n-2q} = c_{2q} \cap \mathfrak{M}\,, \tag{27}$$

where here $\mathfrak{M}$ denotes the fundamental n-cycle of $\mathfrak{M}$, then C_{2n-2q} is *the* $(2n-2q)$*-th characteristic homology class of* CHERN: it is consequently dual to $c_{2q}(\mathfrak{M})$ and of dimension $2(n-q)$.

b) Now let P be a *prime non-singular* divisor on $\mathfrak{M}$, which is supposed to be *algebraic and non-singular*, and let $\mathfrak{B}_n$ be the bundle consisting of all the vectors which are *unitary* and *normal* to P on $\mathfrak{M}$: this is obviously a unitary 1-sphere bundle over P, whose unique non-zero CHERN class, besides the unity class, is the c_2, which, following KODAIRA, we are going to determine.

To this end let f be a meromorphic function on $\mathfrak{M}$ having at P a simple zero-divisor and put

$$D = P - (f)\,.$$

Using f we define on $P - D \cdot P$, the following continuous field of unitary normal vectors n^α:

$$n^\alpha = \sum_\beta g^{\alpha\beta^*}\, \overline{\partial_\beta f} \Big/ \Big(\sum_{\gamma,\beta} g^{\gamma\beta^*}\, \partial_\gamma f\, \overline{\partial_\beta f}\Big)^{1/2}, \quad \partial_\alpha = \partial/\partial z^\alpha\,.$$

By taking at any point $p \in P$ local coordinates such that the equation of P, locally at p, is of the form $z_p^n = 0$, we find $f = z_p^n \cdot g_p$, where $(g_p) = -D$ locally at p. Therefore

$$n^\alpha = (\bar{g}_p / |g_p|) \cdot n_p^\alpha \,, \quad n_p^\alpha = g^{\alpha n^*} / (g^{n n^*})^{1/2} \,.$$

Let us now introduce a simplicial decomposition $\mathfrak{K}$ of P, whose 1-dimensional skeleton does not intersect $D \cdot P$, and such that every simplex of $\mathfrak{K}$ lies in the domain of at least one system of local coordinates.

Then, for each 2-simplex of $\mathfrak{K}$, one may define the continuous mapping

$$\mu : z \to \overline{g_p(z)} / |g_p(Z)| \,,$$

of the boundary ∂T of T onto the unit circle u on the plane of a complex variable. The transform of the 1-cycle ∂T by μ is then homologous to an integral multiple $\nu(T) \cdot u$ of the 1-cycle u. Associating with each simplex T the corresponding integer $\nu(T)$, we find a 2-cocycle which belongs to c_2.

But it is clear that:

$$\nu(T) = \frac{1}{2\pi i} \int_{\partial T} d \log \bar{g}_p = I(D, T) \,,$$

where $I(D, T)$ denotes the KRONECKER index. Hence the homology class dual to that of $c_2(\mathfrak{M})$ is that of the $(n-2)$-cycle $D \cdot P$, or, since $D \equiv P$, of $P^{[2]}$.

Recalling that in our case $\{\tilde{\pi}(\mathfrak{M})\} = \{\pi, \pi^{[2]}, 0, 0, \ldots\}$ {see (VIII, 5b)}, we have at once that, *when P is a non-singular $\mathfrak{M}$-divisor, the* SEGRE *series $\{\tilde{\pi}(\mathfrak{M}), x\}$ and the* CHERN *series $\{c(\mathfrak{B}_n), x\}$ are topologically dual.*

c) We shall now extend the above result to the general case where P is any non-singular subvariety of $\mathfrak{M}$, following the proof given by VESENTINI.

Let $\mathfrak{B}_n$ denote the bundle of the unitary orthogonal vectors to P in $\mathfrak{M}$ and let us assume the existence of a prime non-singular divisor A on $\mathfrak{M}$ passing through P.

We wish to prove that:

(i) *The series $\{\tilde{\pi}(\mathfrak{M}), x\}$ and $\{c(\mathfrak{B}_n), x\}$ are topologically dual on $\mathfrak{M}$.*

Since the theorem is true when $v - p = 1$, we may proceed by induction with respect to $v - p$, supposing our statement true for all positive integers less than $v - p$. Therefore if $\overline{\mathfrak{B}}_n$, $\mathfrak{B}_n^2$ are the respective bundles of the unitary normal vectors to A in M and to P in A, and if we denote by $\mathfrak{B}_n^1$ the restriction of $\overline{\mathfrak{B}}_n$ to P, then $\mathfrak{B}_n$ is evidently the product of $\mathfrak{B}_n^1$ and $\mathfrak{B}_n^2$.

Hence, by the duality theorem of CHERN {see CHERN [3]}, we derive:

$$\{c(\mathfrak{B}_n), x\} = \{c(\mathfrak{B}_n^1), x\} \cdot \{c(\mathfrak{B}_n^2), x\} \,, \tag{28}$$

the product being defined by the *cup-product* between the coefficients.

On the other hand, putting

$$\{\tilde{\alpha}(\mathfrak{M}), x\} \cdot P = \{\tilde{\alpha}'(\mathfrak{M}), x\},$$

we have by (VIII, 6, i) the relation

$$\{\tilde{\pi}(\mathfrak{M}), x\}_P = \{\tilde{\pi}(A), x\} \cdot \{\tilde{\alpha}'(\mathfrak{M}), x\}, \tag{29}$$

where the product is, instead, the *intersection-product* between equivalence classes of $A(P)$. Our inductive hypothesis gives

$$D_P\{c(\mathfrak{B}_n^2), x\} = \{\tilde{\pi}(A), x\}, \tag{30}$$

where D_P is the topological duality on P.

Moreover (b) yields

$$D_A\{c(\overline{\mathfrak{B}}_n), x\} = \{\tilde{\alpha}(\mathfrak{M}), x\},$$

where D_A is the topological duality on A, and hence it easily follows that

$$D_P\{c(\mathfrak{B}_n^1), x\} = \{\tilde{\alpha}'(\mathfrak{M}), x\}. \tag{31}$$

Now from (30) and (31), and taking into account (28) and (29), we derive

$$D_P\{c(\mathfrak{B}_n), x\} = \{\tilde{\pi}(\mathfrak{M}), x\}.$$

Then, if c_{2q}, $q = 1, 2, \ldots, v - p$, is the $2q$-th CHERN class of the unitary $(2v - 2p - 1)$-sphere bundle normal to P in $\mathfrak{M}$ and if $C_{2p-2q} = c_{2q} \cap P$, we have $C_{2p-2q} = \tilde{\pi}_q(\mathfrak{M}), q = 1, 2, \ldots, p$, which proves our statement. The additional hypothesis of the existence of an A as above may be eliminated, as VESENTINI has proved, by modifying suitably the preceding proof {see VESENTINI [4]}.

d) From (i), by recalling the definition of the canonical sequence $\{\tilde{y}^*\}$ of the variety $\mathfrak{M} = V$ {see (VIII, 7)} and observing that the tangent sphere bundle to V is isomorphic to the unitary normal sphere bundle to the diagonal variety of $V \times V$ in this product, we have at once:

(ii) $C_{2v-2i} = (-1)^i\, y_i^*$, *where* C_{2v-2i} *denotes the i-th homology class of* CHERN *for* V *and* y_i^* *the homology class on* $\mathfrak{M}$ *to which belongs the* SEGRE *equivalence class* y_i^*.

This theorem had been conjectured by HODGE {see HODGE [11]} and proved by him when V is a complete intersection of prime non-singular divisor of a projective space or when V admits suitable integrals.

The fact that the characteristic classes of V admit algebraic representatives has also been proved by CHERN by means of an interesting restatement of the definition of such classes {see CHERN [3]}.

X. The Applications of Stack Theory to Algebraic Geometry

1. Complex Line Bundles

{See e. g. KODAIRA [3] and SERRE [3]; the notion of complex line bundle and its association with a divisor is due to A. WEIL in [2].}

Let V be a compact KÄHLER variety of complex dimension n. We denote by F *a complex line bundle over* V, i. e. an analytic fibre bundle whose fibre is a complex line $\mathbf{C}^1$ and whose structure group is the multiplicative group $\mathbf{C}^*$ of complex numbers acting on $\mathbf{C}^1$.

The bundle F may be described as follows: Let $\{U_j\}$ be a sufficiently fine finite covering of V and let π be the canonical projection of F onto V. Then we have $\pi^{-1}(U_j) = U_j \times \mathbf{C}^1$, and $(z, \zeta_j) \in U_j \times \mathbf{C}^1$ is identical with $(z, \zeta_k) \in U_k \times \mathbf{C}^1$ if and only if $\zeta_j = f_{jk}(z) \cdot \zeta_k$, where $f_{jk}(z)$ is a non-vanishing holomorphic function defined in $U_j \cap U_k$.

The bundle F is then said to be defined by the system $\{f_{jk}\}$ of the transition functions f_{jk}, and ζ_j are called the fibre coordinate at the point (z, ζ_j) on F over the neighbourhood U_j. Clearly we have

$$f_{jk}\, f_{kj} = 1,\; f_{jk}\, f_{kl}\, f_{lj} = 1\,. \tag{1}$$

Two bundles F, F' defined, respectively, by f_{jk}, f'_{jk} are said to be *analytically equivalent* if there exist non-vanishing holomorphic functions f_j defined, respectively, in U_j such that

$$f'_{jk} = f_j\, f_k^{-1}\, f_{jk}\,. \tag{2}$$

Then we shall identify F' with F. For any pair of bundles F, G determined, respectively, by f_{jk}, g_{jk}, we define $F + G$ as the bundle determined by $f_{jk}\, g_{jk}^{-1}$. Then the set $\mathfrak{F} = \{F\}$ of all complex line bundles can be regarded as *an additive group*, whose zero element is *the trivial bundle* $\mathbf{C}^1 \times V$.

Given a divisor D on V defined in each U_j by a local meromorphic function $R_j(D) = R_j(z; D)$, the system of functions

$$f_{jk}(D) = R_j(D)/R_k(D)\,, \tag{3}$$

defines a complex line bundle over V which will be denoted by $\{D\}$, two such bundles being equivalent *if and only if the associated divisors D and D' are linearly equivalent.*

Thus each linear equivalence class is an element of $\mathfrak{F}$ and *the divisor class group can be regarded as a subgroup of* $\mathfrak{F}$.

2. The Stack Concept

{The notion of stack *(faisceau)* is due originally to J. LERAY [1, 2], but it has been closely scrutinised, extended and applied by CARTAN and his school; see H. CARTAN [a, b, c]; CARTAN-EILENBERG [a]; CARTAN [2].}

We recall briefly the CARTAN definition of a stack.

A *stack* over V is a set $\mathfrak{S}$ endowed with a projection π of $\mathfrak{S}$ onto V and with the following two structures:

(1) For each $z \in V$, $\pi^{-1}(z) = \mathfrak{S}_z$ is a *module* over the complex field $\mathbf{C}$;

(2) $\mathfrak{S}$ is a topological space such that: (a) The composition laws of $\mathfrak{S}$ related to the $\mathbf{C}$-module structure of $\mathfrak{S}_z$ are *continuous* when they are defined; (b) π is a *local homomorphism* of $\mathfrak{S}$ onto V.

A *section* s of $\mathfrak{S}$ over a neighbourhood $U \subset V$ is a continuous map $U \to \mathfrak{S}$ such that $\pi \circ s = 1$: the set of all sections over U is evidently a $\mathbf{C}$-module denoted by $\Gamma(\mathfrak{S}, U)$ and if $U' \subset U$ there is the restriction homomorphism $\Gamma(\mathfrak{S}, U) \to \Gamma(\mathfrak{S}, U')$. Therefore $\mathfrak{S}_z$ is the *direct limit* of the modules $\Gamma(\mathfrak{S}, U)$ related to the neighbourhoods of z.

Conversely, suppose that with each neighbourhood U of a fundamental set for V is associated a $\mathbf{C}$-module $\mathfrak{S}_U$ endowed with a homomorphism $f_{U'U}: \mathfrak{S}_U \to \mathfrak{S}_{U'}$ for each pair (U, U') with $U' \subset U$, such that $f_{U''U} = f_{U''U'} \circ f_{U'U}$ when $U'' \subset U' \subset U$. Then define $\mathfrak{S}_z$ as the direct limit of $\mathfrak{S}_U$ over the neighbourhoods $U \supset z$ and introduce into the union $\mathfrak{S}$ of the $\mathfrak{S}_z$ the topology determined by the fundamental system of sets everyone of which consists of the transforms onto $\mathfrak{S}_z$, for any $z \in U$, of an element $\alpha \in \mathfrak{S}_U$: the set $\mathfrak{S}$ is then a stack.

This generation may be immediately applied to the case where we take as $\mathbf{C}$-module $\mathfrak{S}_U$: (1) *The set of holomorphic functions on* U; (2) *The set of meromorphic functions on* U. Hence we obtain respectively the stacks denoted by $\mathfrak{O}$ and $\mathfrak{M}$. The modules $\mathfrak{O}_z$ and $\mathfrak{M}_z$ are called *the modules of germs respectively of holomorphic or meromorphic functions.* We shall meet in the sequel with many other such examples.

A stack $\mathfrak{S}$ is said to be *analytic* if $\mathfrak{S}_z$ is an $\mathfrak{O}_z$-module, provided the map $(f, \alpha) \to f\alpha$ be *continuous* whenever $f \in \mathfrak{O}_z$ and $\alpha \in \mathfrak{S}_z$: such are obviously the stack $\mathfrak{O}^p$, $p \geqq 1$, of p germs of holomorphic functions when we define: $f(f_1, f_2, \ldots, f_p) = (ff_1, ff_2, \ldots, ff_p)$, and also any sub-stack $\mathfrak{J}$ of $\mathfrak{O}$ such that $\mathfrak{J}_z$ is an ideal of the ring $\mathfrak{O}_z$ (the latter is a so-called *ideal-stack*).

3. Cohomology Groups over a Stack

{See e. g. CARTAN [2].}

The importance of the stack concept derives largely from the fact that a stack $\mathfrak{S}$ can be used to give local coefficients of cohomology groups on V. We recall the definition of these groups $H^q(V, \mathfrak{S})$ for $q \geqq 0$ ($H^q(V, \mathfrak{S}) = 0$, if $q < 0$).

Let $\mathfrak{U} = \{U_j\}$ be a finite open covering of V, and let $\mathfrak{N}(\mathfrak{U})$ be its nerve. With each simplex $u = (U_{j_0}, U_{j_1}, \ldots, U_{j_q}) \in \mathfrak{N}(\mathfrak{U})$ we associate the open set $X(u) = U_{j_0} \cap U_{j_1} \cap \cdots \cap U_{j_q}$ where $X(u) \neq 0$. Given a stack $\mathfrak{S}$, a cochain f is a function of simplex u such that $f(u) \in \Gamma(\mathfrak{S}, X(u))$, where

$\Gamma(\mathfrak{S}, X(u))$ is the module of sections of $\mathfrak{S}$ over $X(u)$. The support of f is defined as the union of the supports of the values $f(u)$ for all $u \in \mathfrak{N}(\mathfrak{U})$.

The cochains then form a *graded module* with coboundary which we denote by $\mathbf{C}(\mathfrak{U}, S)$.

Let $\mathbf{Z}(\mathfrak{U}, \mathfrak{S})$ be the submodule of *cocycles*, $\mathbf{B}(\mathfrak{U}, \mathfrak{S})$ the submodule of *coboundaries*, and define

$$\mathbf{H}(\mathfrak{U}, \mathfrak{S}) = \mathbf{Z}(\mathfrak{U}, \mathfrak{S})/\mathbf{B}(\mathfrak{U}, \mathfrak{S}) \,. \tag{4}$$

We say that the covering $\mathfrak{V} = \{V_j\}$ *refines* $\mathfrak{U}$ if each V_j is contained in some U_i and we then write $\mathfrak{V} < \mathfrak{U}$. Now $\mathbf{H}^q(V, \mathfrak{S})$ *is defined to be the limit module of* $\mathbf{H}^q(\mathfrak{U}, \mathfrak{S})$ *over all finite open covering of* V. Obviously $\mathbf{H}^0(V, \mathfrak{S})$ remains canonically identified *with the section group* $\Gamma(V, \mathfrak{S})$.

Each homomorphism of stacks over V $\mathfrak{S} \to \mathfrak{S}'$ — i. e. a continuous map of $\mathfrak{S}$ into $\mathfrak{S}'$ inducing a homomorphism $\mathfrak{S}_z \to \mathfrak{S}'_z$ for each $z \in V$ — defines evidently a *homomorphism* $\mathbf{H}^q(V, \mathfrak{S}) \to \mathbf{H}^q(V, \mathfrak{S}')$, or, more precisely, $\mathbf{H}^q(V, \mathfrak{S})$ is a covariant functor of $\mathfrak{S}$.

Now, if $\mathfrak{T}$ is a substack of $\mathfrak{S}$, we can define a *natural homomorphism*

$$\delta^q : \mathbf{H}^q(V, \mathfrak{S}/\mathfrak{T}) \to \mathbf{H}^{q+1}(V, \mathfrak{T}) \,, \tag{5}$$

where $\mathfrak{S}/\mathfrak{T}$, called a *quotient stack*, is the union of the group $\mathfrak{S}_z/\mathfrak{T}_z$ endowed with the quotient topology.

We may now enunciate the fundamental property of the stacks:

(i) *The sequence of groups and homomorphisms:*

$$0 \to \mathbf{H}^0(V, \mathfrak{T}) \to \mathbf{H}^0(V, \mathfrak{S}) \to \mathbf{H}^0(V, \mathfrak{S}/\mathfrak{T}) \xrightarrow{\delta^0} \mathbf{H}^1(V, \mathfrak{T}) \to \cdots \to$$
$$\to \mathbf{H}^q(V, \mathfrak{T}) \to \mathbf{H}^q(V, \mathfrak{S}) \xrightarrow{\delta^q} \mathbf{H}^{q+1}(V, \mathfrak{S}/\mathfrak{T}) \to \cdots$$

is an exact sequence, i. e. the image of each homomorphism is the kernel of the following homomorphism.

4. A Theorem of DOLBEAULT

{This theorem is contained in DOLBEAULT's work [1]; the proof of KODAIRA is given in [3].}

We outline here KODAIRA's method of deriving an important theorem of DOLBEAULT by extending suitable the theory of differential forms on V.

Let F be a complex line bundle over V {see (X, 1)}. By a form φ of type (r, s), of class C^∞, *with coefficients in* F, we shall mean a system $\{\varphi_j\}$ of exterior differential forms φ_j of type (r, s), of class C^∞, defined, respectively, in U_j such that $\varphi_j = f_{jk} \cdot \varphi_k$ in $U_j \cap U_k$: we denote by $\Gamma^{r,s}(F)$ the linear space of these forms. Setting $a_{jk} = |f_{jk}|^2$, we consider the principal bundle over V defined by the system $\{a_{jk}\}$ and with structure group the multiplicative group of positive real numbers.

This bundle is topologically trivial and therefore we can find a system $\{a_j\}$ of positive real functions a_j, of class C^∞, defined, respectively, in U_j, such that

$$|f_{jk}|^2 = a_j/a_k\,, \quad \text{in} \quad U_j \cap U_k\,.$$

Now, for any $\varphi = \{\varphi_j\} \in \Gamma^{r,s}(F)$, we define $\bar\partial\varphi = \{(\bar\partial\varphi)_j\}$ and $\mathfrak{d}\varphi = \{(\mathfrak{d}\,\varphi)_j\}$ by taking

$$\left.\begin{aligned}(\bar\partial\,\varphi)_j &= \bar\partial\,\varphi_j\,,\\ (\mathfrak{d}\,\varphi)_j &= -\,{}^*a_j\,\partial\Big(\frac{1}{a_j}\,{}^*\varphi_j\Big),\end{aligned}\right\} \tag{6}$$

where ∂ or $\bar\partial$ denotes the exterior derivative with respect to the variables z^α or $\bar z^\alpha$. The operators $\bar\partial$ and $\mathfrak{d}$, intrinsically defined, map the group $\Gamma^{r,s}(F)$ respectively into the groups $\Gamma^{r,s+1}(F)$ and $\Gamma^{r,s-1}(F)$.

Next we introduce the complex LAPLACE-BELTRAMI operator

$$\Box = \mathfrak{d}\,\bar\partial + \bar\partial\,\mathfrak{d}\,. \tag{7}$$

By defining conveniently the inner product (φ, ψ), with respect to which $\mathfrak{d}$ is the formal adjoint of $\bar\partial$, and denoting $\mathbf{H}^{r,s}(F)$ the subspace of $\Gamma^{r,s}(F)$ consisting of all solutions of $\Box\,\varphi = 0$, we have the decomposition formula

$$\Gamma^{r,s}(F) = \bar\partial\,\Gamma^{r,s-1}(F) \oplus \mathfrak{d}\,\Gamma^{r,s+1} \oplus \mathbf{H}^{r,s}(F)\,, \tag{8}$$

where $\oplus$ denotes a direct sum of mutually orthogonal linear subspaces: the proof is a generalisation of the usual one for the operators d and δ.

Moreover it is known that $\Box\,\varphi = 0$, which is a partial differential equation of elliptic type with the principal part $-\sum g^{\alpha\beta^*}\,\partial^2/\partial z^\alpha\,\partial\bar z^\beta$ on V, has a finite number of linearly independent solutions. Then $\mathbf{H}^{r,s}(F)$ is *a finite group*, called *the group of harmonic forms of type* (r, s) *with coefficients in* F.

We can now consider the module of p-forms with coefficients in F and holomorphic over a subset U_j of V, such that $\mathfrak{U} = \{U_j\}$ is a finite open covering of V and then apply the process described in (X, 2), so obtaining *the stack* $\Omega^p(F)$ *of germs of holomorphic* p*-forms with coefficients in* F.

From the definitions in (X, 3) and from what has been said above, we have immediately *the isomorphism*

$$\mathbf{H}^q(V, \Omega^p(F)) \cong \mathbf{Z}^{p,q}(F)/\bar\partial\,\Gamma^{p,q-1}(F)\,, \tag{9}$$

where $\mathbf{Z}^{p,q}(F)$ is the subspace of cocycles of $\Gamma^{p,q}(F)$ with respect to the $\bar\partial$-cohomology and $\bar\partial\,\Gamma^{p,q-1}(F)$ the subgroup of coboundaries.

From (8) we infer that $\mathbf{Z}^{p,q}(F) = \bar\partial\,\Gamma^{p,q-1}(F) \oplus \mathbf{H}^{p,q}(F)$. This yields the isomorphism

$$\mathbf{Z}^{p,q}(F)/\bar\partial\,\Gamma^{p,q-1} \cong \mathbf{H}^{p,q}(F)\,, \tag{10}$$

and, finally, we have the following theorem of DOLBEAULT:

(i) *The groups* $\mathbf{H}^q(V; \Omega^p(F))$ *and* $\mathbf{H}^{p,q}(F)$ *are isomorphic.*

The original theorem of DOLBEAULT referred to the case $F = \mathbf{C}^1 \times V$; a different, more intrinsic, proof has been given by SERRE {see SERRE [3], p. 15}.

From (i) it follows that the group $\mathbf{H}^q(V; \Omega^p(F))$ *is finite.* Moreover, by observing that the mapping

$$\{\varphi_j\} \to \left\{\frac{1}{a_j} {}^*\overline{\varphi}_j\right\} \tag{11}$$

maps $\mathbf{H}^{p,q}(F)$ isomorphically onto $\mathbf{H}^{n-p,n-q}(-F)$, we have *the* SERRE *isomorphism* {see SERRE [3], p. 22}

$$\mathbf{H}^q(V; \Omega^p(F)) \cong \mathbf{H}^{n-q}(V; \Omega^{n-p}(-F)) . \tag{12}$$

5. Positive Complex Line Bundles

{This notion is due to KODAIRA [4], where the reader will find further details.}

From the relation (1) it follows that

$$\log f_{jk} + \log f_{kl} + \log f_{lj} = 2\pi i\, c_{jkl}, \quad \text{in} \quad U_j \cap U_k \cap U_l, \tag{13}$$

where c_{kjl} is a constant rational integer. The system of the integers c_{jkl} defines a 2-cocycle on the nerve $\mathfrak{N}(\mathfrak{U})$ and therefore determines an element c_N of the group $\mathbf{H}^2(\mathfrak{N}; \mathbf{Z})$ with integer coefficients $\mathbf{Z}$. Proceeding to the direct limit with respect to the order $<$ between finite open coverings {see (X, 3)} we obtain a well-defined element $c(F) \in \mathbf{H}^2(V; \mathbf{Z})$.

This element is the characteristic class of the principal bundle associated with F — which is obtained by removing the origin $\zeta_j = 0$ of each fibre — and it is also called the characteristic class of F. Clearly the mapping $F \to c(F)$ is a homomorphism of the group F into $\mathbf{H}^2(V; \mathbf{Z})$.

KODAIRA and SPENCER have then shown that:

(i) *The image of this homomorphism is composed of those cohomology classes of* $\mathbf{H}^2(V; \mathbf{Z})$ *which, regarded as classes* $c_{\mathbf{R}}(F)$ *of* $\mathbf{H}^2(V; \mathbf{R})$ ($\mathbf{R}$ *real numbers), contain a closed real* (1, 1)*-form* γ.

Naturally here any cohomology class is regarded, in view of the DE RHAM theorem {see (IX, 4)}, as *a class of d-closed forms.*

A real (1, 1)-form γ on V can be written as

$$\gamma = i \sum \gamma_{\alpha\beta^*}(z, \bar{z})\, dz^\alpha \wedge d\bar{z}^\beta , \tag{14}$$

where $\overline{\gamma}_{\alpha\beta^*} = \gamma_{\beta\alpha^*}$. Such a form γ is said to be *positive,* $\gamma > 0$, if the Hermitian form $\sum \gamma_{\alpha\beta^*}(z, \bar{z})\, u^\alpha \bar{u}^\beta$ in n variables $u^1, u^2, \ldots, u^n$ is positive definite at each point z on V. A complex line bundle F is called *positive,* $F > 0$, if its characteristic class $c_R(F)$ contains a closed real form $\gamma > 0$. Now KODAIRA has proved that

(ii) *If* $F > 0$, *the groups* $\mathbf{H}^q(V, \Omega^n(F))$ *and* $\mathbf{H}^{n-q}(V, \Omega^0(-F))$ *both vanish for* $1 \leq q \leq n$.

Defining the canonical bundle over V as the complex line bundle K associated with the canonical linear equivalence class, i. e. the bundle determined by the system J_{jk} of Jacobians $J_{jk} = \partial(z_k^1, \ldots, z_k^n)/\partial(z_j^1, \ldots, z_j^n)$, where $(z_j^1, \ldots, z_j^n)$ are local coordinates in U_j, we have

$$\Omega^0(F) \cong \Omega^n(F - K) . \tag{15}$$

Therefore from (ii) we deduce that

(iii) *The group* $\mathbf{H}^q(V, \Omega^0(F))$ *vanishes for* $1 \leq q \leq n$, *if* $F - K > 0$, *where now* $\Omega^0(F)$ *is the stack* $\mathfrak{O}$ *of germs of holomorphic functions of* F.

With this we end our introductory account, passing on to outline the principal applications to algebraic geometry.

6. The Picard Variety in Stack Theory

{This section summarises Kodaira-Spencer [2].}

The following elegant concept of the (first) Picard variety is due to Kodaira and Spencer; we shall later prove in (XI, 4) that, for an algebraic variety, it coincides with that defined in Ch. VIII.

a) Let $\mathbf{H}_{1,1}^2(V; \mathbf{Z})$ and $\mathfrak{P}$ be respectively the image and the kernel of the homomorphism $F \to c(F)$ defined in (X, 5). Then we can write:

$$\mathfrak{F}/\mathfrak{P} = \mathbf{H}_{1,1}^2(V; \mathbf{Z}) . \tag{16}$$

Now we wish to determine the structure of the group $\mathfrak{P}$. If $F \in \mathfrak{P}$, we may choose suitable rational integers c_{jk}, so that the functions $h_{jk} = \log f_{jk} + 2\pi i\, c_{jk}$ satisfy:

$$h_{jk} + h_{kl} + h_{lj} = 0 , \quad \text{in} \quad U_j \cap U_k \cap U_l . \tag{17}$$

Therefore the system $\{h_{jk}\}$ defines an element h_N of $\mathbf{H}^1(\mathfrak{N}; \Omega^0)$ where $\Omega^0 = \Omega^0(\mathbf{C} \times V)$, for each covering $\mathfrak{N}(\mathfrak{U})$. Thus we may proceed to obtain, by direct limit (X, 2) a well defined element $h \in \mathbf{H}^1(V; \Omega^0)$. By means of the Hodge isomorphism $\mathbf{H}^1(V) \to \mathbf{H}^1(V; \mathbf{C})$ (IX, 4 b) and the Dolbeault isomorphism of (X, 4) with $F = \mathbf{C}^1 \times V$, we can easily show that each bundle $F \in \mathfrak{P}$ is determined by an element $\bar{\alpha} \in \mathbf{H}^{0,1}(V)$ and that the map $\bar{\alpha} \to F$ is a homomorphism from $\mathbf{H}^{0,1}(V)$ onto $\mathfrak{P}$, whose kernel, i. e. the subset of $\mathbf{H}^{0,1}$ associated with the zero bundle $\mathbf{C}^1 \times V$, is the discrete subgroup $\mathfrak{I}$ of $\mathbf{H}^{0,1}(V)$ defined by

$$\mathfrak{I} = \{\omega | \omega = 2\pi i\, \pi_{0,1} H c , \quad c \in \mathbf{H}^1(V; \mathbf{Z})\} , \tag{18}$$

where $\pi_{0,1} = P_1$ {see (IX, 5)}.

The space $\mathbf{H}^{0,1}(V)$ is a complex vector space of dimension q {see (IX, 8 a)} and $\mathfrak{I}$ is a discrete subgroup of $\mathbf{H}^{0,1}(V)$ generated by $2q$ vectors which are linearly independent in R. Hence we have

$$\mathfrak{P} \cong \mathbf{H}^{0,1}(V)/\mathfrak{I} , \tag{19}$$

where the factor group is a complex torus of dimension q.

Moreover if $\bar{\alpha} = \sum_{\nu=1}^{q} t_\nu \alpha_\nu$, where $\{\alpha_\nu\}$, $\nu = 1, 2, \ldots, q$, is a base for the simple differentials of the first kind on V, it can be immediately seen that the bundle $F \in \mathfrak{P}$ depends analytically on the parameters $(t_1, t_2, \ldots, t_q)$ and thereby we can introduce a complex structure on $\mathfrak{P}$ by identifying $\mathfrak{P}$ with $\mathbf{H}^{0,1}(V)/\mathfrak{J}$. Then $\mathfrak{P}$ is called *the* PICARD *variety attached to the* KÄHLER *variety* V.

b) Up to now we have assumed that V was a compact KÄHLER variety: now let V be algebraic.

If $F = \{D\}$, the map {see (X, 1)}

$$\zeta_j(z) \times z \to \zeta_j(z)/R_j(z; D) \tag{20}$$

maps $\Omega^0(F) = \Omega(\{D\})$ isomorphically onto the stack $\mathfrak{M}_D^0 = M^0(D)$ of germs of meromorphic functions which are multiples of $-D$. Therefore

$$\mathbf{H}^q(V; \Omega(\{D\})) \cong \mathbf{H}^q(V; \mathfrak{M}_D^0)\,, \tag{21}$$

and

$$\dim \mathbf{H}^0(V; \Omega(\{D\})) = \dim|D| + 1\,. \tag{22}$$

If $|D - K| = |S|$, where $|S|$ is the system of hyperplane sections in its ambient space, *the condition for* (X, 5, iii) *is obviously satisfied and hence*

$$\mathbf{H}^q(V; \mathfrak{M}_D^0) = 0\,, \quad \text{for} \quad q \geqq 1\,. \tag{23}$$

Now we are able to describe the proof of the following theorem of KODAIRA-SPENCER which is fundamental for our subject:

(i) *Each element* $F \in \mathfrak{F}$ *is such that* $F = \{D\}$.

The proof is by induction on the dimension n of V. The sequence

$$0 \to \Omega(F - \{S\}) \xrightarrow{i} \Omega(F) \xrightarrow{r} \Omega(F_S) \to 0\,, \tag{24}$$

— where S is a generic hyperplane section, F_S is the restriction of F to S, i and r denote respectively the inclusion and restriction map — is an exact sequence. Hence, setting

$$F_m = F + \{S_m\} = F_{m-1} + \{S\}\,, \tag{25}$$

where $|S_m|$ is the m-ple of $|S|$, we find the exact sequence

$$0 \to \Omega(F_{m-1}) \xrightarrow{i} \Omega(F_m) \xrightarrow{r} \Omega(F_{m,S}) \to 0\,, \tag{26}$$

and hence the corresponding exact cohomology sequence:

$$\begin{aligned} 0 &\to \mathbf{H}^0(V, \Omega(F_{m-1})) \xrightarrow{i^*} \mathbf{H}^0(V, \Omega(F_m)) \xrightarrow{r^*} \mathbf{H}^0(S, \Omega(F_{m,S})) \xrightarrow{\delta^*} \\ &\xrightarrow{\delta^*} \mathbf{H}^1(V, \Omega(F_{m-1})) \xrightarrow{i^*} \mathbf{H}^1(V, \Omega(F_m)) \xrightarrow{r^*} \mathbf{H}^1(S, \Omega(F_{m,S})) \to \cdots\,. \end{aligned} \tag{27}$$

By the inductive hypothesis, there exists a divisor $\bar{D}$ on S such that $\{\bar{D}\} = F_S$ on S and then

$$F_{m,S} = F_S + \{S_m\}_S = \{\bar{D} + S_m \cdot S\}\,, \text{ on } S\,. \tag{28}$$

Now there exists an integer m_0 such that

$$\mathbf{H}^1(S, \Omega(F_{m,S})) = \mathbf{H}^1(S, M^0(\bar{D} + S_m \cdot S)) = 0, \quad m \geqq m_0, \tag{29}$$

being $|\bar{D} + S_m \cdot S|$ sufficiently ample for large m to apply a preceding observation. But the groups $\mathbf{H}^q(V, \Omega(F_m))$ have finite dimension by (X, 4), and therefore we obtain from (27):

$$\dim \mathbf{H}^0(S, \Omega(F_{m,S})) = \dim \mathbf{H}^0(V; \Omega(F_m)) - \dim \mathbf{H}^0(V, \Omega(F_{m-1})) + \\ + \dim \mathbf{H}^1(V, \Omega(F_{m-1})) - \dim \mathbf{H}^1(V, \Omega(F_m)), \quad m \geqq m_0 .$$

From here, by summing both sides with respect to m, we find easily $\dim \mathbf{H}^0(V, \Omega(F_m)) \to +\infty$ for $m \to +\infty$. This shows that, for large m, the bundle F_m has a global analytic section $\zeta_j(z) \times z$ different from the zero-section.

If D_m is the divisor locally defined by $\zeta_j(z) = 0$, it follows that $F_m = \{D_m\}$ and setting $D = D_m - S_m$, we obtain $F = \{D\}$. Moreover for $n = 1$, $\mathbf{H}^0(S, \Omega(F_{m,S})) = \Omega(F_{m,S}) = \Omega(S)$ is a linear space of dimension $s \geqq 1$, where s is the degree of S, and $\mathbf{H}^1(S, \Omega(F_{m,S})) = 0$ for any m: the proof is then obtained by applying the above argument for $n = 1$.

c) The identification of the bundle group $\mathfrak{F}$ with the linear divisor class group $\mathbf{G}/\mathbf{G}_l$ of an algebraic variety V prepares the way for some important facts.

Namely, it can be proved that *the homology class of the V-divisor D is the dual of the class* $c(\{D\})$, and hence, denoting by $\mathbf{G}_0$ the subgroup of $\mathbf{G}$ consisting of all V-divisors integrally homologous to zero, we have $D \in \mathbf{G}_0$ if and only if $\{D\} \in \mathfrak{P}$.

Thus we obtain the theorem:

(ii) *If $\mathfrak{P}$ is the* Picard *variety attached to V in the sense of* (a), *then* $\mathbf{G}_0/\mathbf{G}_l = \mathfrak{P}$.

This statement is known as *the first duality theorem of* Igusa {see Igusa [3]; A. Weil [4]}.

Again, we have the canonical isomorphism

$$\mathbf{H}^2_{1,1}(V; \mathbf{Z}) \cong \mathbf{H}^{1,1}_{2n-2}(V; \mathbf{Z}), \tag{30}$$

where the group on the right is the subgroup of $\mathbf{H}_{2n-2}(V; \mathbf{Z})$ consisting of all homology classes whose harmonic parts are of type (1, 1). Taking into account (a) we get

$$\mathbf{G}/\mathbf{G}_0 = \mathbf{H}^{1,1}_{2n-2}(V; \mathbf{Z}), \tag{31}$$

and this leads immediately to the following criterion for $(2n-2)$-cycles to be algebraic, due to Lefschetz and Hodge {see Hodge [a], p. 213; Hodge [4, 6, 7, 8]; Lefschetz [a], p. 82}:

(iii) *An integral $(2n-2)$-cycle Γ on an algebraic variety V is homologous to an (algebraic) divisor if and only if the closed $(2n-2)$-forms of type $(n, n-2)$ have zero periods on it.*

It follows at once that every torsion cycle is algebraic, or, in other words, denoting by $\mathbf{T}_q$ the q-th torsion group of V, we have

$$\mathbf{T}_{2n-2}(V) \subseteqq \mathbf{H}^{1,1}_{2n-2}(V;\mathbf{Z}) . \tag{32}$$

Therefore

$$\mathbf{G}_r/\mathbf{G}_0 = \mathbf{T}_{2n-2}(V) , \tag{33}$$

where $\mathbf{G}_r$ is the subgroup of $\mathbf{G}$ consisting of all divisors which are rationally homologous to zero. Recalling that $\mathbf{T}_{2n-2}$ is dual to $\mathbf{T}_1$, we arrive at the theorem:

(iv) *The factor group* $\mathbf{G}_r/\mathbf{G}_0$ *is dual to the first torsion group* $\mathbf{T}_1(V)$ *of* V, which is known as the second duality theorem of IGUSA {see IGUSA [3]; A. WEIL [4]}.

We observe that, by (i), the PICARD variety in the present sense will be identifiable with the PICARD variety defined in Ch. VIII, as soon as we have proved that *the groups of algebraic equivalence and of integral topological equivalence coincide on* V.

7. The Theorem of RIEMANN-ROCH for Adjoint Systems

Any statement giving the dimension of a linear system, arbitrary or also somewhat particular, on an algebraic variety V, as a function of some suitable characters of the variety and of the system itself, is called *a theorem of* RIEMANN-ROCH. Here we shall describe how KODAIRA and SPENCER have derived a theorem of this type *for adjoint systems* on V by using stack theory.

First progress in RIEMANN-ROCH's theorems for higher dimensional varieties is due to SEVERI [13] and afterwards to B. SEGRE [5]. Recently these theorems have been thoroughly investigated by KODAIRA [1, 2, 5]; SPENCER [2]; HIRZEBRUCH [1]; HODGE [13]; and by SERRE, whose works on the subject have not been published up to now, except for some hints in [3, 4]. We take the opportunity of observing that in [4] SERRE gives a synthesis of abstract and topological methods by using ZARISKI's topology and stack theory: this is undoubtedly a very useful idea for abstract algebraic geometry, and we regret that the report of this work is exceedingly lengthy.

This section expounds SPENCER [1]: HIRZEBRUCH's work will be referred to in (X, 9). For the notion of POINCARÉ and SEVERI residues see respectively: POINCARÉ [1] and SEVERI [4]. The result of LEFSCHETZ, quoted below, is in [a], pp. 88—91: see also ECKMANN [1, 2] and KODAIRA [5], pp. 91—96.

We begin by introducing *the* EULER *characteristics* $\chi^r = \chi(V, \Omega^r)$ and $\chi^r_D = \chi(V, \mathfrak{M}^r_D)$ relative to the algebraic non-singular variety V, with respect to the stack Ω^r of germs of holomorphic r-forms on V or to the stack $\mathfrak{M}^r_D$ of germs of meromorphic r-forms on V multiples of the divisor $-D$, i. e. such that $R_j(z; D) \cdot \tau_p$ is holomorphic at $p \in U_j$, where τ_p is such a germ at p.

Then $\mathfrak{M}^r_D \cong \Omega^r(\{D\})$ and $\Omega^r = \Omega^r(\mathbf{C}^1 \times V)$ with the symbols of (X, 4), and by (X, 4, i) all the cohomology groups $\mathbf{H}^s(V, \Omega^r)$ and $\mathbf{H}^s(V, \mathfrak{M}^r_D)$ are finite and moreover they vanish if $s < 0$ or if $s > n$.

Therefore we can define

$$\chi^r = \sum_{s=0}^{n} \mathbf{H}^s(V, \Omega^r) , \quad \chi^r_D = \sum_{s=0}^{n} \mathbf{H}^s(V, \mathfrak{M}^r_D) . \tag{34}$$

Suppose that the divisor D is a non-singular divisor S, represented locally at p by $z_p^1 = 0$, where $z_p^1, z_p^2, \ldots, z_p^n$ are local coordinates on V. Setting $z_p^i = \bar{z}_p^{i-n}$, $i = n+1, \ldots, 2n$, we associate with any s-form

$$\varphi = \sum_{i_1 < i_2 < \cdots < i_s} \varphi_{i_1 i_2 \cdots i_s} \, d z_p^{i_1} \wedge \cdots \wedge d z_p^{i_s} , \tag{35}$$

defined locally at p, the $(s-1)$-form

$$\nu_p \cdot \varphi = \sum_{1 < i_2 < \cdots < i_s} \varphi_{1 i_2 \cdots i_s} \, d z_p^{i_2} \wedge \cdots \wedge d z_p^{i_s} . \tag{36}$$

Now let $\tau_p \in \mathfrak{M}^r_S$: then τ_p is an $(r, 0)$-current and a calculation shows that

$$\bar{\partial} \tau_p[\psi] = 2\pi i \int_S \{\nu_p(z_p^1 \tau_p) \cdot \psi + (-1)^r (z_p^1 \tau_p) \cdot \nu_p \psi\} \tag{37}$$

where ψ is of type $(n-r, n-1)$, of class C^∞ and with a compact carrier in $U_p \supset p$. The divisor C on S associated with the bundle on S defined by the restriction h_{pq} to S of the functions z_p^1/z_q^1 in $U_p \cap U_q$, $p, q \in S$, is obviously the characteristic divisor of S in V.

Taking

$$\mathfrak{P}(\tau_p) = (\nu_p \cdot z_p^1 \tau_p)_S , \quad \mathfrak{S}(\tau_p) = (z_p^1 \tau_p)_S / h_p \tag{38}$$

which are known respectively as *the* POINCARÉ and SEVERI *residues* of τ_p, and setting

$$\mathfrak{R}_S(\tau_p) [\psi] = \int_S \{\mathfrak{P}(\tau_p) \cdot \psi + (-1)^r \mathfrak{S}(\tau_p) \, h_p \cdot \nu_p \psi\} \tag{39}$$

where $(h_p) = C$, we find the formula

$$\bar{\partial} \tau_p[\psi] = 2\pi i \, \mathfrak{R}_S(\tau_p) [\psi] . \tag{40}$$

The current $\mathfrak{R}_S(\tau_p)$ is called the total residue of τ_p on S. If $\mathfrak{R}^r_S$ is the residue stack obtained from $\mathfrak{M}^r_S$ by applying $\bar{\partial}$, we have the exact sequence

$$0 \to \Omega^r \xrightarrow{i} \mathfrak{M}^r_S \xrightarrow{\bar{\partial}} \mathfrak{R}^r_S \to 0 , \tag{41}$$

where i is the inclusion map.

Moreover the above formulae easily give the following canonical isomorphism:

$$\mathfrak{R}^r_S \cong \Omega^{r-1}(S) + \mathfrak{M}^r_C(S) . \tag{42}$$

From the exact cohomology sequence associated with (41), which is

$$\cdots \to \mathbf{H}^s(V, \Omega^r) \xrightarrow{i^*} \mathbf{H}^s(V, \mathfrak{M}^r_S) \xrightarrow{\partial^*} \mathbf{H}^s(V, \mathfrak{R}^r_S) \xrightarrow{\delta^*} \mathbf{H}^{s+1}(V, \Omega^r) \to \cdots \tag{43}$$

we obtain

$$\begin{aligned} &\dim \mathbf{H}^s(V, \mathfrak{M}^r_S) \\ &= \dim \mathbf{H}^s(V, \mathfrak{R}^r_S) - \dim \mathbf{H}^{s+1}(V,\Omega^r) + \dim\{\mathbf{H}^s(V,\Omega^r)/\delta^* \mathbf{H}^{s-1}(V,\mathfrak{R}^r_S)\} + \\ &\quad + \dim\{\mathbf{H}^{s+1}(V,\Omega^r)/\delta^* \mathbf{H}^s(V, \mathfrak{R}^r_S)\}\,, \end{aligned} \tag{44}$$

where (42) furnishes

$$\dim \mathbf{H}^s(V, \mathfrak{R}^r_S) = \dim \mathbf{H}^s(S, \Omega^{r-1}) + \dim \mathbf{H}^s(S, \mathfrak{M}^r_C)\,. \tag{45}$$

Taking in (44) $s = 0$, $r = n$ we have

$$\begin{aligned} &\dim \mathbf{H}^0(V, \mathfrak{M}^n_S) \\ &= \dim \mathbf{H}^0(V, \mathfrak{R}^r_S) - \dim \mathbf{H}^1(V, \Omega^n) + \dim \mathbf{H}^0(V, \Omega^n) + \\ &\quad + \dim\{\mathbf{H}^1(V, \Omega^n)/\delta^* \mathbf{H}^0(V, \mathfrak{R}^n_S)\}\,. \end{aligned} \tag{46}$$

Now $\dim \mathbf{H}^0(V, \mathfrak{M}^n_S) = \dim \mathbf{H}^0(V, \mathfrak{M}^0_{K+S}) = \dim |K + S| + 1$, $\dim \mathbf{H}^0(V, \mathfrak{R}^n_S) = \dim \mathbf{H}^0(S, \Omega^{n-1})$ since $\mathfrak{M}^n_C = 0$: hence $\dim \mathbf{H}^0(V, \mathfrak{R}^n_S) = \dim \mathbf{H}^{n-1,0}(S)$. Moreover $\mathbf{H}^{n-1,0}(V)$ is isomorphic to the group of harmonic forms on S of type $(n-1, 0)$, so that $\dim \mathbf{H}^0(V, \mathfrak{R}^n_S) = g_{n-1}(S)$, where $g_{n-1}(S)$ denotes the number of linearly independent q-fold differentials of the first kind on S. Similarly we can find $\dim \mathbf{H}^1(V, \Omega^n) = \dim \mathbf{H}^{n,1}(V) = \dim \mathbf{H}^{n-1,0}(V) = g_{n-1}(V)$ and $\dim \mathbf{H}^0(V, \Omega^n) = g_n(V)$.

Finally, *if* $n \geqq 2$, it can be proved that $\mathbf{H}^1(V, \Omega^n)/\delta^* \mathbf{H}^0(S, \Omega^{n-1})$ is isomorphic to the group of $(n-1)$-fold differentials of the first kind on V which vanish on S: if l is the dimension of this group we obtain, substituting in (46), the following formula

$$\dim |K + S| = g_n(V) - g_{n-1}(V) + g_{n-1}(S) - 1 + l\,. \tag{47}$$

This is precisely *the* RIEMANN-ROCH *theorem for the adjoint system of an arbitrary non-singular divisor.*

When $|S|$ is an *ample linear system*, any k-fold differentials of the first kind, $1 \leqq k \leqq n-1$, which vanish on S *vanish also on* V: this is a classical result of LEFSCHETZ, which may be easily obtained from (IX, 6). Therefore in this case $l = 0$, and (47) yields the simpler relation:

$$\dim |K + S| = g_n(V) - g_{n-1}(V) + g_{n-1}(S) - 1\,. \tag{48}$$

Defining

$$a(V) = \sum_{r=0}^{n-1} (-1)^r g_{n-r}(V)\,, \tag{49}$$

we can restate (48) in the form

$$\dim |K + S| = a(V) + a(S) - 1\,, \tag{50}$$

as soon as we use the fact that $g_k(V) = g_k(S)$, $1 \leqq k \leqq n - 2$; this follows from another theorem of LEFSCHETZ which states that *a k-cycle on S is homologous to zero on S if it so on V* {see (XI, 3)}.

We end by remarking that the process described may be generalised to prove that *if S is a non-singular divisor, T a non-singular divisor belonging to an ample linear system, such that $T \cdot S$ is a prime non-singular S-divisor and $n \geqq 3$, then it is*

$$\dim|K + S + T| = a(V) + a(S) + a(T) + a(T \cdot S) - 1 . \tag{51}$$

8. The Arithmetic Genera

This section is devoted to some additional information concerning arithmetic genera, introduced in Ch. V.

a) Let $S, T, U, \ldots$ be prime non-singular divisor on V. If $n = \dim V^n = 1$, $a(V^1)$ is simply the genus of the curve V^1, so that, when $n = 2$, we derive from the adjunction formula {see (IV, 5)}

$$2\,a(S) - 2 = I(S, K + S) . \tag{52}$$

It is now easily seen, by induction on n, that

$$a(S) = a(T) , \quad \text{if} \quad S \equiv T \quad \text{and} \quad n \geqq 2 , \tag{53}$$

this being true in virtue of (52) when $n = 2$. Moreover if $U \equiv S + T$, and $S \cdot T$ is prime non-singular on T, we have

$$a(U) = a(S) + a(T) + a(S \cdot T) . \tag{54}$$

The proof for $n = 2$ is obvious. Then suppose (54) to be true for $n \leqq m - 1$ and let us prove it for $n = m$.

If $|U|$ and $|S|$ are ample, (54) follows immediately from (50). If, instead, these systems are not ample, let $|E|$ be an ample linear system and denote respectively by E_h or S_h a generic member of $|h\,E|$ or of $|h\,E + S|$, which are certainly both ample, so that E_h and S_h are both prime non-singular divisors on V {see (III, 7, vi)}.

Then we can apply (54) to the ample systems $|K + E_h + U|$, $|K + S_h + T|$, which are of the same dimension since $E_h + U \equiv S_h + T$: hence we find a relation, which, by using our inductive hypothesis, quickly gives the proof.

b) If $|E|$ is the hyperplane section system of V, we have

$$a(E_h) = \sum_{k=1}^{n} \binom{h}{k} \{a(E^{[k]}) + (-1)^{n-k}\} + (-1)^n , \tag{55}$$

where $a(D) =$ degree $D - 1$, when D is a point group. We omit the simple inductive proof resting on (a).

Moreover, if $|U|$ is an ample system, the generic member U_l of the ample system $|U + l\,E|$ is a non-singular prime divisor, so that $a(U_l)$ is defined, and then $a(U_l)$ is *a polynomial in l of degree n, $l \geqq 0$*. Namely, since $U_l \equiv U + E_l$, $l \geqq 1$, we have from (54) $a(U_l) = a(U) + a(E_l) + a(E_l \cdot U)$, and hence the proof follows by (55). For $l = 0$ the fact is obvious.

c) The system $|D + h E|$ is ample for any D, when $h \geqq h_0(D)$, where $h_0(D)$ is a suitable integer depending on D {see (III, 7, vi)}: this allows us to define $a(D_h)$ for $h = l + h_0$ and $l \geqq 0$. Then $D_h \equiv U + l E$, where $U = D_{h_0}$. By (b) there exists a polynomial $a(h; D, V)$ of degree n in h such that

$$a(h; D, V) = a(D_h), \quad h \geqq h_0. \tag{56}$$

It can now be shown that the value of the polynomial $a(h; D, V)$ for $h = 0$ is independent of the choice of the ample system $|E|$, depending only upon D. Thus *we can define*

$$a_V(D) = a(0; D, V) \tag{57}$$

for any D, noticing that if D is prime and non-singular, *then* $a_V(D) = a(D)$. Moreover

$$a(h; D, V) = a_V(D + h E) \tag{58}$$

as follows from the identity $a(l + h; D, V) = a(l; D + h E, V)$ when $l = 0$.

Our definitions give obviously

$$a_V(D) = a_V(D'), \quad \text{if} \quad D \equiv D'. \tag{59}$$

d) If S is a prime non-singular divisor and D *an arbitrary divisor not containing* S, the modular relation (54) extends to the following:

$$a_V(D + S) = a_V(D) + a(S) + a_S(D \cdot S). \tag{60}$$

For the proof consider a generic member U_h of the system $|D + S + h E|$.

Then for large h, U_h and D_h are prime and non-singular. Now (54) applied to $U_h \equiv D_h + S$, and the definition of $a_V(D)$, together justify our assertion.

Now let S be a generic member of the system $|D - K|$ supposed to be ample. The Riemann-Roch theorem (49) gives: $\dim|D| = \dim|K + S| = a(V) + a(S) - 1$. Since $a(S) = a_V(D - K)$, by (59), we find

$$\dim|D| = a(V) + a_V(D - K) - 1. \tag{61}$$

On the other hand the system $|D + h E - K| = |D_h - K|$ is certainly ample for large h and (61), together with (58), yields

$$\dim|D + h E| = a(V) + a(h; D - K, V) - 1, \quad h \geqq h_0. \tag{62}$$

Therefore there exists a polynomial $v(h; D, V)$ in h such that

$$\dim|D + h E| = v(h; D, V), \quad \text{for large } h, \tag{63}$$

where, besides,

$$v(h; D, V) = a(V) + a(h; D - K, V) - 1 . \tag{64}$$

Consequently for $h = 0$ we have

$$v(0; D, V) = a(V) + a_V(D - K) - 1 , \tag{65}$$

and so $v(0; D, V)$ is independent of $|E|$ and may be called the virtual dimension of the system $|D|$. Obviously by (V, 7, i) the present definition of the virtual dimension of $|D|$ coincides with the SEVERI definition given in (V, 7) (as HODGE has proved directly [11]) and hence, using the same symbol, we derive $\delta(D) = v(0; D, V)$. From this and the formula (37) of Ch. V it follows at once that

$$P^a(V) = v(0; K, V) + 1 - (-1)^n . \tag{66}$$

Since $a_V(0) = (-1)^n$, we infer from (49), (65), (66) the following important theorem, conjectured by SEVERI and proved by KODAIRA in the manner described:

(i) *The arithmetic genus $P^a(V)$ of a non-singular algebraic variety defined over the complex field coincides with the functional $a(V) = \sum_{r=0}^{n-1} (-1)^r g_{n-r}(V)$, where $g_s(V)$ is the number of linearly independent s-fold differentials and of the first kind on V.*

e) Moreover, by taking $A \equiv 0$ in (28) of Ch. V and by what has been said above we have

$$p_a(V) = (-1)^n v(0; 0, V) , \tag{67}$$

where the second zero in $v(0; 0, V)$ is the zero divisor on V.

From (65) we infer that

$$p_a(V) = (-1)^n \{a(V) + a_V(-K) - 1\} . \tag{68}$$

Now let S be a prime non-singular divisor and let $K + S$ be a generic member of $|K + S|$. Then $K_S = (K + S) \cdot S$ is a canonical divisor on S {see (IV, 5)}.

By (68) we have

$$p_a(S) = (-1)^{n-1} \{a(S) + a_S(-K_S) - 1\} , \tag{69}$$

and by (60)

$$a_S(-K_S) = a_S((-K - S) \cdot S) = a_V(-K) - a_V(-K - S) - a(S) . \tag{70}$$

Therefore we conclude with the formula

$$p_a(S) = (-1)^{n-1} \{a_V(-K) - a_V(-K - S) - 1\} . \tag{71}$$

This suggests the definition, for any divisor D on V,

$$p_a(D) = (-1)^{n-1} \{a_V(-K) - a_V(-K - D) - 1\} . \tag{72}$$

Then, from (59), it follows easily that $p_a(D) = p_a(D')$ if $D \equiv D'$. We also have *the modular property*, where we suppose that D does not contain the prime non-singular divisor S:

$$p_a(D + S) = p_a(D) + p_a(S) + p_a(D \cdot S) . \tag{73}$$

This shows that $p_a(D)$, on the present definition, is the same as the analogous character defined in (V, 3) following SEVERI and ZARISKI.

f) We shall now prove *the fundamental identity:*

$$P^a(V) = p_a(V) , \tag{74}$$

for any non-singular variety defined over the complex field. We begin by setting:

$$\alpha(V) = \sum_{s=0}^{n} (-1)^s g_s(V) , \tag{75}$$

where $g_s(V)$ is, as before, the number of linearly independent s-ple differentials of the first kind on V.

Then from (49) we have the relation:

$$a(V) = (-1)^n \{\alpha(V) - 1\} , \tag{76}$$

while we define:

$$a_V(-K) = (-1)^{n-1} \{\alpha_V(-K) - 1\} . \tag{77}$$

Moreover from (68), (57) and (66) it follows that:

$$p_a(V) = \alpha(V) - \alpha_V(-K) - (-1)^n , \tag{78}$$

$$P^a(V) = (-1)^n \{\alpha(V) - 1\} . \tag{79}$$

Our statement (74) is thus reduced to proving the equality:

$$\alpha_V(-K) = [1 - (-1)^n] \, \alpha(V) . \tag{80}$$

By multiplying both sides of (44) by $(-1)^s$ and adding from 0 to n, we obtain, for $r = n$ {see (34)}:

$$\chi^n_S(V) = \chi^n(V) + \chi^{n-s}(S) + \chi^n_C(S). \tag{81}$$

Observing that $\chi^n_C(S) = 0$ by (X, 5, ii) and $\dim \mathbf{H}^{n,s}(\mathfrak{M}) = \dim \mathbf{H}^{n-s,0}(\mathfrak{M}) = g_{n-s}(\mathfrak{M})$, for any compact KÄHLER manifold $\mathfrak{M}$ of dimension n, so that $\chi^{n-1}(S) = (-1)^{n-1} \alpha(S)$, we have, by substituting in (81), the relation

$$\alpha_V(S) = (-1)^{n-1} \{\chi^n_S(V) - \chi^n(V)\} , \tag{82}$$

for an arbitrary non-singular divisor of V.

Now we prove, extending (82), that for any V-divisor D it holds:

$$\bar{\alpha}(D) = \alpha_V(D) , \tag{83}$$

where we define:

$$\bar{\alpha}(D) = (-1)^{n-s} \{\chi^n_D(V) - \chi^n(V)\} , \tag{84}$$

which is permissible, since the cohomology groups $\mathbf{H}^s(V, \mathfrak{M}^n_D)$ are finite by (X, 4). We know {since $\mathfrak{M}^r_D \cong \Omega^r(\{D\})$: see (X, 7)} that $D \equiv D'$ implies

$$\mathbf{H}^s(V, \mathfrak{M}^n_D) \cong \mathbf{H}^s(V, \mathfrak{M}^n_{D'}) , \tag{85}$$

and hence

$$\bar{\alpha}(D) = \bar{\alpha}(D') . \tag{86}$$

Let S be a prime non-singular divisor on V, which is not a component of D. Then we have, for $n > 1$, the exact sequence

$$0 \to \mathfrak{M}^n_D \xrightarrow{i} \mathfrak{M}^n_{D+S} \xrightarrow{\mathfrak{P}} \mathfrak{M}^{n-1}_{D\cdot S}(S) \to 0 , \tag{87}$$

where $\mathfrak{P}$ is the POINCARÉ mapping $\tau_p \to \mathfrak{P}(\tau_p)$ defined in (38). For $n = 1$ we have, instead,

$$0 \to \mathfrak{M}^1_D \xrightarrow{i} \mathfrak{M}^1_{D+S} \xrightarrow{\mathfrak{P}} \Omega^0(S) \to 0 , \tag{88}$$

for any divisor $S = \sum_{k=1}^{m} p_k$, with simple components only, where $\Omega^0(S)$ is the linear space of complex-valued functions defined on the set S.

From the corresponding exact cohomology sequence we readily infer that

$$\left.\begin{aligned} \chi^n_{D+S}(V) &= \chi^n_D(V) + \chi^{n-1}_{D\cdot S}(S) , & n > 1 , \\ \chi^1_{D+S}(V) &= \chi^1_D(V) + m, & n = 1 . \end{aligned}\right\} \tag{89}$$

By using (89) and (84) we get

$$\begin{aligned} \bar{\alpha}(D+S) &= (-1)^{n-1}\{\chi^n_{D+S}(V) - \chi^n(V)\} \\ &= (-1)^{n-1}\{\chi^n_D(V) - \chi^n(V) + \chi^{n-1}_{D\cdot S}(S) - \chi^{n-1}(S)\} + (-1)^{n-1}\chi^{n-1}(S) \\ &= \bar{\alpha}(D) + \alpha(S) - \bar{\alpha}(D\cdot S) . \end{aligned} \tag{90}$$

If $n = 1$, we find from (89)

$$\bar{\alpha}(D+S) = \bar{\alpha}(D) + m . \tag{91}$$

Now we proceed by induction with respect to n. If $n = 1$ our statement (83) is obvious, when we take any divisor D as a difference of two positive divisors and apply, after, (86), (91). Therefore, we assume it for any non-singular variety of dimension less than n and prove it when $\dim V = n \geq 2$.

Let $S = E_h$, with the same notation as in (X, 8 b), and let T be a generic member of $|D + S|$. Then by (90) we get

$$\bar{\alpha}(T) = \bar{\alpha}(D) + \alpha(S) - \bar{\alpha}(D\cdot S) ,$$

and by (60), (76)

$$\alpha(T) = \alpha(D) + \alpha(S) - \alpha_S(D\cdot S) .$$

But, since T is non-singular we have $\alpha(T) = \bar{\alpha}(T)$ and, moreover, $\alpha_S(D\cdot S) = \bar{\alpha}(D\cdot S)$, by our inductive hypothesis: hence $\bar{a}(D) = a_V(D)$.

On the other hand, from (X, 7) and (15) it follows that $\Omega^0 \cong \mathfrak{M}^n_{-K}$ and hence $\mathbf{H}^s(V, \mathfrak{M}^n_{-K}) \cong \mathbf{H}^s(V, \Omega^0)$, so that $\chi^n_{-K}(V) = \chi^0(V)$. Therefore

$$a_V(-K) = (-1)^{n-1}\{\chi^n_{-K}(V) - \chi^n(V)\} = (-1)^{n-1}\{\chi^0(V) - \chi^n(V)\}$$
$$= [1 - (-1)^n]\,\alpha(V)\,,$$

which proves the result.

We may observe that the above proof and (49) give a new demonstration of the relative invariance of the arithmetic genus of V, taking into account the analogous character of $g_s(V)$, as VAN DER WAERDEN has proved {see VAN DER WAERDEN [10]; the conjecture of SEVERI is in [13], p. 87}.

9. The RIEMANN-ROCH Theorem

This theorem has recently undergone a deep analysis by SERRE and HIRZEBRUCH. We shall give here an account of HIRZEBRUCH's results, omitting proofs which, at present, have appeared only in outline and which will be expounded by HIRZEBRUCH in his monograph [a].

a) Let $\mathfrak{W}$ be a *complex analytic bundle* over V, algebraic over the complex field and non-singular, with the complex vector space $\mathbf{C}^q$ as fibre and the complex linear q-dimensional group $\mathbf{G}\,L(q, \mathbf{C})$ as structure group.

The bundle $\mathfrak{W}$ is locally isomorphic with $V \times \mathbf{C}^q$, and the changes of the local coordinate systems have to be effected by the invertible holomorphic matrixes of order q. If $s(z)$ denotes a holomorphic section of $\mathfrak{W}$ over an open neighbourhood U of V, and if $f(z)$ is a holomorphic function on U, the product $f(z) \circ s(z)$ is still a holomorphic section of $\mathfrak{W}$ over U, and the sum of two holomorphic section over U is also such a section.

Hence it follows that the stack $\mathfrak{S}(\mathfrak{W})$ of germs of holomorphic sections of $\mathfrak{W}$ possesses the structure of an analytic stack, and, since $\mathfrak{W}$ is locally isomorphic to $V \times \mathbf{C}^q$, $\mathfrak{S}(\mathfrak{W})$ is locally isomorphic to $\mathfrak{O}^q$, i. e. to the product of q stacks $\mathfrak{O}$ {see (X, 2)}.

All this may be easily inverted by proceeding from an analytic stack $\mathfrak{T}$ locally isomorphic to $\mathfrak{O}^q$ and arriving at an analytic bundle of the same type as $\mathfrak{W}$, and such that $\mathfrak{T} \cong \mathfrak{S}(\mathfrak{W})$.

Therefore there exists a one-to-one correspondence between the analytic stacks locally isomorphic to $\mathfrak{O}$ and our analytic fibre bundle of type $\mathfrak{W}$.

We remark that this correspondence, in particular, associates: (1) The bundle of the *covariant p-vectors* with the stack of germs of *holomorphic p-forms Ω^p*; (2) The complex line bundle F related to the divisor D considered in (X, 1) with the stack $\Omega^0(\{D\}) \cong \mathfrak{M}^0_D$; (3) More particularly, the trivial bundle $F = \mathbf{C}^1 \times V$ with the stack Ω^0.

b) Let c_i be the CHERN classes of the tangent bundle over V and so of V {see (IX, 9)}, $c_i \in \mathbf{H}^{2i}(V, \mathbf{Z})$, $0 \leq i \leq n$, $c_0 = 1$, and let d_j be the CHERN classes of $\mathfrak{W}$, $d_j \in \mathbf{H}^{2i}(V, \mathbf{Z})$, $0 \leq j \leq q$, $d_0 = 1$.

We introduce the formal roots γ_i, δ_j of the respective CHERN polynomials, defined by writing

$$\sum_{i=0}^{n} c_i\, x^i = \prod_{i=1}^{n} (1 + \gamma_i\, x)\,, \quad \sum_{j=0}^{q} d_j\, x^j = \prod_{j=1}^{q} (1 + \delta_j\, x)\,. \tag{92}$$

Each formal power series which is symmetric in the γ_i as well in the δ_j can be considered as a power series in the CHERN classes c_i and d_j, and, therefore, as an element of the cohomology ring $\mathbf{H}^*(V, Q)$ of V where Q denotes the rationals. If $u \in \mathbf{H}^*(V, Q)$, we shall put $k_n[u] = \bar{u} \cap V$ where $\bar{u}$ denotes the topological $2n$-dimensional component of u and V the fundamental cycle of V: $k_n[u]$ is always a rational number.

c) The fundamental statement of HIRZEBRUCH is now the following:

(i) *Putting*

$$\chi(V, \mathfrak{S}(\mathfrak{W})) = \sum_{i=0}^{n} (-1)^i \dim \mathbf{H}^i(V, \mathfrak{S}(\mathfrak{W}))\,, \tag{93}$$

we have

$$\chi(V, \mathfrak{S}(\mathfrak{W})) = k_n\left[\sum_{j=1}^{q} e^{\delta_j} \cdot \prod_{i=1}^{n} \frac{-\gamma_i}{e^{-\gamma_i} - 1}\right] = k_n\left[e^{c_1/2} \cdot \sum_{j=1}^{q} e^{\delta_j} \cdot \prod_{i=1}^{n} \frac{\gamma_i/2}{\sinh \gamma_i/2}\right] \tag{94}$$

As a first consequence we see that $\chi(V, \mathfrak{S}(\mathfrak{W}))$ may be considered as a polynomial in c_1 and in the CHERN classes of V and W: moreover, if we define as PONTRJAGIN *classes* of V the classes $p_i \in \mathbf{H}^{4i}(V, \mathbf{Z})$ given by $p_0 = 1$ and by the relations

$$\sum_{i=0}^{[n/2]} (-1)^i\, p_i = \sum_{i=0}^{n} (-1)^i\, c_i \sum_{i=0}^{n} c_i\,, \quad i = 0, 1, \ldots, [n/2]\,, \tag{95}$$

and if we observe that the classes $(\gamma_i)^2$ may be considered as the formal roots of the PONTRJAGIN polynomial $\sum_{i=0}^{[n/2]} p_i \times x^i$, and that the series $x/\sinh x$ depends only upon x^2, then we may more strongly affirm that $\chi(V, \mathfrak{S}(\mathfrak{W}))$ *is a polynomial in* c_1, *in the* PONTRJAGIN *classes of* V *and in the* CHERN *classes of* $\mathfrak{W}$.

d) We now take $\mathfrak{W} = V \times \mathbf{C}^1$ and remark that, from (75), (76) and the equality $p_a(V) = a(V)$ proved in (X, 8, i) and (X, 8 f), we obtain:

$$\alpha(V) = (-1)^n p_a(V) + 1 = \sum_{i=0}^{n} (-1)^i g^i = \chi(V, \mathfrak{S}(\mathbf{C}^1 \times V)) = \chi(V, \Omega^0)\,. \tag{96}$$

Then, by (i), we find ($q = 1$, $\delta_1 = 0$) that

$$\alpha(V) = k_n\left[\prod_{i=1}^{n} \frac{-\gamma_i}{e^{-\gamma_i} - 1}\right]. \tag{97}$$

This is an expression for the character $\alpha(V)$, called *the* TODD *genus of* V, *in terms of the characteristic classes of* V *and thence, in terms of the canonical classes of* SEGRE-TODD {see (IX, 9)}: obviously, then, *the arithmetic genus of V can also be expressed as a function of such classes.*

In particular, by a simple calculation, we find for $n = 1, 2, 3$ the following formulae:

$$\alpha(V^1) = 1/2\, c_1[V]\,, \quad \alpha(V^2) = 1/12\,(c_1^2 + c_2)\,[V^2]\,, \quad \alpha(V^3) = 1/24\, c_1\, c_2[V^3]\,, \tag{98}$$

which, apart the form, are all well known.

In effect for $n = 1$ we have $c_1[V^1] = 2 - 2p$, and then $p_a(V^1) = p$, where p is the genus of our curve. For $n = 2$, we have $c_1^2[V^2] = K^{[2]}$ and $c_2[V^2] = \chi(V^2)$, and hence: $p_a(V^2) = 1/12\,(K^{[2]} + \chi) - 1$, which is a well-known formula of surface theory {recall that $\chi = I + 4$, where I is the ZEUTHE-SEGRE invariant and see ZARISKI [a], p. 62, 113}.

Finally the third formula may similarly be written in the form $24\,(p_a(V^3) - 1) = K\,C$, where C is the canonical curve of V: this formula is due to TODD {see TODD [9], p. 215}.

We mention here that *the* TODD *genus of a non-singular subvariety* A^{n-r} *of V, which is a complete intersection of the V-divisors* $U_1, U_2, \ldots, U_r$ of cohomology classes $u_1, u_2, \ldots, u_r$, $u_i \in \mathbf{H}^2(V, \mathbf{Z})$ *is given by the formula*

$$\alpha(A^{n-r}) = k_n\left[\prod_{i=1}^{r}(1 - e^{-u_i}) \cdot \sum_{j=0}^{n-r} \alpha_j(c_1, c_2, \ldots, c_j)\right], \tag{99}$$

where

$$\alpha_j(c_1, c_2, \ldots, c_j) = k_j\left[\prod_{i=1}^{j} \frac{-\gamma_i}{e^{-\gamma_i} - 1}\right]. \tag{100}$$

e) Let us suppose, more generally, that $\mathfrak{W} = F$, $q = 1$, and let $\delta_1 = f = c(F)$, $f \in \mathbf{H}^2(V, \mathbf{Z})$, be the characteristic class of F. From (i) it follows that

$$\chi(V, \mathfrak{S}(F)) = k_n\left[e^{f + c_1/2} \prod_{i=1}^{n} \frac{\gamma_i/2}{\sinh \gamma_i/2}\right], \tag{101}$$

which is a polynomial in $f + c_1/2$ and in the PONTRJAGIN classes of V.

By (X, 6 b) $\mathfrak{S}(F) \cong \Omega^0(\{D\})$, since $F = \{D\}$ {see (X, 6, i)}: therefore, recalling (12), (15) and (22) which furnish: $\dim|K - D| + 1 = \dim \mathbf{H}^n(V, \Omega^0(\{D\}))$, we have the formula

$$\dim|D| + 1 + (-1)^n(\dim|K - D| + 1) = k_n\left[e^{f + c_1/2} \prod_{i=1}^{n} \frac{\gamma_i/2}{\sinh \gamma_i/2}\right] + \delta, \tag{102}$$

where we have put

$$\delta = +\sum_{i=1}^{n-1} (-1)^{i-1} \dim \mathbf{H}^i(V, \Omega^0(\{D\}))\,. \tag{103}$$

The dimensional formulae (102) and (103) give *a more comprehensive expression for the* RIEMANN-ROCH *theorem.*

We give, with HODGE, an alternative derivation of the RIEMANN-ROCH theorem, together with an important complement which states the geometrical significance of the terms appearing in (102) and which is due also to HODGE {see HODGE [13]}. By applying (76), (77), (82) we deduce that

$$a(V) + a_V(D - K) = \chi^n_{D-K}(V)\,, \tag{104}$$

or, by SERRE's duality theorem (X, 4) and by recalling (X, 8 d) that

$$\delta(D) = \chi^0_D - 1\,, \tag{105}$$

where $\delta(D)$ is the virtual dimension of SEVERI. Moreover, by (22) we have that

$$\dim|D| + 1 = \dim \mathbf{H}^0(V, \mathfrak{M}^0_D)\,, \tag{106}$$

which together with (105), yields

$$\dim|D| + 1 = \delta(D) + \sum_{i=1}^{n} (-1)^{i-1}\, \mathbf{H}^i(V, \mathfrak{M}^0_D)\,. \tag{107}$$

If we take into account HIRZEBRUCH's result (102) we have the following cohomological interpretation of SEVERI's virtual dimension:

$$\delta(D) = k_n \left[e^{f + c_1/2} \prod_{i=1}^{n} \frac{\gamma_i/2}{\sinh \gamma_i/2} \right] - 1\,. \tag{108}$$

Moreover, if $|X|$ is an ample linear system on V such that the system $|X + D - K|$ is ample, then, HODGE has proved,

$$\operatorname{def}|D + i\,X| \cdot X^{[i]} = \dim \mathbf{H}^i(V, \mathfrak{M}^0_D)\,, \quad i = 1, 2, \ldots, n - 1\,. \tag{109}$$

Thus we obtain the fundamental formula

$$\dim|D| = \delta(D) + \sum_{i=1}^{n-1} (-1)^{i-1} \operatorname{def}|D + i\,X| \cdot X^{[i]} + (-1)^{n-1} s\,, \tag{110}$$

where

$$s = \dim|K - D| + 1 \tag{111}$$

is the so-called index of speciality of $|D|$ and where $\operatorname{def}|D + i\,X| \cdot X^{[i]}$, which is independent of $|X|$ by (109), may be called, with HODGE, the i-th index of irregularity of $|D|$ on V.

The character $\delta = \sum_{i=1}^{n-1} (-1)^{i-1} \operatorname{def}|D + i\,X| \cdot X^{[i]}$ is called the *superabundance* of $|D|$. When $|D - K|$ is ample we have by (23) $\delta = s = 0$, and thus we reobtain the previous results of SEVERI, ZARISKI and KODAIRA.

Finally we remark that by taking $D = 0$ or $D = K$, we obtain, on the same hypotheses as above, the following expressions for the number of independent i-ple differentials of the first kind on V {see (X, 7)}:

$$g_i = \operatorname{def}|i\,X| \cdot X^{[i]}\,, \qquad i = 1, 2, \ldots, n - 1\,, \tag{112}$$

$$g_{n-i} = \operatorname{def}|K + i\,X| \cdot X^{[i]}\,, \quad i = 1, 2, \ldots, n - 1\,. \tag{113}$$

f) We shall now apply (102) to the particular cases $n = 1, 2, 3$. When $n = 1$ we have immediately from (102):

$$\dim|D| - \dim|K - D| = (f + c_1/2)\,[V^1]\,, \tag{114}$$

where now $f[V^1] = \deg D = m$ and $c_1[V^1] = \chi(V^1) = 2 - 2p$. Hence and from (104) follows the usual formula:

$$\dim|D| = m - p + s\,. \tag{115}$$

g) When $n = 2$ we have

$$\begin{aligned} k_2\left[e^{f+c_1/2}\prod_{i=1}^{2}\frac{\gamma_i/2}{\sinh\gamma_i/2}\right] &= \frac{1}{2}(f^2 + f c_1)\,[V^2] + \alpha(V^2)\\ &= \frac{1}{2}(D^{[2]} - D\cdot K) + p_a(V^2) + 1, \end{aligned} \tag{116}$$

and taking $D^{[2]} = m = \deg|D|$ and recalling the formula $D \cdot K = -m + 2\pi - 2$, where π is the genus of D, we obtain finally:

$$\dim|D| = n - \pi + p_a(V) + 1 - s + \delta\,. \tag{117}$$

Moreover we have here the classical complement {see ZARISKI [a], p. 66} to the effect that $\delta \geqq 0$, since in this case (103) furnishes: $\delta = \mathbf{H}^1(V, \Omega^0(\{D\}))$.

h) When $n = 3$ we have

$$\begin{aligned} g = k_3&\left[e^{f+c_1/2}\prod_{i=1}^{3}\frac{\gamma_i/2}{\sinh\gamma_i/2}\right]\\ &= \left[\frac{1}{6}f^3 + \frac{1}{4}f^2 c_1 + \frac{1}{12}f(c_1^2 + c_2) + \frac{1}{24}c_1 c_2\right][V^3]\,. \end{aligned} \tag{118}$$

Observing that in this case c_1 and c_2 are respectively dual of $-K$ and C, the canonical curve of V^3, and that f is dual of D, we obtain:

$$\begin{aligned} g &= \frac{1}{6}D^{[3]} - \frac{1}{4}K D^{[2]} + \frac{1}{12}D(K^{[2]} + C) - \frac{1}{24}K C\\ &= \frac{1}{6}D^{[3]} - \frac{1}{4}K D^{[2]} + \frac{1}{12}D(K^{[2]} + C) - p_a(V^3) + 1\,, \end{aligned} \tag{119}$$

where we have used (98). On the other hand, as we may at once calculate by (d), we have:

$$p_a(D) = \alpha(D) - 1 = \frac{1}{6}D^{[3]} + \frac{1}{4}K D^{[2]} + \frac{1}{12}(K^{[2]} + C)\,D - 1\,, \tag{120}$$

and

$$p_a(D^{[2]}) = -\,\alpha(D^{[2]}) + 1 = D^{[3]} + \frac{1}{2}K D^{[2]} + 1\,. \tag{121}$$

Therefore we finally obtain:

$$\dim|D| = D^{[3]} - p_a(D^{[2]}) + p_a(D) - p_a(V^3) + 2 + s + \delta\,. \tag{122}$$

Now

$$\delta = \dim\mathbf{H}^1(V, \Omega^0(\{D\})) - \dim\mathbf{H}^2(V, \Omega^0(\{D\})) \gtreqless 0\,, \tag{123}$$

as is proved by simple examples. If $\dim|S| \geqq 1$ and $|D| = |K + S|$ then $i = 0$ and one may show, with SEVERI {see SEVERI [13], p. 63; KODAIRA [2], p. 323}, that then $\delta \geqq 0$. If $|S|$ is ample then $\delta = 0$ also, and we again have by a simple calculation, the result of (X, 7) in a more explicit form.

10. Miscellaneous Results

We state here without proof the following theorem of KODAIRA {see KODAIRA [6]; the quoted proof of CASTELNUOVO-ENRIQUES is in [1], Ch. III: see also ZARISKI [a], p. 72; for th. (v) below see also KODAIRA [5], p. 128 and CHOW-KODAIRA [1]: it has been conjectured by A. WEIL in [4]}:

(i) *A compact complex analytic variety V carrying a* HODGE *metric, i. e. a* KÄHLER *metric* $2\sum g_{\alpha\beta^*}(dz^\alpha\, d\bar{z}^\beta)$ *such that the associated exterior form* $\omega = i\sum g_{\alpha\beta^*}\, dz^\alpha \wedge d\bar{z}^\beta$ *belongs to the cohomology class of an integral 2-cocycle {see* (IX, 6)*}, is biregularly equivalent to a non-singular algebraic variety embedded in a projective space.*

The proof rests on stack theory, and in particular on the results of (X, 5), on some property of local quadratic transformations and finally on an embedding argument inspired by a classical proof of CASTELNUOVO-ENRIQUES in surface theory. This theorem is very useful as a criterion for deciding whether a given variety is algebraic or otherwise. KODAIRA has proved in this way that the following classes of varieties possess non-singular projective algebraic models:

(ii) *Complex toruses endowed with a* RIEMANN *metric.*

(iii) *Compact complex varieties with a Hermitian metric and a* RICCI *curvature which is positive (or negative) definite everywhere.*

(iv) *The factor spaces* $\mathfrak{B}/\Delta$ *where* $\mathfrak{B}$ *is a bounded domain in the space* $(z^1, z^2, \ldots, z^n)$ *and* Δ *is a discontinous group of automorphisms of* $\mathfrak{B}$ *such that* $\mathfrak{B}$ *is compact relatively to* Δ, *which is supposed to be without fixed points in* $\mathfrak{B}$, *except the identity.*

(v) *Manifolds which are finite unramified coverings of an algebraic variety.*

(vi) *Projective bundles over algebraic varieties having as fibre a complex projective space and as structure group the group of projective transformations.*

XI. The Superficial Irregularity and Continuous Systems

1. The Deficiency of a Linear System

We outline here an exposition of the transcendental work of KODAIRA {see KODAIRA [5], pp. 108—128}, which extends to higher varieties some classical results for surfaces, in particular those concerning the deficiencies of linear systems, the irregularity q of a variety and the completeness, under suitable conditions, of the characteristic series of continuous systems. The chief tool is the theory of currents {see (IX, 3)}, though some of the topics could be treated by the methods of the preceding chapter {see KODAIRA-SPENCER [3]}.

a) Let S be a prime non-singular divisor on an algebraic non-singular variety V, defined over the complex field, and let $\mathfrak{A}(S)$ be the linear space of all *additive meromorphic functions F on V*, which are *multiples of* $-S$, i. e. of those which are additively altered by a constant c_γ by analytic continuation along a closed continuous curve γ and are such that $(F) + S \geqq 0$: then, clearly, $\mathfrak{A}(S)$ is *the linear space of* PICARD *integrals of the first or of the second kind* {see (IX, 8 a)}.

Let F_0 be a meromorphic function on V having S as *simple polar divisor*, and let $D = (F_0) + S$. Then, for any $F \in \mathfrak{A}(S)$, the ratio F/F_0 induces on S *a single valued meromorphic function* $f = (F/F_0)_S$, multiple of the divisor $-\Gamma = -D \cdot S$ cut by D on S; f is called *the* SEVERI *residue* {see (X, 7)} of F on its polar divisor S {see also (X, 7)}.

Let $f(\Gamma)$ be the linear space of the meromorphic functions on S, which are multiples of $-\Gamma$, and set, for each point $p \in S$, $S = (R_p)$ and $\Gamma = (h_p)$. We also denote by L a *generic spatial 1-section* of V and by $A_{\nu L}$ the differential induced on L by the simple differential of the first kind A_ν, $\nu = 1, 2, \ldots, g_1$, on V.

It may now be proved, by the technique of harmonic integrals {see KODAIRA [5], p. 110}, that the mapping $F \to f$ of $\mathfrak{A}(S)$ into $f(\Gamma)$ is *onto* $f(\Gamma)$ and that *the image into $f(\Gamma)$ of the subspace $\mathfrak{M}(S)$ of $\mathfrak{A}(S)$ consisting of the single-valued functions is made up of those functions $f \in f(\Gamma)$ for which the system*

$$\sum_{p \in S \cap L} a_{\nu p}\, \xi_p = 0, \quad \nu = 1, 2, \ldots, g_1, \tag{1}$$

admits the solution $\xi_p = (f\, h_p)_p$, where we have put

$$a_{\nu, p} = \mathrm{Res}_p(A_{\nu L}/R_p)\,, \tag{2}$$

b) Now let D be an *arbitrary* divisor on V not containing S as a component and let $\mathfrak{F}(D)$ be the space of the meromorphic functions on V multiples of $-D$ and $f(\Gamma)$ the space of meromorphic functions on S multiples of $-\Gamma = -D \cdot S$.

The result (a) furnishes an answer to the question *if there exists in $\mathfrak{F}(D)$ a function F which induces on S a given function $f = F_S \in f(\Gamma)$, at least when $S \equiv D$, or when the system $|S - D|$ contains a connected member T not containing S as a component.* Putting $(F_0) = D + T - S$ and $h_p = (R_p F_0)_S$ for each point $p \in S$, *the condition of existence for such a function is that the system* (1) *should have the solution* $\xi_p = (f\, h_p)_p$.

In fact, if $D \equiv S$, and then $T \equiv 0$, the mapping $F \to F/F_0$ furnishes a one-to-one correspondence between $\mathfrak{F}(S)$ and $\mathfrak{F}(D)$, while the relation

$f = (F/F_0)_S$ is equivalent to the other $f = F_S$, so that we are in the case considered in (*a*).

In the other case, $T \not\equiv 0$, we put $D_1 = D + T$, $\Gamma_1 = D_1 \cdot S$. Then $D_1 \equiv S$ on V and $\mathfrak{F}(D) \subseteq \mathfrak{F}(D_1)$, $f(\Gamma) \subseteq f(\Gamma_1)$. If $f_0(\Gamma_1)$ is the subspace of $f(\Gamma_1)$ satisfying the condition in (a), we have simply to prove that if $f \in f_0(\Gamma_1) \cap f(\Gamma)$, then there exists a function $F \in \mathfrak{F}(D)$, such that $F_S = f$, which easily follows from the hypothesis that T is connected.

c) On the same hypotheses as before for D and S, the mapping $F \to F_S$ with $F \in \mathfrak{F}(D)$ is a homomorphism of $\mathfrak{F}(D)$ into $f(D \cdot S)$ of kernel $\mathfrak{F}(D - S)$.

The linear system $|D| \cdot S$, consisting of the S-divisors $(M_S) + D \cdot S$, with $M \in \mathfrak{F}(D) - \mathfrak{F}(D - S)$, is cut on S by the system $|D|$, and we have

$$\dim |D| \cdot S = \dim |D| - \dim |D - S| - 1 \,. \tag{3}$$

We define *the deficiency* $\operatorname{def}(D/S)$ *of the V-divisor D on S by the expression*

$$\operatorname{def}(D/S) = \dim |D \cdot S|_S - \dim (|D| \cdot S) \,. \tag{4}$$

If $D \equiv D'$ we have $\operatorname{def}(D/S) = \operatorname{def}(D'/S)$, so that $\operatorname{def}(D/S)$ can also be considered as *the deficiency of the system* $|D|$ on the divisor S, at least when the system $|D| \cdot S$ is not empty.

Moreover we may remove the restriction that D should not contain S as a component by replacing D, if necessary, by another divisor D' satisfying the condition and such that $D' \equiv D$. In particular the number $\operatorname{def}(S/S)$ is called *the characteristic deficiency* of S.

Now putting $F(D)_S = \{F_S \mid F \in \mathfrak{F}(D)\}$, we deduce from the definitions that

$$\operatorname{def}(D/S) = \dim f(D \cdot S) - \dim \mathfrak{F}(D)_S. \tag{5}$$

Moreover from (b) it follows that $\mathfrak{F}(D)$ consists of all functions f satisfying the condition stated in (a). Therefore, when D and S satisfy the additional condition of (b), $\operatorname{def}(D/S)$ *is equal to the number of linearly independent solutions of the system* (1), i. e. to $g_1(V) - j_s(D)$, where $j_s(D)$ is the number of linearly independent differentials A of the first kind satisfying the system

$$\sum_{p \in L \cap S} (f\, h_p)_p \operatorname{Res}_p(A_L/R_p) = 0 \,, \quad \text{for any} \quad f \in f(D \cdot S) \,, \tag{6}$$

which is the transposed system of (1). In conclusion we have the important result

$$\operatorname{def}(D/S) = g_1(V) - j_s(D) \,. \tag{7}$$

We remark the immediate consequence that

$$\operatorname{def}(S/S) \leqq g_1(V) \,. \tag{8}$$

d) If $|S - D|$ is ample, it may be considered as the system of hyperplane sections of some model, birational and biregular, of V, which we shall denote by V. Then (7) is valid since the generic element T of $|D-S|$ is a non-singular variety. Moreover we may suppose that the curve L lies on T and that $(h_p) = D \cdot S + T \cdot S$, locally at $p \in S$.

Therefore, if $f \in f(D \cdot S)$, $f h_p$ vanishes on $T \cdot S$, and hence $(f h_p)_p = 0$ for each $p \in S \cap L$, since $S \cap L \subset T \cdot S$. Then $j_s(D) = b^1(V)$, and thereby $\operatorname{def}(D/S) = 0$; we conclude with the theorem

(i) *If S is a non-singular divisor such that the system $|S - D|$ is ample, where D is any divisor, then the deficiency of D on S is zero, i. e. the linear system $|D|$ cuts a complete system on S.*

e) From this follows immediately the following lemma of Enriques-Severi-Zariski {see Enriques [a], p. 129; Severi [a], p. 372; Zariski [a], p. 67; Zariski's proof is in Zariski [21], pp. 570—578}, which, as Zariski has proved, is also valid when V is *defined and normal over a field k* of any characteristic:

(ii) *For any complete system $|D|$ on V, there exists an integer $h(D)$, depending upon D, such that [for $h \geqq h(D)$] $|D|$ cuts a complete linear system on the generic member E_h of the h-ple of the system of the hyperplane sections of V.*

f) The number r of *simple* Picard *integrals of the second kind* on V is obviously given by $r = \dim \mathfrak{A}(S) - \dim \mathfrak{F}(S)$. From (a) it follows that

$$\dim \mathfrak{A}(S) = \dim f(\Gamma) + g_1(V) + 1\,, \tag{9}$$

while from (c) we derive

$$\operatorname{def}(S/S) = \dim f(\Gamma) - \dim \mathfrak{F}(S) + 1\,. \tag{10}$$

We thus have the noteworthy formula

$$r = g_1(V) + \operatorname{def}(S/S)\,, \tag{11}$$

where S is a non-singular prime divisor on V.

2. The Poincaré Families

We have already defined the first Picard variety $\mathfrak{P}$ and the second Picard variety $\mathfrak{A}$ attached to a non-singular algebraic variety V from the abstract viewpoint in Ch. VII, and we have again, independently, defined the first Picard variety attached to V, when this is defined over the complex field, by using the theory of stacks.

Here, leaving aside the question of proving the identity, over the complex field, of both definitions, we give, following Kodaira's exposition {see Kodaira [5], p. 115}, the transcendental derivation of the variety $\mathfrak{A}$ attached to V, and we describe the more salient related facts, which are partly classical and partly due to Igusa, Weil and Chow {see Igusa [3]; A. Weil [4]; Chow [6] and also Andreotti [1, 2]; the reader interested in the recent developments of the notion of Abelian varieties must also take into account the important works [29] of Severi and [13] of Roth; see also for these questions Roth [b], p. 106}.

Later, we shall establish the uniqueness of all these definitions {see (XI, 4)}.

a) Let $\gamma_1, \gamma_2, \ldots, \gamma_{2q}$, $2q = b^1(V)$, be a base for 1-cycles with integral coefficients on V and let $A_1, A_2, \ldots, A_q$, be a bases for simple differentials of the first kind on V, such that $\int\limits_V A_\lambda \cdot {}^*\bar{A}_\nu = \delta_{\lambda\nu}$, $\delta_{\lambda\nu} = 0$, if $\lambda \neq \nu$; $\delta_{\lambda\nu} = 1$, if $\lambda = \nu$: moreover we put $\omega_{\nu j} = \int\limits_{\gamma_j} A_\nu$.

The $2q$ vectors $\omega_j = (\omega_{1j}, \omega_{2j}, \ldots, \omega_{\nu j}, \ldots, \omega_{qj})$ generate a discrete subgroup $\Delta = \{\sum m_j \omega_j \,|\, m_j = 0, \ \pm 1, \ \pm 2, \ldots\}$ of the q-dimensional complex vector space $\{u\}$, so that the factor group $\mathfrak{A} = \{u\}/\Delta$ becomes a complex torus with which is associated the RIEMANN matrix $\omega = (\omega_{\nu j})$ {see (IX, 8 b)}.

By (X, 10, ii) $\mathfrak{A}$ is an Abelian variety which may be realised by an analytic biregular mapping onto a non-singular algebraic projective model: *$\mathfrak{A}$ is called the second* PICARD *variety of V.*

It is clear that *V can be applied holomorphically into $\mathfrak{A}$* by the homomorphism $\varphi: z \to [P(z)]$, where $[u]$ denotes the point of $\mathfrak{A}$ which arises from the point (u) of $\{u\}$, and where $P(z) = (P_1(z), P_2(z), \ldots, P_q(z))$, with $P_\nu(z) = \int\limits_{z_0}^{z} A_\nu$, being z_0 an arbitrary fixed point of V.

b) A many-valued meromorphic function on V, not everywhere 0 or ∞, is said to be *multiplicative* if the absolute value $|F(z)|$ is one-valued on V.

By analytic continuation along a closed continuous curve γ, the function $F(z)$ is then multiplied by a constant factor $\chi(\gamma)$ with $|\chi(\gamma)| = 1$ depending only on the homology class of γ on V. If G is the fundamental group of V and G' the commutator group of G, it is well known that $H = G/G'$ is the one-dimensional homology group of V with integral coefficients.

Let $\hat{V}$ be the covering manifold of V belonging to the subgroup G' of G; every element σ of H determines an automorphism of $\hat{V}$, transforming each point $\hat{M} \in \hat{V}$ into a point $\sigma \hat{M}$ lying over the same point M of V as $\hat{M}$: then F is multiplicative if $|F(\sigma \hat{M})| = |F(\hat{M})|$ for every $\sigma \in H$. One can now define *the divisor (F) on V of any meromorphic multiplicative function F* as follows {see KODAIRA-DE RHAM [a], p. 110}.

If $\hat{P}$ is a point lying over P on $\hat{V}$, the local coordinates $(z_1, z_2, \ldots, z_n)$ on V at P can be used as local coordinates on $\hat{V}$ at $\hat{P}$, and F can be represented as a product $F = F(z) \cdot \prod\limits_j F_j(z)^{m_j}$ of a unit factor $F(z)$ in the ring of holomorphic functions locally at P and of powers of irreducible holomorphic functions $F_j(z)$ in the ring of holomorphic functions at P. The $F_j(z)$ or m_j are not affected by replacing $\hat{P}$ by $\sigma \hat{P}$.

The divisor of F on V will then be defined locally at P, as $\sum m_j \cdot W_j$ where W_j is the irreducible algebroid variety, of complex dimension $n-1$, defined by $F_j(z)=0$. Then the divisor (F) of F will be defined globally, by means of any suitable finite open covering of V.

c) Now let $\mathbf{G}_0$ be the group of the V-divisors which are rationally homologous to zero on $V(D \sim 0)$: if $D \in \mathbf{G}_0$ we can prove that *there is a meromorphic multiplicative function F on V, determined to within a multiplicative constant, such that $(F)=D$, and conversely.* This function is altered by analytic continuation along a closed continuous curve by the factor $\chi(\gamma, D) = \exp 2\pi i \int_Q H\gamma$, where Q is a $(2n-1)$-chain with integral coefficients and with the boundary $\partial Q = D$.

This statement, as may be easily proved by using STOKES's theorem, is equivalent to LEFSCHETZ's theorem which affirms *the existence of a* PICARD *differential of the third kind on V with the given "residue" D, if and only if $D \sim 0$ on V* {see LEFSCHETZ [a], p. 146; A. WEIL [1]; see also ATIYAH-HODGE [1], p. 66}: as a corollary to a more general existence theorem it is deduced in this form by A. WEIL.

Taking

$$\eta_j(D) \equiv \int_Q H\,\gamma_j \pmod 1\,, \quad \partial Q = D\,, \tag{12}$$

we can associate with every divisor $D \in \mathbf{G}_0$ the point $[\eta(D)]$ belonging to the $2q$-dimensional real torus $\mathfrak{P}$, quotient of a vectorial real space with $2q$ dimensions by its subgroup of vectors with rational components.

So we obtain a homomorphism of $\mathbf{G}_0$ into $\mathfrak{P}$, whose kernel is evidently the subgroup $\mathbf{G}_l$ of $\mathbf{G}_0$ consisting of the V-divisors linearly null: in fact only for these we have $\eta_j(D) \equiv 0 \pmod 1$. Hence this yields *an isomorphism of $\mathbf{G}_0/\mathbf{G}_l$ into $\mathfrak{P}$.*

We can now define the matrix $\pi = (\pi_{\nu j})$, of type $q \times 2q$, by the relations $\pi\,\bar\omega_{-1} = i$, $\pi\,\omega_{-1} = 0$, and, with its help, we can introduce a complex structure $\{\eta'\}$ in the space $\{\eta\}$ by putting $\eta'_\nu = \sum_{j=1}^{2q} \pi_{\nu j}\,\eta_j$, $\nu = 1, 2, \ldots, q$. Consequently $\mathfrak{P}$ becomes an *Abelian variety having as model a complex algebraic torus.*

d) It is well known from function theory that, since the matrix ω is Riemannian, we can associate with ω a holomorphic function $\vartheta(u)$, called a *theta function,* which does not vanish identically and which satisfies the identity

$$\vartheta(u+\omega_j) = \exp\left\{\sum_\nu \left(\sum_\lambda h_\nu\,\lambda\,\bar\omega\,\lambda_j\right) u_\nu\right\} \cdot \vartheta(u)\,, \quad j = 1, 2, \ldots, 2q\,, \tag{13}$$

where $h = (h_{\nu\lambda}) = (i\,\bar\omega\,J^{-1}\,\omega_{-1})^{-1}$, where J is a $(2q \times 2q)$ skew-symmetric integral matrix related to ω.

Moreover, for any complex vector $t \in \{u\}$, we can find function such that the function $\vartheta(P(z)+t)$ of (z) does not vanish everywhere on V and the divisor $\Delta_t = (\vartheta(P(z) + t))$ is a non-singular algebraic subset of V with $\Delta_t - \Delta_0 \sim 0$ for any vector t. Then the mapping $[t] \to [\eta(\Delta_t - \Delta_0)]$ is *an analytic homomorphism λ representing the variety $\mathfrak{A}$ on the variety $\mathfrak{P}$* described in (a).

This obviously implies that the homomorphism ψ of (a) is *onto*; incidentally we thus have a new proof of (X, 6, ii) and moreover the actual variety $\mathfrak{P}$, which here appears as the dual in the sense of WEIL {see e. g. A. WEIL [4], p. 866} of the variety $\mathfrak{A}$, is the same as that defined in (X, 6).

Finally we remark that the system of divisors $\Delta_t - \Delta_0$, $t \in \{u\}$ is *a* POINCARÉ *family* {see (VII, 3)} *with respect to the rational homology on V, in the sense that every class of the group $\mathbf{G}_0/\mathbf{G}_t$ has in such a system one and only one representative.*

3. The Superficial Irregularity

We have proved {see (8) and (IX, 8 a)} that, if S is a prime non-singular divisor on V, then $\operatorname{def}(S/S) \leqq q$. The following theorem, which is classical for surfaces, gives a fundamental geometrical characterisation, due to KODAIRA {see KODAIRA [5], p. 119}, of the number $q = g_1$ of linearly independent simple differentials of the first kind on V, which, as we have seen in (XI, 2) is the same as the dimension of the varieties $\mathfrak{P}$ and $\mathfrak{A}$ transcendentally attached to V and which, as we shall see later in (XI,4), is also the same as the superficial irregularity defined in (VII,8).

(i) *The irregularity q of an algebraic non-singular variety is equal to the maximum of the characteristic deficiencies* $\operatorname{def}(S/S)$ *of non-singular prime divisors S on V.*

We sketch here KODAIRA's proof.

a) To begin with KODAIRA proves that the arithmetic genus of a divisor D depends *only upon the homology class of D.*

The more interesting part of the demonstration, to which we may easily reduce the rest by applying the modular property of the arithmetic genus, consists in proving that $p_a(D) = (-1)^n$ if $D \sim 0$. This may be seen by induction with respect to the dimension of V, by using the two facts: (1) If $D \sim 0$, there is a divisor $\Delta_t - \Delta_0 \sim D$ by (XI, 2 d); (2) The numbers $p_a(\Delta_t - \Delta_0)$ are all less than a constant with respect to $t \in \{u\}$.

We remark that this proof is substantially the same in its objective as MATSUSAKA's proof in (VII, 5) regarding the invariance of the arithmetic genus with respect to algebraic equivalence {see (XI, 4)}.

b) Our statement (i) may be demonstrated by proving that there exist $2q$ meromorphic additive functions multiple of $-S$: in fact, since then $\operatorname{def}(S/S) \geqq q$, we must necessarily have $\operatorname{def}(S/S) = q$. The construction of such functions is effected by KODAIRA in the following manner.

Suppose both the systems $|S - K + \Delta_a - \Delta_t|$ and $|S + \Delta_a - \Delta_t|$ to be ample, where a is a fixed vector of $\{u\}$, t belongs to a neighbourhood $U(a)$ of a on $\{u\}$, and, Δ_a and Δ_t belong to the divisor system on V described in (XI, 2 d). Then the system $|S + \Delta_a|$ is also ample and hence it can be identified with the system $|E_\sigma|$ cut on a suitable birational and biregular model of V, which we still denote by V, by the hyperplanes of the ambient space $\mathbf{P}^d$, where we suppose V to be properly embedded.

Let $\sigma \cdot \zeta = \sum_{i=0}^{d} \sigma^i \zeta_i$ be the equation of σ in $\mathbf{P}^d$ and let $\overline{\mathbf{P}}^d = \{\sigma^0, \ldots, \sigma^d\}$ be the projective space whose points represent biunivocally the hyperplanes of $\mathbf{P}^d$. It is easily seen that the system $|E - \Delta_t|$ consists of all positive or zero divisors of the type $E_\sigma - \Delta_t$. If $\overline{\mathbf{P}}_t^r$ is the subspace of $\overline{\mathbf{P}}^d$ representing the hyperplanes σ passing through Δ_t, then we have:

$$r = \dim|E - \Delta_t| = p_a(V) + p_a(E - K - \Delta_t) - 1\,, \tag{14}$$

and since $\Delta_t \sim \Delta_a$ we have also by (a):

$$r = p_a(V) + p_a(E - K - \Delta_a) - 1\,. \tag{15}$$

It is then proved that $\overline{\mathbf{P}}_t^r$ depends holomorphically on t, in an obvious sense: therefore we can say, roughly speaking, that *the system* $|S + + \Delta_a - \Delta_t|$ *depends holomorphically on t in a suitable neighbourhood* $U(a)$, *of a on* $\{u\}$ *and that its dimension is constant with respect to these values of t.*

c) Since $S \in |E - \Delta_a|$, it will be $S = E_\alpha - \Delta_a$, with $\alpha \in \overline{\mathbf{P}}_a^r$. Now let $\overline{\mathbf{P}}^{d-r}$ be a generic subspace of $\overline{\mathbf{P}}^d$ containing α. If $\sigma(t)$ is the intersection point of $\overline{\mathbf{P}}_t^r$ with $\overline{\mathbf{P}}^{d-r}$, and if we normalise the α^i and the $\sigma^i(t)$ by setting $\alpha^0 = \sigma^0(t) = 1$, then every $\sigma^j(t)$ is holomorphic in t, $t \in U(a)$, and we have $\sigma^j(a) = \alpha^j$.

Under obvious conditions, which may always be assumed, the expression $\sigma(t) \cdot \zeta(z) = \sum_{i=0}^{d} \sigma^i(t)\, \zeta_i(z)$, where $\zeta_j = \zeta_j(z)$ is a holomorphic function of (t, z) for $t \in U(a)$, $z \in U(p)$, where $U(p)$ is a neighbourhood on V of a given point p arbitrary on E_α.

For each fixed t we have $(\sigma(t) \cdot \zeta(z)) = E_{\sigma(t)}$ and the divisor $S_t = E_{\sigma(t)} - \Delta_t$ is a positive divisor with $S_a = S$. Hence, setting

$$\vartheta(P(z) + t) \cdot R_p(z, t) = \sigma(t) \cdot \zeta(z)\,, \tag{16}$$

we see that $R_p(z, t)$ is holomorphic in (t, z) when $t \in U(a)$, $z \in U(p)$.

Moreover, for each t fixed in $U(a)$, we have

$$(R_p(z, t)) = S_t\,, \quad \text{in} \quad U(p)\,. \tag{17}$$

On the other hand the function

$$F_\nu(z) = \partial_\nu \log \vartheta(P(z) + a)\,, \quad \partial_\nu = \partial/\partial t_\nu\,, \tag{18}$$

is, as one easily sees, a meromorphic additive function multiple of $-\Delta_a$, while

$$G_\nu(z) = \partial_\nu \log \sum \alpha^j \zeta_j(z) \tag{19}$$

is a single-valued meromorphic function on V with

$$(G_\nu) \geqq -E_\alpha = -S_a - \Delta_a . \tag{20}$$

Consequently $F_\nu - G_\nu$ is *a meromorphic additive function on* V, with $(F_\nu - G_\nu) \geqq -S_a - \Delta_a$. Deriving (16) with respect to t, we obtain

$$F_\nu(z) - G_\nu(z) = -\partial_\nu \log R_p(z, a) , \tag{21}$$

for any p in $E_\alpha = S_a + \Delta_a$. Since $(R_p(z, a)) = S_a$ in $U(p)$, we find that $F_\nu - G_\nu$ is multiple of $-S_a = -S$.

Finally, by calculating the periods, we can prove that the q functions $F_\nu - G_\nu$ and the q independent simple differentials of the first kind on V are all independent, which completes KODAIRA's proof.

c) Now it is clear that the system $|h\,E|$, i. e. a suitable multiple of the system of the hyperplane sections, satisfies all the hypotheses stated above: therefore there exist linear systems on V which have characteristic deficiency zero: this proves the statement (i).

We notice again that this theorem, combined with (11), gives at once the important consequence:

(ii) *On the non-singular variety V there exist exactly $2\,q$ independent simple* PICARD *integrals of the second kind, where q is the superficial irregularity of V.*

We give here some comments on the theory of the superficial irregularity. The reader will perhaps find some interest in a comparison with the classical situation.

Let C be a non-singular divisor of a surface F, over the complex field, whose genera are denoted by p_g, p_a, and for which the numbers of linearly independent integrals respectively of the first and of the second kind are g_1 and g_1^*. Moreover let us define as irregularity q of F the character $p_g - p_a$.

Then the classical facts are briefly as follows, where we shall omit the references which may be found in ZARISKI [a].

(a) $b^1(F) = g_1^*$; this was proved by transcendental methods by PICARD. Recently a proof based on stack theory and valid for any dimension has been given by ATIYAH and HODGE in [1], p. 64.

(b) $\operatorname{def} C/C \leqq q$; this was proved by a purely algebraic proof by CASTELNUOVO with later improvements due to SEVERI.

(c) $q = g_1, 2q = g_1^*$: the proof of such statements was given by the following stages:

(c_1) $g_1^* - g_1 \leqq q$; this was obtained by SEVERI by means of (b) and some considerations of the residues (of SEVERI) of the simple integrals of the second kind.

(c_2) $g_1 \leqq q$, $g_1^* \leqq 2\,q$; this was proved by means of (a) and a direct construction by SEVERI and by appealing to the properties of reducible Abelian integrals due to PICARD.

(c_3) $g_1 \geqq q$; this was derived by assuming the existence of a POINCARÉ family and by showing that there are at least q integrals of the first kind (CASTELNUOVO uses for this end the PICARD variety of F and SEVERI, more simply, a transcendental criterion for linear equivalence).

We point out that the equalities $g_1 = q$, $g_1^* = 2q$ have also received a direct transcendental proof by LEFSCHETZ on the following bases: the theorem of POINCARÉ on reducible integrals and a classical property of the so-called normal functions of POINCARÉ {see for all ZARISKI [a], Ch. VII}.

Furthermore if V is a non-singular variety of any dimension over the complex field, we have the statement:

(d) The number of independent differentials of the first kind on V is the same of the same number for any non-singular planar 2-section of V.

The first proof was given by CASTELNUOVO-ENRIQUES in [1]. We may therefore conclude with the following classical results:

(e) There is a character q associated with any non-singular surface F, which admits the following characterisations:

(e_1) $q = p_g - p_a$,

(e_2) $q = g_1$,

(e_3) $q = 1/2\, b^1(F)$,

(e_4) $q = 1/2\, g_1^*$,

(e_5) q is the maximum of the characteristic deficiencies of the non-singular divisors on F.

Now KODAIRA assumes (e_2) as definitions of the irregularity for any non-singular variety V over the complex field and he proves directly (e_4) and (e_5). The tools, as we have seen in the text, are: the theorem (i) in (XI, 1) which solves an existence problem on simple integrals of the second kind, the invariance of the arithmetic genus of divisors by homology, the dimensional formula of SEVERI-ZARISKI for ample linear systems expressed by means of arithmetic genera, and finally the existence of POINCARÉ families, which is assured by the previous considerations of (XI, 2).

This proof is an effective improvement of classical proofs, even for surfaces, by reason of its simplicity and compactness, though, obviously, in this case, it somewhat resembles the transcendental analysis of SEVERI and LEFSCHETZ.

Furthermore we remark that (e_1), when V is a surface, becomes in KODAIRA's treatment a theorem whose proof remains that of (b) which gives immediately the assertion (e_1) by means of (e_5).

When V is not a surface we wish to establish (d). The best available proof starts from (e_3), which has been again proved by HODGE in the elegant manner of (IX, 8a): it either uses well known topological properties of the BETTI number $b^1(V)$ or may also be derived from the equality $g_1(V) = g_1(F)$, where F is a non-singular planar 2-section of V.

This equality may in its turn be immediately derived from a statement of LEFSCHETZ {[a], pp. 88—91} to the effect that a 1-cycle on F is homologous to zero on F if it is so on V or may also be proved directly, following KODAIRA {[5], p. 93}, by means of the theory of HODGE-ECKMANN in (IX, 6), which is again equivalent to deriving it from another classical theorem of LEFSCHETZ to the effect that an arbitrary 1-cycle with rational coefficients on V is rationally homologous to a 1-cycle with rational coefficients on F {[a], pp. 88—91}.

This is in effect a satisfactory situation for our transcendental-topological theory, but if we go over to abstract fields, we are somewhat worse off, in spite of the important contributions of Ch. VII. Obviously we can no longer speak of integrals, but in virtue of the abstract constructions of the first and second PICARD variety, we may well replace the character $g_1(V)$ by the dimension q^* of both such varieties. Then (d) remains true in the new connotation by MATSUSAKA's statement in (VII, 8); we have also the fact that the characteristic deficiencies on a surface are bounded, which is a pure algebraic statement (b).

Nevertheless it remains to prove that q^* concides with $p_g - p_a$, which will be achieved only if one proves that q^* satisfies (e_5): this is an important question which probably derives its solution from the theory of A. WEIL for abstract Abelian varieties.

We recall the works [3, 4, 5] of MUHLY on these subjects, and we remark that an interesting, strictly algebraic, characterisation of the irregularity q of a threefold has been given by B. SEGRE in [5], p. 76.

SEVERI has recently extended very considerably the theory of superficial irregularity {see SEVERI [42, 43]}. For any non-singular variety V^r we define the character $q_r = p_g(V) - p_a(V)$ as the *last* or *r-dimensional* irregularity: here we may have $q_r \gtreqless 0$ {see also SEVERI [13]}.

A non-singular sub-variety V^h of $V^r (2 \leqq h \leqq r-1)$ is called *ordinary* when it possesses exactly the same number of linearly independent differential forms of every degree $s (1 \leqq s \leqq h-1)$ as V^r itself. It may be shown that *the last irregularity q^h of such a V^h is independent of V^h*, and is absolutely invariant under birational transformations of V^r: it is called the *h-irregularity of V^r*.

We thus have $r-1$ irregularities $q_2, q_3, \ldots, q_r$ of V_r, and in addition, an irregularity q_1 (always zero) of any algebraic curve of V^r. V^r is called *h-irregular* if $q_h \neq 0$ and *completely regular* if $q_h = 0 (h = 1, 2, \ldots, r)$.

SEVERI has proved that *the number of linearly independent differential forms of degree $s (1 \leqq s \leqq r-1)$ attached to V^r is equal to $q_s + q_{s+1} (q_1 = 0)$*. For $s = r$ this number is of course $p_g(V^r)$ {see (IX, 7)}.

We quote also the interesting theorem: *a n. a. s. c. that a superficially irregular variety V^r should contain an involutory system $\omega^{r-h} (1 \leqq h \leqq r-1)$ of superficial irregularity $\geqq h+1$ (a pencil of genus $\geqq 1$ of divisors if $h = 1$) is that V^r should possess $h+1$ linearly independent forms of the first species, whose product is zero and such that none of the products, taken h at a time, vanishes* {see SEVERI [42], where many consequences of this theorem are given}.

4. Characteristic Systems of Complete Continuous Systems

The results of this section, all due to KODAIRA {see KODAIRA [5], p. 123}, are of outstanding importance for algebraic geometry over the complex field and generalise to higher varieties the deepest results obtained in the surface theory by the combined efforts of the French and Italian schools.

a) Let $|D_0|$ be a complete linear system on V, such that $|D_0 - K + \Delta_0 - \Delta_t|$ and $|D_0 + \Delta_0 - \Delta_t|$ are both ample for any t. Then $|E| = |D_0 + \Delta_0|$ is still ample and the dimension $r = \dim |E - \Delta_t|$, is independent of t {see (XI, 3 b)}. If $\mathfrak{C}$ is the system of all divisors on V, positive and rationally homologous to D_0, it is clear that this system contains all linear systems $|E - \Delta_t|$. By using the homomorphism λ of $\mathfrak{A}$ onto $\mathfrak{P}$ {see (XI, 2 d)} we deduce at once that the mapping $\mu: D \to [\eta(D - D_0)]$ maps $\mathfrak{C}$ onto $\mathfrak{P}$.

We now wish to determine the structure of $\mathfrak{C}$.

To this end let α be an arbitrary point on $\mathfrak{P}$ and $\mathfrak{a}$ a point of $\mathfrak{A}$ such that $\alpha = \lambda(\mathfrak{a})$: moreover we denote by $U(\mathfrak{a})$ a neighbourhood of $\mathfrak{a}$ on $\mathfrak{A}$ biregularly mapped by λ onto a neighbourhood $U(\alpha)$ of α on $\mathfrak{P}$. We can suppose that, for a suitable $\vartheta^{(\alpha)}(u)$ the divisor Δ'_a, where $\mathfrak{a} = [a]$, is prime and non-singular: then the divisor $\Delta_\tau^{(\alpha)} = \Delta_t^{(\alpha)}$ with $\tau = \lambda(t)$ is also prime and non-singular, and depends holomorphically on t and therefore on τ when $\tau \in U(\alpha)$, where $U(\alpha)$ is suitably selected.

If we consider the subsystem $\mathfrak{C}_\alpha$ of $\mathfrak{C}$ consisting of the divisors $D = \mu^{-1}(\tau), \tau \in U(\alpha)$, we have for each such divisor $\mu(D) = [\eta(\Delta_0 - \Delta_\tau^{(\alpha)})]$ and hence $D \in |E - \Delta_\tau^{(\alpha)}|$. We thus obtain the result:

(i) *If V has superficial irregularity q there exist systems of divisor on V consisting, locally, of ∞^q linear systems.*

b) Now proceeding similarly to (XI, 3 c) and using analogous symbols we deduce that each divisor of $\mathfrak{C}_\alpha$ is of the type $E_\sigma - \Delta_\tau^{(\alpha)}$, where $\sigma \in \overline{\mathbf{P}}_{\alpha,t}^r$, and $\tau \in U(\alpha)$. By setting

$$\sigma_{\alpha k}(\tau) = \mathbf{P}_{\alpha,t}^r \cdot \overline{\mathbf{P}}_k^{d-r}, \quad k = 0, 1, \ldots, r+1, \tag{22}$$

we may introduce into $\overline{\mathbf{P}}_{\alpha,t}^r$ homogeneous projective coordinates $(\xi_\alpha, \zeta_\alpha, \ldots, \xi_\alpha^r)$ with respect to the points $(\sigma_{\alpha k}(\tau))$ as fundamental points of reference: we then normalise these coordinates by putting

$$\sum_{k=0}^{r+1} \sigma_{\alpha k}^j(\tau) = 0. \tag{23}$$

Now, if $\mathbf{P}^r$ is a projective space $(\xi^0, \xi^1, \ldots, \xi^r)$, we obtain clearly a one-to-one correspondence between $\mathbf{P}^r \times U(\alpha)$ and $\mathfrak{C}_\alpha$, and a simple calculation proves that, if α, β belong to $\mathfrak{P}$, with $U(\alpha) \cap U(\beta) \neq 0$, then each member $D \in \mathfrak{C}_\alpha \cap \mathfrak{C}_\beta$ has two representatives: one, $\xi_\alpha \times \tau$, is in $\mathbf{P}^r \times U(\alpha)$ and the other, $\xi_\beta \times \tau$, is in $\mathbf{P}^r \times U(\beta)$, which correspond to each other by a homography in $\mathbf{P}^r$, holomorphically depending on τ.

By identifying such elements, the system $\{\mathbf{P}^r \times U(\alpha)\}$ with $\alpha \in \mathfrak{P}$, becomes an analytic bundle Λ over $\mathfrak{P}$, whose fibre is the projective space $\mathbf{P}^r$ and whose structure group is the projective group in $\mathbf{P}^r$. Hence and from (X, 10, vi) it follows that Λ may be considered as *an algebraic projective non-singular variety.*

This means that $\mathfrak{C}$ is an algebraic system: in effect let p be a point of V and $\lambda_0 = 0_\alpha \times \alpha$ a point of Λ and let us set

$$R_p(z, \lambda) = \sum_{k=0}^{r} \xi_\alpha^k R_{p,k}^{(\alpha)}(z, t), \quad \lambda = \xi_\alpha \times \tau \text{ in } U(\lambda_0) \supset \lambda_0, \tag{24}$$

where the functions $R_{p,k}^{(\alpha)}(z, t)$, holomorphic in (z, t), are defined as in (XI, 3 c) and where we suppose for example $\xi_\alpha = 1$.

Then $R_p(z, \lambda)$ is a holomorphic function of (z, λ) in $U(p) \times U(\lambda_0)$ on $V \times \Lambda$, and, if λ is fixed, we have $(R_p(z, \lambda)) = D_\lambda$, being D_λ the divisor of $\mathfrak{C}$ associated with λ. Now if $\mathfrak{R} = (R_p(z, \lambda))$, we have, in the neighbourhood of each point $p \times \lambda_0$ of V,

$$D_\lambda \times \lambda = \mathfrak{R} \cdot (V \times \lambda). \tag{25}$$

Therefore $\mathfrak{C}$ is parametrised by the divisor $\mathfrak{R}$ of $V \times \Lambda$ which, by Chow's theorem (IX, 1), is algebraic and hence $\mathfrak{C}$ also is algebraic. Obviously the system $\mathfrak{C}$ is complete whether as a divisor system or as a system of complete linear systems.

Hence follows immediately *the identification of algebraic equivalence on V with topological equivalence over the integrals*, which is a well known result of Lefschetz for surfaces.

Consequently the groups $\mathbf{G}_0$ and $\mathbf{G}_a$ {see (X,6)} coincide and all results valid for the former are also valid for the latter, and conversely: in particular the definitions of Picard variety, in the strict algebraic sense of Ch. VIII or in the transcendental sense of this Ch. XI or also in that of Ch. X are all equivalent. Such definitions also coincide for the dual concept of the second Picard variety; and, finally, one and the same character q acquires all the meanings attached to it in (XI, 3).

c) We shall now introduce the notion of characteristic linear system of the algebraic system $\mathfrak{C}$ on any one of its prime non-singular members. Let $S = D_{\lambda_0}$ be such a member of $\mathfrak{C}$ and let us denote by $\partial_j R_p(z, \lambda_0)_S$ the holomorphic function induced on S by the partial derivative

$$\partial_j R_p(z, \lambda_0) = [\partial R_p(z, \lambda)/\partial \lambda^j]_{\lambda = \lambda_0}, \tag{26}$$

where $(\lambda^1, \lambda^2, \ldots, \lambda^{r+q})$ are local coordinates in a neighbourhood $U(\lambda_0)$ of λ_0 on Λ.

Then it is immediately seen that the divisor on S, given by the expression

$$\overline{D}_\mu = \Big(\sum_j \mu^j\, \partial_j\, R_p(z, \lambda_0)_S\Big),$$

is well defined for any value of the quantities μ^j, provided that the function on the right does not vanish identically on S.

So we evidently obtain a linear system on S, which is precisely the so-called *characteristic linear system* $\{\overline{D}_\mu\}$ *of* $\mathfrak{C}$ *on* S: it is determined only by $\mathfrak{C}$ and S.

d) We shall now describe Kodaira's proof of the completeness of the system $\{\overline{D}_\mu\}$.

Let $\lambda_0 = 0_\alpha \times \alpha$, $0_\alpha = (1, 0, \ldots, 0)$ with reference to the symbols used in (b) and, if $\lambda = \xi_\alpha \times \tau$, with $\xi_\alpha = (1, \xi^1, \ldots, \xi^r)$ and $\tau = \varphi(t)$ assume as local coordinates of (λ) the $\lambda^1 = \xi^1, \ldots, \lambda^r = \xi^r, \lambda^{r+1} = t^1, \ldots, \lambda^{r+q} = t^q$. Then from the definition of the function $R_p(z, \lambda)$ in (b) we deduce that

$$\sum_{j=1}^{r+q} \mu^j\, \partial_j R_p(z, \lambda_0) = \sum_{j=1}^{r} \mu^j\, R_{pj}^{(\alpha)}(z, \alpha) + \sum_{\nu=1}^{q} \mu^{r+\nu}\, \partial_\nu R_{p0}^{(\alpha)}(z, \alpha)\,, \tag{27}$$

where

$$\partial_\nu = [\partial/\partial t_\nu\, R_{p0}^{(\alpha)}(z, \tau)]_{\tau = \alpha}\,. \tag{28}$$

Since $S = (R^{(\alpha)}_{p0}(z, \alpha))$, it is easily seen that the function

$$M(z) = \left[\sum_{k=1}^{r} \mu^k R^{(\alpha)}_{pk}(z, \alpha) : R_{p,0}(z, \alpha)\right] \tag{29}$$

is meromorphic and single-valued on V with $(M) \geqq -S$, while, by (XI, 3 c), each $M_\nu = -\partial_\nu \log R^{(\alpha)}_{p0}$ is meromorphic additive on V with $(M_\nu) \geqq -S$.

Now if $\sum \mu^j \partial_j R_p(z, \lambda_0)_S$ vanishes identically on S, from (27) it follows that the function $M - \sum_{\nu=1}^{q} \mu^{r+\nu} M_\nu$ is everywhere holomorphic on V, so that this function is a Picard integral of the first kind on V, which is absurd unless it is constant and the $\mu^{r+1}, \ldots, \mu^{r+q}$ all vanish, since $M_1, M_2, \ldots, M_q, P_1, P_2, \ldots, P_q$ are independent, by (XI, 3c). Therefore $M = c_0$: then the definition of M shows that all the μ's and also c_0 vanish. It follows that $\dim\{\overline{D}_\mu\} = r + q - 1$.

On the other hand, it is obvious from the definitions that $\{\overline{D}_\mu\} \geqq |S| \cdot S$ and hence, if $T \in |S| \cdot S$, we have $\{\overline{D}_\mu\} \subseteqq |T|_S$ and since $\dim|T|_S = \dim|S| - 1 + \operatorname{def}(S/S)$, with $\dim|S| = r$ and $\operatorname{def}(S/S) \leqq q$ by (8), we have $\dim|T|_S \leqq \dim\{\overline{D}_\mu\}$. Therefore $\{\overline{D}_\mu\} = |T|_S$ and hence the characteristic system of $\mathfrak{C}$ is complete. We may conclude with the following theorem of Kodaira:

(ii) *Let $|D_0|$ be a complete system on V such that $|D_0 - K + \Delta_0 - \Delta_t|$ and $|D_0 + \Delta_0 - \Delta_t|$ are both ample for any t, and let $\mathfrak{C}$ be the algebraic system consisting of all the positive V-divisors $D \sim D_0$ on V, and let S be a prime non-singular divisor belonging to $\mathfrak{C}$. Then the characteristic linear system on S of $\mathfrak{C}$ is complete.*

We remark that, if D_1 is an arbitrary V-divisor, then the hypotheses of our theorem are certainly satisfied by the linear system $|D_0| = |D_1 + h E|$, for large h.

We recall briefly the classical question of the completeness of the characteristic system of continuous complete systems.

We know that, on any prime non-singular divisor S of a non-singular variety V, which we suppose defined over the complex field, the complete linear system $|S^{[2]}|$ is defined {see (VI, 9)}, independently of the existence or otherwise of a linear or algebraic system of divisors containing S. We ask now whether it is possible to construct a maximal family $\{S\}$ on V which contains S and whose characteristic system Γ_S on S (defined analogously to (XI, 4)) is $|S^{[2]}|$: when V is a surface F the characteristic series has been stated by Severi to answer this question.

In such a case it was known that, if F is regular (i. e. $p_g = p_a$), then $\{S\}$ is necessarily a linear complete system with a complete characteristic series (Enriques, Castelnuovo; see Zariski [a], Ch. IV). Successive proofs of Enriques and Severi to the effect that, if $\{S\}$ is complete, then so is also Γ_S, were not correct, as Severi himself pointed out in [17] {see criticism in Zariski [a], Ch. V}. Moreover the result itself is not invariably true, as was proved by examples {see Zappa [1, 2, 3]}.

After this mathematicians retired to more cautious positions, which had at least the merit of saving the applications already made {as e. g. the theory of irregularity; see (XI, 3)}. A useful tool for this purpose was the theorem proved by Poincaré in [4] by means of his normal functions and successively improved by LEFSCHETZ in [a] and by SEVERI in [17]: it affirms the existence on F of a q-dimensional algebraic family whose curves are all isolated with respect to linear equivalence, where q is the irregularity of F.

Such a family, as we have seen in Ch. VII, has been called a POINCARÉ family {see also (XI, 2)}. Hence SEVERI deduced, by means of the RIEMANN-ROCH theorem, the existence of complete continuous families with complete characteristic series: such is in particular any non-special total family in the sense of (VII, 6), which is all compounded of arithmetically effective curves in the sense of SEVERI {see (VII, 6) and ZARISKI [a], p. 86}.

The present treatment of KODAIRA {see also, for surfaces, ANDREOTTI [1], p. 17} constitutes an improvement of POINCARÉ's proof: it holds for any dimension and settles together with our question that of proving the coincidence of algebraic equivalence with integral homology, which for surfaces is the so-called theorem of LEFSCHETZ {see LEFSCHETZ [a], pp. 80—82; ZARISKI [a], p. 143}. This has been made possible by the beautiful th. (vi) of KODAIRA in (X, 10). The last part of the proof is partly classical: see e. g. the researches of B. SEGRE in [5]. The problem of finding a strictly algebraic proof of these results remains outstanding for abstract algebraic geometry.

5. Miscellaneous Results

a) The identification of algebraic equivalence with integral homology on V gives the result that the subgroup of $\mathbf{G}$, consisting of all divisors rationally homologous to zero on V coincides with the subgroup $\mathbf{G}_s$ of the divisors D which divide the zero of algebraic equivalence, i. e. are such that $\lambda D \equiv 0$, with λ integral.

Hence *the group* $\mathbf{G}_s/\mathbf{G}_a$, *which is the so-called division group or* SEVERI *group on* V {this group has been introduced by SEVERI in [12, 14]; see also ZARISKI [a], p. 95}, *coincides with the group* $\mathbf{G}_r/\mathbf{G}_0$ *of* (X, 6 c) *and this, by* (X, 6, iv), *with the torsion group* $\mathrm{T}_{2n-2}(V)$ *of* V.

Interesting examples of varieties endowed with torsion have been given by ROTH {see ROTH [10]; the preceding theorem has been proved by LEFSCHETZ for surfaces in [a], pp. 80—82; see also ZARISKI [a], p.143}.

b) A V-divisor is said to be *arithmetically equivalent to zero if it has zero intersection with each* 1-*cycle algebraic on* V {see SEVERI [22], p. 247}.

It is possible to prove that $\mathbf{G}_r(V)$ *coincides with the system of* V-*divisors which are arithmetically equivalent to zero, so that the group* $\mathbf{G}/\mathbf{G}_r$ *becomes the group of the arithmetic equivalence on* V.

The proof is immediate by expressing D in the form $D \sim \sum_{i=1}^{\varrho} a_i D_i$, mod $\mathbf{Q}$, where $\{D_i\}$ is a basis of the group $\mathbf{G}(V)/\mathbf{G}_r(V)$ and $\mathbf{Q}$ denotes the rationals: then observe that if D is arithmetically zero, so is also $\sum_{i=1}^{\varrho} a_i D_i$. Hence $\sum_{i=1}^{\varrho} a_i(D_i, \gamma_k) = 0$ for any algebraic 1-cycle γ_k defined by $\gamma_k = L \cdot D_k$ where L is a surface obtained by a generic linear section of V. Since $\|(D_i, \gamma_k)\| \neq 0$, it follows that $a_i = 0$, and hence $D \sim 0$, mod $\mathbf{Q}$ {see IGUSA [3], p. 20}.

We remark that this is equivalent to affirming, in virtue of well known topological results that, if D has a zero algebraic intersection number with the algebraic curves of V, the same is true *for the transcendental 2-cycles on* V {see SEVERI [22], p. 248}.

When V is a *surface* SEVERI has proved that, *if two curves A and B on V cut the same number of points on a 1-cycle C with $C^{[2]}$ of degree zero, then A and B are pseudo-algebraically equivalent:* i. e. $\lambda A \equiv \lambda B$.

The identification of algebraic equivalence with topological equivalence with division allowed may also be founded on this purely algebraic criterion, following ALBANESE and SEVERI {see SEVERI [34]; the first case considered by SEVERI was restricted by projective conditions: see e. g. ZARISKI [a], p. 90; see also TODD [12]}.

c) The rank of the group $\mathbf{G}(V)/\mathbf{G}_r(V)$ coincides with the BETTI number of *the algebraic module* of the V-divisors, i. e. with the so-called *algebraic base number of* V*:* this number ϱ is known as the number of PICARD of V and, by (XI, 2 c), it admits a transcendental characterisation by means of the simple differentials of the third kind on V, entirely analogous to the one, due to PICARD, for surfaces {see e. g. ZARISKI [a], p. 114 and the note 13 above}.

It is sufficient to observe that the linear combination of $k+1$ V-divisors on V is always homologous to zero, and then algebraically zero, if $k \geqq \varrho$ and it is not generally so if $k < \varrho$.

d) It is obvious that $\varrho \leqq b^{2n-2}(V)$. The number $\varrho_0 = b^{2n-2}(V) - \varrho$ may be called the LEFSCHETZ number of V, and it clearly represents *the least number of transcendental cycles on V which are linearly independent on V.*

The *absolute* invariance of this number may be easily guessed, since birational transformations affect only algebraic cycles, and it may be formally proved by using the LEFSCHETZ-HODGE criterion for algebraic cycles in (X, 6, iii).

It follows that the number ϱ is a relative invariant of the same type as BETTI's number. On the other hand we notice that *the order of the torsion group, or* SEVERI's *number σ is, instead, an absolute invariant* {see ZARISKI [a], p. 93}.

The number ϱ_0 coincides with the number of independent 2-forms of the second kind on V: this is a classical theorem of LEFSCHETZ, which has been recently reviewed in the framework of stack theory by ATIYAH and HODGE in [1]. This research constitutes actually a deep analysis of the integrals attached to an algebraic variety which, unfortunately, cannot be briefly summarised.

e) The concept of arithmetic equivalence, which is of great importance in enumerative geometry, can be extended to the k-cycles on V, $k = 0, 1, \ldots, n$, by saying that *a k-cycle Y^k is arithmetically equivalent to zero if and only if the intersection product $Y^k \cdot Z^{n-k}$ is equal to zero when it is defined, where Z^{n-k} is any $(n-k)$-cycle on V.*

The existence of *a finite base* for the associated group of equivalence follows at once, as SEVERI observes, from the fact that if $Y \sim 0$, mod $\mathbf{Q}$, then it is also arithmetically zero. From this SEVERI has proved that the k-cycles and the $(n-k)$-cycles of V have *equal base numbers, with non-vanishing intersection matrix* with respect to the dual bases {see SEVERI [22], pp. 248—249}. We recall that, instead, the existence of a finite base for the algebraic equivalence of any dimension on V, which has been conjectured by SEVERI, has been reduced by this author to assuming the possibility of stating a finite system of conditions in some suitable characters of V, for a cycle X on V to be positive: see SEVERI [22], p. 251}.

f) We state here two theorems of CHOW {see CHOW [5]} concerning the behaviour of the fundamental group of V or POINCARÉ group, which are of importance in many applications.

(i) *Let V be non-singular and let M be a member of an irreducible algebraic system $\sum$ having V as support. If $f(z)$ is a continuous mapping of the unity interval I into V with $f(0) = f(1) = P_0 \in M$, then a finite k-th power of f is homotopic rel.* 0, 1, *to a continuous mapping of I into M. If $\sum$ is involutory, then $k = 1$.*

(ii) *Let φ be a birational transformation of a non-singular variety V onto a non-singular variety V', with* $\dim V = n$, $\dim V' = m$, *which is such that:* (1) *$\varphi^{-1}(\eta')$ is a variety if η' is a generic point on V';* (2) *The system H' of the points on V' which are either fundamental for φ^{-1} or else such that the corresponding set on V has some multiple component, has a dimension less than $m - 2$.*

Then there exists a homomorphism λ of $\mathfrak{G}(V)$ onto $\mathfrak{G}(V')$, where $\mathfrak{G}(V)$ and $\mathfrak{G}(V')$ denote respectively the fundamental groups of V or V', whose kernel is the subgroup $\mathfrak{G}_V(\varphi^{-1}(P_0'))$ of $\mathfrak{G}(V)$ consisting of the fundamental group relative to $\varphi^{-1}(P_0')$, where P_0' is a sufficiently general point on V', embedded in $\mathfrak{G}(V)$ by the inclusion map $\varphi^{-1}(P_0') \subset V$.

We remark that (i) generalises partially a transcendental criterion of SEVERI hitherto restricted to surfaces, and that (ii) contains as a particular case the theorem, also classical {see e. g. ZARISKI [a], p. 144}, that there exist precisely $2p$ independent 1-cycles on a surface which are not homologous to some 1-cycle belonging to the generic curve of a pencil of genus p. In fact it is sufficient to consider as variety V' the CHOW variety of the curves of the pencil and to apply after (ii).

We notice that if φ is a birational transformation we obtain a well known theorem of EHRESMANN contained in his thesis, which affirms that the groups $\mathfrak{G}(V)$ and $\mathfrak{G}(V')$ are then isomorphic: a result which we have implicitly applied before in (d) {see EHRESMANN [1]}.

Bibliography

Treatises, Monographs and Reports

BERTINI, E.: [a] Introduzione alla geometria proiettiva degli iperspazi, Principato. Messina 1923; [b] Einführung in die projektive Geometrie mehrdimensionaler Räume. Wien 1924.

BOCHNER, S., and W. T. MARTIN: [a] Several complex variables. Princeton: Princeton Press 1948.

CARTAN, H.: [a] Séminaire E. N. S. 1950—1951; [b] Séminaire E. N. S. 1951—1952; [c] Séminaire E. N. S. 1953—1954.

— and S. EILENBERG: [a] Homological algebra. Princeton: University Press 1955.

CASTELNUOVO, G.: [a] Memorie Scelte. Bologna: Zanichelli 1937.

— and F. ENRIQUES: [a] Die algebraischen Flächen vom Gesichtspunkte der birationalen Transformationen aus. Enzyklop. d. math. Wiss. III 6 b, 1914.

CONFORTO, F.: [a] Lo stato attuale della teoria dei sistemi di equivalenza e delle corrispondenze algebriche tra varietà. Atti Congresso U. M. I., pp. 49—83. Roma 1942; [b] Funzioni abeliane e matrici di RIEMANN. Roma: Libreria Università 1942.

DE RHAM, G.: [a] Variétés différentiables. Act. Sci. Ind., 1222. Paris: Hermann 1955.

— and K. KODAIRA: [a] Harmonic Integrals. Princeton: Inst. for Advanced Study 1950.

DUBREIL, P.: [a] Quelques propriétés des variétés algébriques. Act. Sci. Ind. **210**. Paris: Hermann 1935.

GAETA, M. F.: [a] Quelques progrès récents dans la classification des variétés algébriques d'un espace projectif. Coll. de Géom. Alg., pp. 145—183. Liège 1952.

GODEAUX, L.: [a] Correspondances entre deux courbes algébriques. Mémor. Sci. Math. 111. Paris: Gauthier-Villars 1949.

GRÖBNER, W.: [a] Moderne algebraische Geometrie. Wien: Springer 1949.

HILBERT, D.: [a] Gesammelte Abhandlungen, 2 Bd. Berlin: Julius Springer 1933.

HIRZEBRUCH, F.: [a] Arithmetische Geschlechter und der Satz von RIEMANN-ROCH. Erg. Math. Berlin: Julius Springer 1956.

HODGE, W. V. D.: [a] The theory and applications of harmonic integrals. Cambridge: University Press 1941.

— and D. PEDOE: [a] Methods of algebraic geometry, I, II, III. Cambridge: University Press 1952—1954 (quoted as a_1, a_2, a_3).

KRULL, W.: [a] Idealtheorie. Erg. Math. Berlin: Julius Springer 1935.

LEFSCHETZ, S.: [a] L'Analysis situs et la géométrie algébrique. Paris: Gauthier-Villars 1924; [b] Hyperelliptic functions and Abelian varieties. Bull. Nat. Res. Counc. 63. Washington 1928; [c] Topology. Coll. Publ. New York 1930; [d] Algebraic Geometry. Princeton: University Press 1953.

LICHNEROWICZ, A.: [a] Gruppo d'olonomia e omologia. Pubbl. Istituto Mat. Roma 1954.

MACULAY, F. S.: [a] The algebraic theory of modular systems. Cambridge Tracts, 19, Cambridge 1916.

NORTHCOTT, D. G.: [a] Ideal theory. Cambridge: University Press 1953.

ROTH, L.: [a] Algebraic threefolds. Rend. mat. Roma (5), **10**, 1—50 (1951); [b] Algebraic threefolds (with special regard to problems of rationality). Erg. Math. Berlin: Julius Springer 1955.

SAMUEL, P.: [a] Commutative algebra. Cornell University, 1953; [b] Algèbre locale. Mém. Sci. Math. 123. Paris: Gauthier-Villars 1953; [c] Méthodes d'algèbre abstraite en géométrie algébrique. Erg. Math. Berlin: Julius Springer 1955.

SCHIFFER, M., and D. C. SPENCER: [a] Functionals of finite RIEMANN surfaces. Princeton: University Press 1953.

SCHUBERT, F.: [a] Kalkül der abzählenden Geometrie. Leipzig: B. G. Teubner 1879.

SCHWARTZ, L.: [a] Théorie des distributions, I, II. Act. Sci. Ind., 1091, 1122. Paris: Hermann 1950—1951.

SEGRE, B.: [a] Geometria algebrica nei paesi anglosassoni (dal 1939 al 1945). Relat. Pont. Acad. Sci. **11**, 1—52 (1946); [b] Questions arithmétiques sur les variétés algébriques. Coll. Int. d'algèbre et théorie des nombres, pp. 83—94, Paris 1949; [c] Forme differenziali e loro integrali. Roma: Docet 1951.

SEGRE, C.: [a] Mehrdimensionale Räume. Enzyklop. d. math. Wiss. III, **7**, 1912, 1921.

SEVERI, F.: [a] Serie, sistemi d'equivalenza e corrispondenze algebriche sulle varietà algebriche. I. Roma: Cremonese 1942; [b] Fondamenti di geometria algebrica. Cedam, Padova, 1948; [c] Introduzione alla geometria algebrica (Geometria numerativa), I, II. Roma: Docet 1948—1949; [d] La géométrie algébrique Italienne. Coll. de Géom. Alg., pp. 9—55. Liège 1949; [e] Memorie scelte. Bologna: Zuffi 1950.

STEENROD, N.: [a] The topology of fibre bundles. Princeton: University Press 1951.

WAERDEN, B. L. VAN DER: [a] Moderne Algebra, I, II. (quoted as a_1, a_2). Berlin: Julius Springer 1955; [b] Einführung in die algebraische Geometrie. Berlin: Julius Springer 1939.

WEIL, A.: [a] Foundations of algebraic geometry. Colloquium Publ. New York 1946; [b] Sur les courbes algébriques et les variétés qui s'en déduisent. Act. Sci. Ind. 1041. Paris: Hermann 1948; [c] Variétés abéliennes et courbes algébriques. Act. Sci. Ind., 1064. Paris: Hermann 1948.

WU WEN-TSUN: [a] Sur les classes charactéristiques des structures fibrées sphériques. Act. Sci. Ind. 1183. Paris: Hermann 1952.

ZARISKI, O.: [a] Algebraic surfaces. Erg. Math. Berlin: Julius Springer 1935; [b] The fundamental ideas of abstract algebraic geometry. Proc. of the Int. Congress of Math., Cambridge, Mass., II, 1950, pp. 77—89.

List of Papers

ABELLANAS, P. F.: [1] Dimension of an algebraic variety. Rev. Mat. Hisp.-Amer. (4), **2**, 13—21 (1942); [2] On the geometrical theory of algebraic surfaces for a perfect coefficient field of characteristic p. Rev. Accad. Ci. Madrid **36**, 482—499 (1942); [3] Théorie arithmétique des correspondances algébriques. Rev. Mat. Hisp.-Amer. (4), **9**, 175—233 (1949); (4), **11**, 159—179 (1951); [4] Fundamental subvariety for an algebraic correspondence. Rev. Mat. Hisp.-Amer. (4), **10**, 207—232 (1950); [5] Orientation of algebraic varieties. Rev. Mat. Hisp.-Amer. (4), **12**, 79—101 (1952); [6] Primals of an algebraic variety. Rev. Mat. Hisp.-Amer. (4), **13**, 255—282 (1953).

AKIZUKI, Y.: [1] Theorems of BERTINI on linear systems. J. Math. Soc. Jap. **3**, 170—180 (1951).

ALBANESE, G.: [1] Intorno ad alcuni concetti e teoremi fondamentali sui sistemi algebrici di curve d'una superficie algebrica. Ann. di Mat. (3), **24**, 159—233 (1915); [2] Sul genere aritmetico delle varietà a quattro dimensioni. Rend. Accad. Linc. (**5**), 33, 179—182 (1924); [3] Invarianza del genere P_a d'una varietà algebrica a quattro dimensioni. Rend. Accad. Linc. (5), **33**, 210—214 (1924); [4] Formule fondamentali della geometria sopra una varietà algebrica.

Ann. di Mat. **4**, 154—184 (1927); [5] Sul teorema fondamentale della base per la totalità delle curve algebriche d'una superficie algebrica. Rend. Accad. Linc. (6), **5**, 481—488 (1927); [6] Corrispondenze algebriche fra i punti di due superficie algebriche. Boll. Un. Mat. Ital. **11**, 131—138 (1932).

ANDREOTTI, A.: [1] Recherches sur les surfaces irrégulières. Acad. Roy. belg. **27**, n. 4, 1—56 (1952); [2] Recherches sur les surfaces irrégulières. Acad. Roy. belg. **27**, n. 7, 1—36 (1952).

ARCHBOLD, J. A.: [1] Multiple intersections on an algebraic V_4. Proc. London Math. Soc. (2), **47**, 101—122 (1941).

ATIYAH, M. F., and W. V. D. HODGE: [1] Integrals of the second kind on an algebraic variety. Ann. of Math. (2), **62**, 56—91 (1955).

BARKER, C. C. H.: [1] Intersections and contacts of surfaces on a V_3. J. London Math. Soc. **26**, 125—131 (1951); [2] Contacts of surfaces on an algebraic fourfold. J. London Math. Soc. **30**, 343—350 (1955).

BARSOTTI, I.: [1] Algebraic correspondences between algebraic varieties. Ann. of Math. (2), **52**, 427—464 (1950); [2] Local properties of algebraic correspondences. Trans. Amer. Math. Soc. **71**, 349—378 (1951); [3] Intersection theory for cycles of an algebraic variety. Pacific J. Math. **2**, 473—521 (1952); [4] A note on Abelian varieties. Rend. Circ. Mat. Palermo (2), **2**, 236—257 (1953); [5] Structure theorems for group-varieties. Ann. di Mat. (4), **38**, 77—119 (1955).

BEHRENS, E. A.: [1] Zur Schnittmultiplizität uneigentlicher Komponenten in der algebraischen Geometrie. Math. Z. **55**, 199—215 (1952).

BIDAL, P., and G. DE RHAM: [1] Les formes différentielles harmoniques. Comm. Math. Helvet. **19**, 1—49 (1946).

BOCHNER, S.: [1] Vectors fields and RICCI curvature. Bull. Amer. Math. Soc. **52**, 776—797 (1946); [2] Curvature and BETTI numbers. Ann. of Math. **49**, 379—390 (1948); **50**, 587—665 (1949).

BRAUER, R.: [1] A note on HILBERT's Nullstellensatz. Bull. Amer. Math. Soc. **54**, 894—896 (1948).

CARTAN, H.: [1] Problèmes globaux dans la théorie des fonctions analytiques de plusieurs variables complexes. Proc. Int. Congress of Math. Cambridge, Mass. **1**, 152—164 (1950); [2] Variétés analytiques complexes et cohomologie. Coll. de Bruxelles **1953**, 41—55.

— and J. P. SERRE: [1] Un théorème de finitude concernant les variétés analytiques compactes. C. r. Acad. Sci. (Paris) **237**, 128—130 (1953).

CASTELNUOVO, G.: [1] Alcune proprietà fondamentali dei sistemi lineari di curve tracciate sopra una superficie algebrica. Ann. di Mat. (2), **25**, 235—318 (1897); [2] Sugli integrali semplici appartenenti ad una superficie irregolare. Rend. Accad. Linc. (5), **14** (1905) (3 notes).

— and F. ENRIQUES: [1] Sopra alcune questioni fondamentali nella teoria delle superficie algebriche. Ann. di Mat. (3), **6**, 165—225 (1901); [2] Sur les intégrales simples de première espèce d'une surface ou d'une variété algébrique à plusieurs dimensions. Ann. Ec. Norm. Sup. (5), **23**, 339—366 (1906).

CHERN, S. S.: [1] Characteristic classes of Hermitian manifolds. Ann. of Math. **47**, 85—121 (1948); [2] On the multiplication in the characteristic ring of a sphere bundle. Ann. of Math. **49**, 362—372 (1948); [3] On the characteristic classes of complex sphere bundles and algebraic varieties. Amer. J. Math. **75**, 565—597 (1953).

CHEVALLEY, C.: [1] On the theory of local rings. Ann. of Math. **44**, 690—708 (1943); [2] On the notion of the ring of quotients of a prime ideal. Bull. Amer. Math. Soc. **50**, 93—97 (1944); [3] Intersections of algebraic and algebroid varieties. Trans. Amer. Math. Soc. **57**, 1—85 (1945).

CHOW, W. L.: [1] On the genus of curves of an algebraic system. Trans. Amer. Math. Soc. **65**, 137—140 (1949); [2] On compact complex analytic varieties. Amer. J. Math. **71**, 893—914 (1949); [3] Algebraic systems of positive cycles in an algebraic variety. Amer. J. Math. **72**, 247—274 (1950); [4] On the defining field of a divisor in an algebraic variety. Proc. Amer. Math. Soc. **1**, 797—799 (1950); [5] On the fundamental group of an algebraic variety. Amer. J. Math. **74**, 726—736 (1952); [6] On PICARD varieties. Amer. J. Math. **74**, 895—909 (1952); [7] On the quotient variety of an Abelian variety. Proc. Nat. Acad. Sci. USA. **38**, 1039—1044 (1952); [8] The Jacobian variety of an algebraic curve. Amer. J. Math. **76**, 453—476 (1954); [9] Abelian varieties over function fields. Trans. Amer. Math. Soc. **78**, 253—275 (1955).

— and K. KODAIRA: [1] On analytic surfaces with two independent meromorphic functions. Proc. Nat. Acad. Sci. USA **38**, 319—325 (1952).

— and B. L. VAN DER WAERDEN: [1] Zur algebraischen Geometrie, IX. Math. Ann. **113**, 692—704 (1937).

COHEN, I. S.: [1] On the structure and ideal theory of complete local rings. Trans. Amer. Math. Soc. **59**, 54—106 (1946); [2] Commutative rings with restricted minimum condition. Duke Math. J. **17**, 27—42 (1950).

— and A. SEIDENBERG: [1] Prime ideals and integral dependence. Bull. Amer. Math. Soc. **52**, 252—261 (1946).

COMESSATTI, A.: [1] Sulla serie canonica d'una superficie algebrica. Rend. Accad. Linc. (6), **16**, 555—560 (1932).

DANTONI, G.: [1] Sulle singolarità della Jacobiana e su quelle della varietà delle ipersuperficie con punto doppio di un generico sistema lineare ∞^r di V_{r-1} di S^r. Ann. Sci. Norm. Sup. Pisa (3) **3**, 1—17 (1949).

DE BAER, J. H.: [1] The relation between the CAYLEY form and the BARSOTTI form of an algebraic chain. Ind. Math. **15**, 158—161 (1953).

DEDEKIND, R., and H. WEBER: [1] Theorie der algebraischen Funktionen einer Veränderlichen. J. f. Math. **92**, 181—290 (1882).

DE RHAM, G.: [1] Sur la théorie des formes différentielles harmoniques. Ann. Univ. Grenoble **22**, 135—152 (1946); [2] Sur la division de formes et de courants par une forme linéaire. Comm. Math. Helvet. **28**, 346—352 (1954).

DERWIDUÉ, L.: [1] Essai sur le problème générale de la réduction des singularitiés d'une variété algébrique. Acad. roy. Belg. (5) **34**, 399—412 and 432—444 (1948); [2] Le problème de la réduction des singularités d'une variété algébrique. Math. Ann. **123**, 302—330 (1951); [3] Sur la réduction des singularités d'une variété algébrique. Mém. Soc. roy. Sci. Liège (4) **13**, 1—41 (1953).

DOLBEAULT, P.: [1] Sur la cohomologie des variétés analytiques complexes. C. r. Acad. Sci. (Paris) **236**, 175—177 (1953).

DUBREIL, P.: [1] Variétés arithmétiquement normales et variétés de première espèce. C. r. Acad. Sci. (Paris) **226**, 548—550 (1948); [2] La fonction charactéristique de HILBERT. Coll. Int. d'algèbre et théorie des nombres, Paris, **1950**, 109—114.

— and M. L. JACOTIN: [1] Divers types d'anneaux intervenant en géométrie algébrique. Coll. de Géom. Alg. Liège **1949**, 57—78.

DU VAL, P.: [1] Removal of singular points from an algebraic surface. Coll. Memoirs, Istanbul **1948**, 21—25; [2] On absolute and non-absolute singularities of algebraic surfaces. Rev. Fac. Sci. Univ. Istanbul (A) **11**, 159—215 (1944).

ECKMANN, B.: [1] Quelques propriétés globales des variétés Kähleriennes. C. r. Acad. Sci. (Paris) **229**, 577—579 (1949); [2] Sur les variétés presque complexes. Proc. Int. Congress of Math., Cambridge, Mass. **1**, 412—419 (1950).

— and H. GUGGENHEIMER: [1] Formes différentielles et métrique hermitienne sans torsion. C. r. Acad. Sci. (Paris) **229**, 464—466 et 489—491 (1949).

EGER, M.: [1] Determinazione del gruppo di punti doppi acquisiti da una forma dell' S_{2k} passante per una data V_k. Boll. Un. Mat. Ital. (1) **15**, 56—61 (1936); [2] Sur les systèmes canoniques d'une variété algébrique. C. r. Acad. Sci. (Paris) **204**, 92—94 et 217—219 (1937); [3] Sur la Jacobienne d'un système de PFAFF. C. r. Acad. Sci. (Paris) **209**, 82—84 (1939); [4] Les systèmes canoniques d'une variété algébrique à plusieurs dimensions. Ann. Sci. Ec. Norm. Sup. (3) **60**, 143—172 (1943).

EHRESMANN, C.: [1] Sur la topologie de certains espaces homogènes (Thesis). Ann. of Math. (2) **35**, 396—443 (1934).

FERNANDEZ BIARGE, J.: [1] Arithmetical investigation of linear systems of divisors of an algebraic variety. Mem. Inst. "Jorge Juan" **11**, 1—80 (1950).

FRANCHIS, B. DE: [1] Intorno al significato di alcuni caratteri delle varietà algebriche. Rend. Circ. Mat. Palermo **56**, 223—227 (1932); **60**, 161—168 (1936); [2] I sistemi canonici e pluricanonici e le forme algebrico-differenziali di prima specie. Ann. di Mat. (4) **19**, 243—249 (1940).

GAETA, F.: [1] Sulle curve algebriche di residuale finito. Ann. di Mat. (4) **27**, 177—241 (1948); [2] Nuove ricerche sulle curve sghembe algebriche di residuale finito. Ann. di Mat. (4) **31**, 1—64 (1950); [3] Sur la limite inférieure l_0 des valeurs de l pour la validité de la postulation régulière d'une variété algébrique. C. r. Acad. Sci. (Paris) **234**, 1121—1123 (1952); [4] Complementi alla teoria delle varietà algebriche V_{r-2} di residuale finito in S_r. Rend. Accad. Linc. (8) **12**, 270—273 (1952); [5] Sui sistemi lineari appartenenti al prodotto di più varietà algebriche. Ann. di Mat. (4) **33**, 91—118 (1952); [6] Sopra un aspetto proiettivamente invariante del metodo di eliminazione di KRONECKER. Rend. Accad. Linc. (8) **18**, 148—150 (1955).

GARABEDIAN, P. R., and D. C. SPENCER: [1] Complex boundary value problems. Office of Naval Research, Technical Report 16 (1951); [2] A complex tensor calculus for KÄHLER manifolds. Acta Math. **89**, 279—331 (1953).

GODEAUX, L.: [1] Sur les surfaces algébriques intersections complètes d'hypersurfaces. Rev. de Tucuman **2**, 211—216 (1941).

GODDARD, L. S..: [1] Bases for the prime ideals associated with certain classes of algebraic varieties. Proc. Cambridge Phil. Soc. **39**, 35—48 (1943); [2] Prime ideals and postulation formulae. Proc. Cambridge Phil. Soc. **44**, 43—49 (1948).

GOLDMAN, O.: [1] HILBERT rings and the HILBERT Nullstellensatz. Math. Z. **54**, 136—140 (1951).

GRÖBNER, W.: [1] Über den Multiplizitätsbegriff in der algebraischen Geometrie. Math. Nachr. **4**, 193—201 (1951); [2] L'ideale aggiunto di una varietà algebrica. Rend. Mat. Roma (5) **10**, 57—63 (1951); [3] Die birationalen Transformationen der Polynomideale. Mh. Math. **58**, 266—286 (1954).

GUGGENHEIMER, H.: [1] Über komplex-analytische Mannigfaltigkeiten mit KÄHLERscher Metrik. Comm. Math. Helvet. **25**, 257—297 (1951); [2] Interpretazione topologica dei covarianti di B. SEGRE. Rend. Acc. Linc. (8) **16**, 331—334 (1954); [3] Omologia delle dilatazioni. Rend. Acc. Linc. (8) **18**, 13—15 (1955).

HILBERT, D.: [1] Über die Theorie der algebraischen Formen. Math. Ann. **36**, 473—534 (1890), or "Gesammelte Abhandlungen", 2. Bd., 199—258.

HIRZEBRUCH, F.: [1] Arithmetic genera and the theorem of RIEMANN-ROCH, for algebraic varieties. Proc. Nat. Acad. Sci. USA **40**, 110—114 (1954); [2] Some problems on differentiable and complex manifolds. Ann. of Math. **60**, 213—236 (1954); [3] Über vierdimensionale RIEMANNsche Flächen mehrdeutiger analytischer Funktionen von zwei komplexen Veränderlichen. Math. Ann. **126**, 1—22 (1953).

HODGE, W. V. D.: [1] Further properties of Abelian integrals attached to algebraic varieties. Proc. Nat. Acad. Sci. USA **17**, 643—650 (1931); [2] On the stationary points of integrals attached to algebraic varieties. J. London Math. Soc. **58**, 280—290 (1937); [3] Note on the degeneration of algebraic varieties. Proc. Cambridge Phil. Soc. **38**, 217—219 (1942); [4] Note on the condition for a p-cycle of an algebraic manifold to be of rank k. Proc. Cambridge Phil. Soc. **43**, 577—580 (1947); [5] A new set of relative birational invariants of algebraic varieties. Accad. d'Ital., Fondaz. Volta 9 (1943); [6] Harmonic integrals on algebraic varieties. Proc. Cambridge Phil. Soc. **44**, 37—42 (1948); [7] On the topology of three-folds whose hyperplane sections have geometric genus zero. Ann. di Mat. (4) **29**, 115—119 (1949); [8] The topological invariants of algebraic varieties. Proc. Int. Congress of Math., Cambridge, Mass. **1**, 182—192 (1950); [9] A special type of KÄHLER manifold. Proc. London Math. Soc. **1**, 104—117 (1951); [10] Differential forms on KÄHLER manifolds. Proc. Cambridge Phil. Soc. **47**, 504—517 (1951); [11] The characteristic classes on algebraic varieties. Proc. London Math. Soc. (3) **1**, 138—151 (1951); [12] Tangent sphere-bundles and canonical models of algebraic varieties. J. London Math. Soc. **27**, 152—159 (1952); [13] A note on the RIEMANN-ROCH theorem. J. London Math. Soc. **30**, 291—296 (1955).

HOPF, H.: [1] Zur Topologie der komplexen Mannigfaltigkeiten. Studies and Essays pr. to R. Courant, 1948, 167—185; [2] Über komplex-analytische Mannigfaltigkeiten. Rend. Mat. Roma **10**, 169—182 (1951).

IGUSA, J.: [1] On the algebraic geometry of CHEVALLEY and WEIL. J. Math. Soc. Japan **1**, 198—201 (1949); [2] Algebraic correspondences between algebraic varieties. J. Math. Soc. Japan **3**, 215—219 (1951); [3] On the PICARD variety attached to algebraic varieties. Amer. J. Math. **74**, 1—22 (1952); [4] Some remarks on the theory of PICARD varieties. J. Math. Soc. Japan **3**, 345—348 (1952); [5] Normal point and tangent cone of an algebraic variety. Mem. Coll. Sci. Univ. Kyoto, A **27**, 189—201 (1951).

IWASAWA, K.: [1] Der BEZOUTsche Satz in zweifach projektiven Räumen. Proc. Japan Acad. **21**, 213—222 (1949); [2] Zur Theorie der algebraischen Korrespondenzen. Proc. Japan Acad. **21**, 204—212 and 411—418 (1949).

KÄHLER, E.: [1] Sui periodi degli integrali multipli sopra una varietà algebrica. Rend. Circ. Mat. Palermo **56**, 69—74 (1932); [2] Forme differenziali e funzioni algebriche. Mem. Acc. Ital. **3**, 1—19 (1932); [3] Über eine bemerkenswerte Hermitische Metrik. Abh. Math. Sem. Hansischen Univ. **9**, 173—186 (1933); [4] Tensori razionali di prima specie sopra una varietà algebrica. Rend. Acc. Linc. (8) **18**, 151—154 (1955).

KAWAHARA, Y.: [1] On the differential forms on algebraic varieties. Nagoya Math. J. **4**, 73—78 (1952).

KNESER, H.: [1] Die Integralen erster Gattung einer algebraischen Mannigfaltigkeit. Math. Ann. **107**, 83—86 (1932); [2] Analytische Mannigfaltigkeiten im komplexen projektiven Raum. Math. Nachr. **4**, 382—391 (1951).

KODAIRA, K.: [1] The theorem of RIEMANN-ROCH on compact analytic surfaces Amer. J. Math. 73, 813—875 (1951); [2] The theorem of RIEMANN-ROCH for adjoint systems on threedimensional algebraic varieties. Ann. of Math. **56**, 288—342 (1952); [3] On cohomology groups of compact analytic varieties with coefficients in some analytic faisceaux. Proc. Nat. Acad. Sci. USA **39**, 865—868 (1953); [4] On a differential-geometric method in the theory of analytic stacks. Proc. Nat. Acad. USA 39, 1268—1273 (1953); [5] Some results in the transcendental theory of algebraic varieties. Ann. of Math. **53**, 86—134 (1954); [6] On KÄHLER varieties of restricted type. Ann. of Math. **60**, 28—48 (1954).

KODAIRA, K. and D. C. SPENCER: [1] On arithmetic genera of algebraic varieties. Proc. Nat. Acad. Sci. USA **39**, 641—649 (1953); [2] Groups of complex line bundles over compact KÄHLER varieties. Divisor class groups on algebraic varieties. Proc. Nat. Acad. Sci. USA **39**, 868—877 (1953); [3] On a theorem of LEFSCHETZ and the lemma of ENRIQUES-SEVERI-ZARISKI. Proc. Nat. Acad. Sci. USA **39**, 1273—1278 (1953).

KOIZUMI, S.: [1] On the differential forms of the first kind on algebraic varieties. J. Math. Soc. Japan **1**, 273—280 (1949).

KRONECKER, L.: [1] Grundzüge einer arithmetischen Theorie der algebraischen Größen. J. f. Math. **92** (1881).

KRULL, W.: [1] Allgemeine Bewertungstheorie. J. f. Math. **167**, 160—196 (1931); [2] Dimensionstheorie in Stellenringen. J. f. Math. **179**, 204—226 (1938).

KUNDERT, E. G.: [1] Über Schnittflächen in speziellen Faserungen und Felder reeller und complexer Linienelemente. Ann. of Math. **54**, 215—246 (1951); [2] A relation between poles and zeros of a simple meromorphic differential form and a calculation of CHERN's characteristic classes of an algebraic variety. Proc. Nat. Acad. Sci. USA **38**, 893—895 (1952).

LEFSCHETZ, S.: [1] The arithmetic genus of an algebraic manifold immersed in another. Ann. of Math. **17**, 197—212 (1916); [2] On certain numerical invariants of algebraic varieties with applications to Abelian varieties. Trans. Amer. Math. Soc. **22**, 327—482 (1921).

LERAY, J.: [1] L'anneau spectral et l'anneau filtré d'homologie d'un espace localement compact. J. de Math. **29**, 1—139 (1950); [2] L'homologie d'un espace fibré, dont la fibre est connexe. J. de Math. **29**, 169—213 (1950).

LEVI, B.: [1] Risoluzione delle singolarità puntuali delle superficie algebriche. Atti Accad. Sci. Torino **33**, 66—86 (1897).

LONGHI, A.: [1] Sulla intersezione di due o più varietà algebriche. Comm. Math. Helvet. **18**, 45—51 (1946).

MACPHERSON, R. E.: [1] Canonical systems on a reducible variety. Proc. Cambridge Phil. Soc. **35**, 389—393 (1939).

MARTINELLI, E.: [1] Sulla varietà delle faccette p-dimensionali di S_r. Mem. Accad. Ital. **16**, 917—943 (1941); [2] Sulla immagine proiettiva delle serie e dei sistemi di equivalenza elementari sopra una varietà. Comm. Pont. Acad. Sci. **6**, 147—151 (1942); [3] Geometria algebrica e geometria riemanniana. Rend. Mat. Roma (5) **8**, 1—25 (1950); [4] Qualche proprietà geometrica nelle varietà a struttura complessa. Atti Accad. Ligure **9**, 1—10 (1952); [5] Alla ricerca di nuovi integrali invarianti sulle varietà algebriche. Atti IV Congresso U. M. I., pp. 1—9. Taormina 1951.

MATSUSAKA, T.: [1] Specialisation of cycles on a projective model. Mem. Coll. Sci. Univ. Kyoto, A **26**, 167—173 (1951); [2] The theorem of BERTINI on linear systems in modular fields. Mem. Coll. Sci. Univ. Kyoto, A **26**, 51—62 (1951); [3] On a generating curve of an Abelian variety. Nat. Sci. Rep. Ochanomizu Univ. **3**, 1—4 (1951); [4] On the algebraic construction of the PICARD variety. Jap. J. Math. (I) **21**, 217—236 (1951); (II) **22**, 51—62 (1952); [5] On algebraic families of positive divisors. J. Math. Soc. Jap. **5**, 113—136 (1953); [6] Some theorems on Abelian varieties. Nat. Rep. Ochanomizu Univ. **4**, 22—35 (1953); [7] On the theorem of CASTELNUOVO-ENRIQUES. Nat. Rep. Ochanomizu Univ. **4**, 164—171 (1953); [8] A remark on my paper (6). Nat. Rep. Ochanomizu Univ. **4**, 172—174 (1953).

MUHLY, H. T.: [1] A remark on normal varieties. Ann. of Math. **42**, 921—925 (1941); [2] Valuations and infinitely near loci. Amer. J. Math. **64**, 457—487 (1942); [3] Independent integral bases and a characterisation of regular surfaces. Trans. Amer. Math. Soc. **54**, 340—360 (1943); [4] The irregularity of an algebraic surface and a theorem on regular surfaces. Bull. Amer. Math. Soc. **55**, 940—947 (1949); [5] Integral bases and varieties multiply of the first species. Proc. Amer. Math. Soc. **2**, 576—580 (1951).

— and O. Zariski: [1] Hilbert's characteristic function and the arithmetic genus of an algebraic variety. Trans. Amer. Math. Soc. **69**, 78—88 (1950).

Nakai, Y.: [1] Note on the intersection of an algebraic variety with the generic hyperplane. Mem. Coll. Sci. Univ. Kyoto, A **26**, 185—187 (1951).

Néron, A.: [1] Problèmes arithmétiques et géométriques rattachés à la notion de rang d'une courbe algébrique dans un corps. Bull. Soc. Math. France **80**, 101—166 (1952); [2] La théorie de la base pour les diviseurs sur les variétés algébriques. Coll. Géom. Alg. Liège **1952**, 119—128.

Noether, E.: [1] Eliminationstheorie und allgemeine Idealtheorie. Math. Ann. **90**, 229—261 (1923).

Noether, M.: [1] Zur Theorie des eindeutigen Entsprechens algebraischer Gebilde von beliebig vielen Dimensionen. Math. Ann. **2**, 293—316 (1870); **8**, 495—533 (1875).

Northcott, D. G.: [1] The number of analytic branches of a variety. J. London Math. Soc. **25**, 275—279 (1950); [2] An application of local uniformisation to the theory of divisors. Proc. Cambridge Phil. Soc. **47**, 279—285 (1951); [3] Some properties of analytically irreducible geometric quotient rings. Proc. Cambridge Phil. Soc. 42, 662—667 (1951); [4] Specialisations over a local domain. Proc. London Math. Soc. (3) **1**, 129—137 (1951); [5] On integrally closed geometric quotient rings and their extensions. Proc. London Math. Soc. (3) **2**, 385—405 (1952); [6] Some results concerning the local analytic branches of an algebraic variety. Proc. Cambridge Phil. Soc. **49**, 386—396 (1953); [7] On the local cone of a point of an algebraic variety. J. London Math. Soc. **29**, 326—333 (1954).

Oehlert, M.: [1] Über die Definition der Zeuthen-Segreschen Invariante. J. f. Math. **171**, 42—54 (1934).

Okugawo, K.: [1] Linear conditions at a point. Mem. Coll. Sci. Univ. Kyoto, A **25**, 99—102 (1949); [2] Base conditions for hypersurfaces at a point. J. Math. Soc. Japan **1**, 242—250 (1950).

Ostrowski, A.: [1] Über ein algebraisches Übertragungsprinzip. Abh. Math. Sem. Univ. Hamburg **1**, 281—326 (1922).

Pannelli, M.: [1] Sopra alcuni caratteri di una varietà algebrica a tre dimensioni. Rend. Accad. Linc. (5) **15**, 483 (1906); [2] Sopra gli invarianti birazionali. Rend. Accad. Linc. (5) **15**, 619 (1906).

Pedoe, D.: [1] On a new analytical representation of curves in spaces. Proc. Cambridge Phil. Soc. **43**, 455—458 (1947).

Perron, O.: [1] Studien über den Vielfachkeitsbegriff und den Bezoutschen Satz. Math. Z. **49**, 654—680 (1944).

Poincaré, H.: [1] Sur la réduction des intégrales abéliennes. Bull. Soc. Math. France **12**, 124—143 (1884); [2] Sur les fonctions abéliennes. Amer. J. Math. **8**, 289—342 (1886); [3] Sur les résidus des intégrales doubles. Acta math. **9**, 321—380 (1886—1887); [4] Sur les courbes tracées sur les surfaces algébriques. Ann. Ec. Norm. Sup. (3) **27**, 55—108 (1910).

Rabinowitsch, S.: [1] Zum Hilbertschen Nullstellensatz. Math. Ann. **102**, 520—520 (1929).

Rosenlichit, M.: [1] Equivalence relations on algebraic curves. Ann. of Math. **56**, 169—191 (1952); [2] Differentials of the second kind for algebraic function fields of one variable. Ann. of Math. (2) **57**, 517—523 (1953); [3] Generalized Jacobian varieties. Ann. of Math. **59**, 505—530 (1954).

Roth, L.: [1] Some formulae for primals in four dimensions. Proc. London Math. Soc. (2) **35**, 540—550 (1933); [2] Adjoint primals in four dimensions. Proc.

London Math. Soc. (2) **40**, 217—234 (1936); [3] Projective characters and invariants of algebraic varieties. Proc. Cambridge Phil. Soc. **33**, 188—198 (1937); [4] Sulle forme che contengono una data varietà algebrica. Rend. Accad. Linc. (8) **3**, 541—545 (1947); [5] Some arithmetical questions in the theory of the base. Ann. di Mat. (4) **27**, 115—134 (1948); [6] Sulle V_3 algebriche su cui l'aggiunzione si estingue. Rend. Accad. Linc. (8) **9**, 246—250 (1950); [7] Sugli invarianti d'una varietà algebrica a tre dimensioni. Rend. Accad. Linc. (8) **10**, 468—472 (1951); [8] Sulle V_3 algebriche che possiedono un sistema anticanonico. Atti IV Congr. U. M. I., Taormina **1951 II**, 434—439; [9] Some threefolds on which adjunction terminates. Proc. Cambridge Phil. Soc. **48**, 233—242 (1952); [10] Alcune V_3 irrazionali a generi nulli. Rend. Accad. Linc. (8) **12**, 265—269 (1952); [11] On threefolds of linear genus unity. Ann. di Mat. (4) **34**, 247—276 (1953); [12] Sull'estensione di un teorema di CASTELNUOVO-HUMBERT. Rend. Accad. Linc. (8) **15**, 376—380 (1953); [13] Some properties of pseudo-Abelian varieties. Ann. di Mat. (4), **38**, 281—302 (1955); [14] Irregular threefolds which possess anticanonical systems. Proc. Cambridge Phil. Soc. **52** (1956).

SAMUEL, P.: [1] La notion de multiplicités en algèbre et en géométrie algébrique. J. Math. (9) **30**, 159—205, 207—274 (1951); [2] Sur les variétés algébroïdes. Ann. Inst. Fourier Grenoble **2**, 147—160 (1950); [3] Singularités des variétés algébriques. Bull. Soc. Math. France **79**, 121—129 (1951).

SCOTT, D. B.: [1] Point-curve correspondences. Proc. Cambridge Phil. Soc. **41**, 135—145 (1945); **42**, 229—239 (1946); **45**, 342—353 (1949); [2] The united curve of a point-curve correspondence on an algebraic surface. Proc. London Math. Soc. (2) **51**, 308—324 (1950); [3] On the fundamental theorem for point-point correspondences with valency on an algebraic surface. Pont. Acad. Sci. Acta **14**, 57—61 (1950); [4] Correspondences of dimension two and three between algebraic surfaces. Proc. London Math. Soc. (3) **2**, 1—21 (1952).

SEGRE, B.: [1] Sulla serie caratteristica d'una superficie sopra una varietà algebrica a quattro dimensioni. Rend. Accad. Linc. (6) **17**, 917—918 (1933); [2] Nuovi contributi alla geometria sulle varietà algebriche. Mem. Accad. Ital. **5**, 479—576 (1934); [3] Intorno alla parità di alcuni caratteri di una varietà di dimensione dispari. Boll. Un. Mat. Ital. (1) **13**, 93—95 (1934); [4] Un problema di geometria numerativa. Boll. Un. Mat. Ital. (1) **15**, 49—55 (1936); [5] Quelques résultats nouveaux dans la géométrie sur une V_3 algébrique. Mem. Acad. Roy. Belgique (2) **14**, 3—99 (1936); [6] Un teorema fondamentale della geometria sulle superficie algebriche ed il principio di spezzamento. Ann. di Mat. (4) **27**, 107—126 (1938); [7] The postulation of a multiple curve. Proc. Cambridge Phil. Soc. **38**, 368—377 (1942); [8] On limits of algebraic varieties, in particular of their intersections and tangential forms. Proc. London Math. Soc. (2) **47**, 351—403 (1942); [9] Sui sistemi continui di ipersuperficie algebriche. Rend. Accad. Linc. (8) **1**, 564—570 (1946); [10] BERTINI forms and Hessian matrices. J. London Math. Soc. **26**, 164—176 (1951); [11] Sullo scioglimento delle singolarità delle varietà algebriche. Ann. di Mat. (4) **33**, 5—48 (1952); [12] Nuovi metodi e risultati nella geometria sulle varietà algebriche. Ann. di Mat. (4) **35**, 1—128 (1953); [13] Dilatazioni e comportamenti associati nel campo analitico. Rend. Circ. Mat. Palermo (2) **1**, 373—379 (1952); [14] Dilatazioni e varietà canoniche nelle varietà algebriche. Ann. di Mat. (4) **37**, 139—155 (1954).

SEGRE, C.: [1] Intorno ad un carattere delle superfie e delle varietà algebriche superiori. Atti Accad. Sci. Torino **31**, 485—501 (1896).

SEIDENBERG, A.: [1] Valuation ideals in polynomial rings. Trans. Amer. Math. Soc. **57**, 387—425 (1945); [2] The hyperplane sections of normal varieties. Trans. Amer. Math. Soc. **69**, 357—386 (1950).

SERRE, J. P.: [1] Quelques problèmes globaux relatifs aux variétés de STEIN. Coll. sur les fonctions de plusieurs variables, Bruxelles **1953**, 57—68; [2] Groups d'homotopie et classes de groups abéliennes. Ann. of Math. **58**, 258—294 (1953); [3] Un theorème de dualité. Comm. Math. Helvet. **29**, 9—26 (1955); [4] Faisceaux algébriques cohérents. Ann. of Math. **61**, 197—278 (1955).

SEVERI, F.: [1] Sulle intersezioni delle varietà algebriche e sopra i loro caratteri e singolarità proiettive. Mem. Accad. Sci. Torino (2) **52**, 61—118 (1902); [2] Su alcune questioni di postulazione. Rend. Circ. Mat. Palermo **17**, 73—103 (1903); [3] Sulla deficienza della serie caratteristica di un sistema lineare di curve appartenente ad una superficie algebrica. Rend. Accad. Linc. (5) **12**, 250—257 (1903); [4] Sulle superficie algebriche che posseggono integrali di PICARD della seconda specie. Math. Ann. **61**, 20—49 (1905); [5] Sulla differenza fra i numeri degli integrali di PICARD della prima e della seconda specie appartenenti ad una superficie irregolare. Atti Accad. Sci. Torino **40**, 288—296 (1905); [6] Il teorema d'ABEL sulle superficie algebriche. Ann. di Mat. (3) **12**, 55—79 (1905); [7] Sul teorema di RIEMANN-ROCH e sulle serie continue di curve appartenenti ad una superficie algebrica. Rend. Ist. Lombardo Sci. (2) **38**, 859—865 (1905); [8] Sulle curve algebriche virtuali appartenenti ad una superficie algebrica. Rend. Ist. Lombardo (2) **38**, 859—865 (1905); [9] Una proprietà delle forme algebriche prive di punti multipli. Rend. Accad. Linc. (5) **15**, 691—696 (1906); [10] Sulla totalità di curve algebriche tracciate sopra una superficie algebrica. Math. Ann. **62**, 194—225 (1906); [11] Osservazioni varie di geometria sopra una superficie algebrica e sopra una varietà. Atti Ist. Veneto Sci. **65**, 625—643 (1906); [12] La base minima pour la totalité des courbes tracées sur une surface algébrique. Ann. Ec. Norm. Sup. Paris (3) **25**, 449—468 (1908); [13] Fondamenti per la geometria sulle varietà algebriche. Rend. Circ. Mat. Palermo **28**, 33—87 (1909); [14] Complementi alla teoria della base per la totalità delle curve di una superficie algebrica. Rend. Circ. Mat. Palermo **30**, 265—288 (1910); [15] Sul principio della conservazione del numero. Rend. Circ. Mat. Palermo **33**, 313—327 (1912); [16] Sulla varietà che rappresenta gli spazi subordinati di data dimensione immersi in uno spazio lineare. Ann. di Mat. (3) **24**, 89—120 (1915); [17] Sulla teoria degli integrali semplici di prima specie appartenenti ad una superficie algebrica. Rend. Accad. Naz. Linc. (5) **30**, 163—167, 204—208, 231—235, 273—280, 296—301, 328—332, 365—367 (1921); [18] La serie canonica e la teoria delle serie principali di gruppi di punti sopra una superficie algebrica. Comm. Math. Helvet. **4**, 268—326 (1932); [19] Un nuovo campo di ricerche nella geometria sopra una superficie e sopra una varietà algebrica. Mem. Accad. Ital. **3**, 1—52 (1932); [20] La teoria delle serie di equivalenza e delle corrispondenza a valenza sopra una superficie algebrica. Rend. Accad. Linc. (6) **17**, 419—425, 491—497, 597—600, 681—685, 759—764, 869—876, 876—881 (1933); [21] Über die Grundlagen der algebraischen Geometrie. Abh. Math. Sem Hans. Univ. **9**, 335—364 (1933); [22] La base per le varietà algebriche di dimensione qualunque contenute in una data e la teoria generale delle corrispondenze fra i punti di due superficie algebriche. Mem. Accad. Ital. **5**, 239—283 (1934); [23] Caratterizzazione geometrica, topologica e trascendente delle serie di equivalenza. Rend. Accad. Linc. **20**, 287—293, 395—397 (1934); [24] La teoria generale delle corrispondenze fra varietà algebriche. Rend. Accad. Linc. (6) **23**, 818—823, 921—925 (1936); [25] Complementi alla teoria generale delle corrispondenze tra varietà algebriche. Rend. Accad. Linc. (6) **24**, 493—497 (1936); (6) **25**, 3—9 (1937); [26] La teoria generale delle corrispondenze tra varietà algebriche e i sistemi di equivalenza. Abh. Math. Sem. Hans. Univ. **13**, 101—112 (1939); [27] Sulla irregolarità superficiale d'una varietà algebrica. Rend. Accad. Ital. (7) **3**, 547—555 (1942); [28] Ulteriori sviluppi della teoria delle serie di equivalenza sulle superficie algebriche. Comm. Pont. Acad. Sci. **6**, 977—1029 (1942); [29] Funzioni quasi abeliane. Pont. Accad. Sci., Scripta varia, no. **4**, 327 (1947); [30] Il concetto generale di molteplicità delle soluzioni pei sistemi di equazioni algebriche e la teoria dell'eliminazione. Ann. di Mat. (4) **26**, 221—270 (1947); [31] Il punto di vista gruppale nei vari tipi di equivalenza sulle varietà algebriche. Comm. Math. Helvet. **21**, 189—224 (1948); [32] Grundlagen der abzählenden Geometrie. S. 1—123. Wolfenbütteler Verlagsanstalt 1949; [33] Sulla molteplicità d'intersezione delle varietà algebriche ed analitiche. Math. Z. **52**, 827—851 (1950); [34] Legami tra certe proprietà aritmetiche delle superficie e la teoria della base. Rend. Mat. Roma (5) **9**, 59—69 (1950); [35] Fondamenti per la geometria sulle varietà algebriche. Ann. di Mat. (4) **32**, 31—81 (1951); [36] Ulteriori complementi alla teoria della base. Rend. Circ. Mat. Palermo (2) **1**, 71—87 (1952); [37] Fondamenti per una teoria generale dei connessi. Acta Salamanticensia, no. 3, 28 pp. (1950); [38] Le diverse concezioni di varietà nella geometria algebrica. Mem. Accad. Naz. IXL (4) **2**, 1—27 (1951); [39] Sugli antigeneri d'una varietà algebrica. Rend. Accad. Linc. (8) **18**, 131—140 (1955); [40] Le equivalenze algebriche. Rend. Accad. Linc. (8) **18**, 357—361 (1955); [41] Le equivalenze razionali. Rend. Accad. Linc. (8) **18**, 443—451 (1955); [42] Contributi alla teoria delle irregolarità d'una varietà algebrica. Rend. Accad. Linc. (8) **20**, 7—16 (1956); [43] Fondamenti per la geometria sulle varietà algebriche Ann. di Mat. (4) **41**, 161—199 (1956).

SPENCER, D. C.: [1] Cohomologie and the RIEMANN-ROCH theorem. Proc. Nat. Acad. Sci. USA **39**, 660—669 (1953).

SPERNER, E.: [1] Über einen kombinatorischen Satz von MACULAY. Abh. Hamburg. Univ. **7**, 149—163 (1930).

THULLEN, P.: [1] Determinazione della serie di equivalenza individuata dal gruppo dei punti doppi impropri d'una superficie dell' S_4. Rend. Circ. Mat. Palermo **59**, 256—260 (1935).

TODD, J. A.: [1] The arithmetic genus of a V_3 in S_4. J. London Math. Soc. **9**, 205—210 (1934); [2] Some group-theoretic considerations in algebraic geometry. Ann. of Math. **35**, 702—704 (1934); [3] Algebraic correspondences between algebraic varieties. Ann. of Math. **36**, 325—335 (1935); [4] On double integrals of the first kind attached to an algebraic variety. J. London Math. Soc. **11**, 35—37 (1936); [5] The geometrical invariants of algebraic loci. Proc. London Math. Soc. (2) **43**, 127—138 (1937); [6] Note on the canonical series of a V_d. Proc. London Math. Soc. (2) **43**, 139—141 (1937); [7] Birational transformations with isolated fundamental points. Proc. Edinburgh Math. Soc. (2) **2**, 117—124 (1937); [8] Intersection of loci on an algebraic V_3. Proc. Cambridge Phil. Soc. **33**, 425—437 (1937); [9] The arithmetical invariants of algebraic loci. Proc. London Math. Soc. (2) **43**, 190—225 (1937); [10] Birational transformations possessing fundamental curves. Proc. Cambridge Phil. Soc. **34**, 144—155 (1938); [11] The geometrical invariants of algebraic loci. Proc. London Math. Soc. (2) **45**, 410—424 (1939); [12] A remark on a theorem of SEVERI. Proc. Cambridge Phil. Soc. **35**, 516—517 (1939); [13] The postulation of a multiple variety. Proc. Cambridge Phil. Soc. **36**, 27—33 (1940); [14] Invariant and covariant systems on an algebraic variety. Proc. London Math. Soc. (2) **46**, 199—230 (1940); [15] Birational transformations with a fundamental surface. Proc. London Math. Soc. (2) **47**, 81—100 (1941); [16] On algebraic curve branches. J. London Math. Soc. **21**, 233—240 (1946); [17] On the overlap of an algebraic surface. J. London Math. Soc. **26**, 73—74 (1951); [18] On the invariants of the canonical system of a V_d. Proc. Cambridge Phil. Soc. **49**, 410 to 412 (1953).

— and E. A. MAXWELL: [1] Note on the invariants of the canonical systems of an algebraic variety. Proc. Cambridge Phil. Soc. **33**, 438—443 (1937).

TORELLI, L.: [1] Sulla postulazione di una varietà e sui moduli di forme algebriche. Ann. di Mat. (3) **18**, 81—98 (1911).

VESENTINI, E.: [1] Classi caratteristiche e varietà covarianti d'immersione. Rend. Accad. Linc. (8) **16**, 199—204 (1954); [2] Ancora sulle classi caratteristiche e sulle varietà covarianti d'immersione. Rend. Accad. Linc. (8) **17**, 196—203 (1954); [3] Campi di vettori dotati di peso sopra una varietà complessa compatta. Rend. Mat. Roma (5) **14**, 1—17 (1955); [4] Un'osservazione sul teorema dell'appartenenza. Rend. Accad. Linc. (8) **18**, 38—43 (1955); [5] Sui punti stazionari di forme differenziali meromorfe sopra una varietà complessa compatta. Rend. Accad. Linc. (8) **18**, 486—494 (1955).

WAERDEN, B. L. VAN DER: [1] Zur Nullstellentheorie der Polynomideale. Math. Ann. **96**, 183—208 (1925); [2] Der Multiplizitätsbegriff der algebraischen Geometrie. Math. Ann. **97**, 756—774 (1926); [3] On HILBERT's fonction, series of composition of ideals. Proc. Kon. Acad. Amsterdam **31**, 749—770 (1928); [4] Topologische Begründung des Kalküls der abzählenden Geometrie. Math. Ann. **102**, 337—362 (1929); [5] Zur algebraischen Geometrie. Notes in Math. Ann.: [I] **108**, 113—125 (1933); [III] **108**, 694—698 (1933); [V] **110**, 128—133 (1934); [VI] **110**, 134—160 (1934); [VII] **111**, 432—437 (1935); [X] **113**, 705 to 712 (1937); [XI] **114**, 683—699 (1937); [XII] **115**, 310—332 (1938); [XIII] **115**, 359—378 (1938); [XIV] **115**, 619—642 (1938); [6] The foundation of the invariant theory of linear systems of curves on an algebraic surface. Indag. Math. **8**, 120—123 (1946); [7] Birational invariants of algebraic manifolds. Acta Salamanticensia, no. 2, 1—57 (1947); [8] Über einfache Punkte von algebraischen Mannigfaltigkeiten. Math. Z. **51**, 497—501 (1948); [9] Divisorenklassen in algebraischen Funktionenkörpern. Comm. Math. Helvet. **20**, 68—109 (1947); [10] Birationale Transformation von linearen Scharen auf algebraischen Mannigfaltigkeiten. Math. Z. **51**, 502—523 (1948); [11] Zur algebraischen Geometrie. Math. Ann. [XVI] **125**, 314—324 (1952); [XVII] **128**, 128—134 (1954); [XVIII] **128**, 135—137 (1954); [12] Zur Konstruktion des Resultantensystems für homogene Gleichungen. Arch. Math. **5**, 371—375 (1954).

WALKER, R. J.: [1] Reduction of singularities of an algebraic surface. Ann. of Math. (2) **36**, 336—365 (1935).

WEIL, A.: [1] Sur la théorie des formes différentielles attachées a une variété analytique complexe. Comm. Math. Helvet. **20**, 110—116 (1947); [2] Fibrespaces in algebraic geometry. Univ. of Chicago, Notes by A. Wallace, 1949; [3] Variétés abéliennes. Coll. Int. d'Alg. et théorie des nombres, pp. 125—127. Paris 1950; [4] On PICARD varieties. Amer. J. Math. **74**, 865—894 (1952); [5] Criteria for linear equivalence. Proc. Nat. Acad. Sci. USA **38**, 258—260 (1952); [6] Sur les critères d'équivalence en géométrie algébrique. Math. Ann. **128**, 95—127 (1954).

YOXALL, A. L.: [1] Note on a paper by J. A. TODD. Proc. Cambridge Phil. Soc. **35**, 125—126 (1939).

ZAPPA, G.: [1] Sull'esistenza di curve algebriche non isolate a serie caratteristica non completa sopra una rigata algebrica. Pont. Accad. Sci. Acta **7**, 1—4 (1943); [2] Su alcuni contributi alla conoscenza della struttura delle superficie algebriche. Pont. Accad. Sci. Acta **7**, 4—8 (1943); [3] Sui sistemi continui di curve sopra una rigata algebrica. Giorn. Mat. Battaglini (4) **77**, 172—183 (1947); [4] Alla ricerca di nuovi significati topologici dei generi geometrico ed aritmetico di una superficie algebrica. Ann. di Mat. (4) **30**, 123—146 (1949).

ZARISKI, O.: [1] Polynomial ideals defined by infinitely near base points. Amer. J. Math. **60**, 151—204 (1938); [2] Some results in the arithmetic theory of algebraic varieties. Amer. J. Math. **61**, 249—294 (1939); [3] The reduction of the singularities of an algebraic surface. Ann. of Math. **40**, 639—689 (1939);

[4] Algebraic varieties over a groundfield of characteristic zero. Amer. J. Math. **62**, 187—221 (1940); [5] Local uniformisation. Ann. of Math. (2) **41**, 852—896 (1940); [6] Pencils on an algebraic variety and a new proof of a theorem of BERTINI. Trans. Amer. Math. Soc. **50**, 48—70 (1941); [7] A simplified proof for the resolution of singularities of an algebraic surface. Ann. of Math. **43**, 583 to 593 (1942); [8] Foundations of a general theory of birational correspondences. Trans. Amer. Math. Soc. **53**, 490—542 (1943); [9] Reduction of the singularities of algebraic three-dimensional varieties. Ann. of Math. **45**, 472—542 (1944); [10] The theorems of BERTINI on the variables singular points of a linear system of varieties. Trans. Amer. Math. Soc. **56**, 130—140 (1944); [11] The compactness of the RIEMANN manifold of an abstract field of algebraic functions. Bull. Amer. Math. Soc. **50**, 683—691 (1944); [12] Generalized semi-local rings. Summa Bras. Math. **1**, 169—195 (1946); [13] The concept of a simple point of an abstract algebraic variety. Trans. Amer. Math. Soc. **62**, 1—52 (1947); [14] A new proof of HILBERT's Nullstellensatz. Bull. Amer. Math. Soc. **53**, 362—368 (1947); [15] Analytical irreducibility of normal varieties. Ann. of Math. **49**, 352—361 (1948); [16] A simple analytical proof of a fundamental property of birational transformations. Proc. Nat. Acad. USA **35**, 62—66 (1949); [17] A fundamental lemma from the theory of holomorphic functions on an algebraic variety. Ann. di Mat. (4) **29**, 187—198 (1949); [18] Quelques questions concernant la théorie des fonctions holomorphes sur une variété algébrique. Coll. Int. Alg. et théorie des nombres, pp. 129—133. Paris 1950; [19] Sur la normalité analytique des variétés normales. Ann. Inst. Fourier Grenoble **2**, 161—164 (1951); [20] Theory and applications of holomorphic functions on algebraic varieties over arbitrary groundfields. Mem. Amer. Math. Soc. no. 5, 1—90 (1951); [21] Complete linear systems on normal varieties and a generalisation of a lemma of ENRIQUES-SEVERI. Ann. of Math. **55**, 552—592 (1952); [22] Le problème de la réduction des singularités d'une variété algébrique. Bull. Sci. Math. France (2) **78**, 31—40 (1954).

Index

For each reference the index gives the Chap., section, subsection (if any) and page, thus: (X, 8c) 153, i. e. Ch. X, section 8, subsection *c*, p. 153. Only the principal definitions, theorems and notations are quoted.